Communications in Computer and Information Science 279

T0183680

For further volumes:
http://www.springer.com/series/7899

Vladimir Vishnevsky · Dmitry Kozyrev
Andrey Larionov (Eds.)

Distributed Computer and Communication Networks

17th International Conference, DCCN 2013
Moscow, Russia, October 7–10, 2013
Revised Selected Papers

 Springer

Editors
Vladimir Vishnevsky
Dmitry Kozyrev
Andrey Larionov
Russian Academy of Sciences
Moscow
Russia

ISSN 1865-0929   ISSN 1865-0937   (electronic)
ISBN 978-3-319-05208-3   ISBN 978-3-319-05209-0   (eBook)
DOI 10.1007/978-3-319-05209-0
Springer Cham Heidelberg New York Dordrecht London

Library of Congress Control Number: 2014934172

Printed on acid-free paper

Springer is part of Springer Science+Business Media (www.springer.com)

Preface

This book contains the proceedings of the 17th International Conference on Distributed Computer and Communication Networks (DCCN 2013). The conference is a continuation of traditional international conferences of the DCCN series, which took place in Bulgaria (Sofia, 1995, 2005, 2006, 2008, 2009), Israel (Tel Aviv, 1996, 1997, 1999, 2001), and Russia (Moscow, 1998, 2000, 2003, 2007, 2010, 2011) in the last 16 years. The main idea of the conference is to assemble researchers from various countries working in the area of theory and applications of distributed computer and communication networks, to exchange the expertise, and to discuss the perspectives of development and collaboration in this area. The content of this volume is related to the following subjects:

1. Computer and communication networks architecture optimization
2. Control in computer and communication networks
3. Performance and QoS evaluation in wireless networks
4. Modeling and simulation of network protocols
5. Queueing theory
6. Wireless IEEE 802.11, IEEE 802.15, IEEE 802.16 and UMTS (LTE) networks
7. FRID technology and its application in intellectual transportation networks
8. Protocols design (MAC, Routing) for centimeter and millimeter wave mesh networks
9. Internet and Web applications and services
10. Application integration in distributed information systems

All the papers selected for the proceedings are given in the form presented by authors. These papers are of interest to everyone working in the field of computer and communication networks.

October 2013   Vladimir Vishnevsky

Organization

DCCN 2013 was organized by Russian Academy of Sciences (RAS), V.A. Trapeznikov Institute of Control Sciences of RAS, the Research and Development Company "Information and Networking Technologies," and the Institute of Information and Communication Technologies (Bulgarian Academy of Sciences).

Program Committee

Conference Co-chair	S.N. Vasilyev (ICS RAS, Russia)
Conference Co-chair	V.M. Vishnevsky (ICS RAS, Russia)
Publicity Chair	T. Atanasova (IICT-BAS, Bulgaria)
Publicity Chair	O.V. Semenova (ICS RAS, Russia)

Reviewers

A.M. Andronov	Riga Technical University, Latvia
L.I. Abrosimov	Moscow Power Engineering Institute, Russia
G.P. Basharin	Peoples Friendship University of Russia, Russia
A.S.Bugaev	Moscow Institute of Physics and Technology, Russia
T. Czachorski	Institute of Informatics of Polish Academy of Sciences, Poland
E. Gelenbe	University of Paris, France
A. Gelman	IEEE Communications Society, USA
A.N. Dudin	Belarusian State University, Belarus
V.V. Devyatkov	Bauman University, Russia
D.-J. Deng	National Changhua University of Education, Taiwan
A.S. Ermakov	Kazakh National Technical University after K.I. Satpayev, Kazakhstan
M.A. Fedotkin	Lobachevsky National Research University of Nizhni Novgorod, Russia
V.A. Ivnitsky	Moscow State University of Railway Engineering, Russia
G. Kotsis	Johannes Kepler University Linz, Austria
E.A. Koucheryavy	Tampere University of Technology, Finland
L. Lakatosh	Budapest University, Hungary
E. Levner	Holon Institute of Technology, Israel
A.N. Latkov	Riga Technical University, Latvia
P.P. Malcev	Institute of Microwave Semiconduct Electronics of RAS, Russia
S.D. Margenov	Institute of Information and Communication Technologies at the Bulgarian Academy of Sciences (IICT-BAS), Bulgaria
E.V. Morozov	Petrozavodsk State University, Russia
G.K. Mishkoy	Academy of Sciences of Moldova, Moldavia
A.N. Nazarov	OJSC Intellect Telecom, Russia

Support and Sponsors

Contents

X   Contents

Next-Generation Internet Projects

John Geske[1] and Peter Stanchev[1,2](✉)

[1] Kettering University, Flint, USA
[2] Institute of Mathematics and Informatics, Bulgarian Academy of Sciences,
Sofia, Bulgaria
{geske,pstanche}@kettering.edu

Abstract. The paper gives an introduction to some projects and initiatives that are connected with the new-generation Internet. US-Ignite partnerships, the European commission Future Internet Program, the European Fire projects, the Global Environment for Network Innovation (GENI), and WiMAX, are presented. New technologies and protocols, such as Software-defined networking (SDN), OpenFlow, a route configuration mechanism and the constituent protocols to add redundant packet forwarding capabilities that will provide high reliability communication for critical applications, are described. The goal of the paper is to understand intricacies, and nuances of some of these techniques and show some of the possibilities of next-generation high-speed networking and their possible applications in the field of the Internet and the applications to education, libraries and museums.

Keywords: US-Ignite initiatives · The European commission Future Internet Program · The European Fire projects · Internet of things · Software-defined networking · OpenFlow · The Global Environment for Network Innovation (GENI) · WiMAX

1 Introduction

The Digital Agenda for Europa states [1]: "The Internet is called on to perform increasingly many tasks - from online banking to tsunami monitoring. The Internet of tomorrow needs to be more powerful, connected and intuitive responding to our needs at home, work or on the go." In Sect. 2, we give an introduction to some projects and initiative connected with the new-generation Internet and the technologies that are linked with them. In Sect. 3, we study new technologies and protocols. In Sect. 4, we show some of the possibilities of next-generation high-speed networking and their possible applications in the field of the Internet of Things and the applications to education, libraries and museums. Our previous analysis of this issue can be found in [2,3].

V. Vishnevsky et al. (Eds.): DCCN 2013, CCIS 279, pp. 1–10, 2014.
DOI: 10.1007/978-3-319-05209-0_1, © Springer International Publishing Switzerland 2014

2   Next-Generation Internet Projects and Initiatives

2.1   US Ignite Partnership

US Ignite is a public-private nonprofit partnership of nationwide scope initiated by White House Office of Science and Technology Policy to "accelerate the development of applications that can take advantage of ultra-high-speed programmable broadband to bring innovative new products and services to the American people." The primary goal of the US Ignite Partnership is to catalyze approximately 60 advanced, next-gen applications over the next five years in six areas of national priority: education and workforce development, advanced manufacturing, health, transportation, public safety, and clean energy. Responsibilities of the Partnership include connecting, convening, and supporting startups, local and state government, universities, industry leaders, federal agencies, foundations, and community and carrier initiatives in conceptualizing and building new applications. The resulting new applications should have a significant impact on the U.S. economy, including providing a broad range of job and investment opportunities.

US Ignite is seeking applications with high societal impact using next generation, high-speed networking. It includes the "programmable broadband", high-speed Internet (1 Gbs+), a networking infrastructure to research, develop, test, prototype, and deploy, next-generation Software Defined Networking applications; a consortium of potential diverse partners. The four most important technical parts of the US Ignite technology include [4]:

- High symmetric bandwidth allows for uncompressed high definition video transmission, which has huge advantages over the IP-based transmission because it minimizes delays in video conferencing. For truly interactive experiences, uncompressed video with its high bandwidth requirements is best, and a number of Ignite applications use uncompressed video particularly in multiple areas, such as healthcare and education.
- The next-generation Internet will take advantage of Software Defined Networks (SDN), which takes the "intelligence" of routing data out of the switches and routers on shelves, and puts more of it into the cloud. SDN tricks servers into thinking that they have the network gear all to themselves, configured exactly the way they like it, when they are really sharing that gear with other servers. More servers can share less network gear, and they can also be moved around easier a big plus for applications such as cloud computing.
- Distributed Cloud Resources (e.g., US Ignite racks) are a kind of cloud computing in which the cloud is itself distributed throughout the network. This has distinct advantages, including pre-staging information where it is needed, processing data traffic more locally, and dramatically improving responsiveness while reducing latency.
- Virtual Networks are tailored to match specific advanced applications, as well as provide unique Access to Advanced Resources, such as advanced computational, sensor, storage and data resources provided by the owners and operators of new technology. The collection of network, distributed, and advanced

resources available in a virtual network to an application is called a "slice." Slices are an important concept because they can be thought of as the delivery mechanism for an application.

2.2 The European Commission Future Internet Program

The European Commission has launched the Future Internet Public-Private Partnership Program [5].

MAIN GOAL: To advance a shared vision for harmonized European-scale technology platforms and their implementation, as well as the integration and harmonization of the relevant policy, legal, political and regulatory frameworks.

PROGRAME AIMS: Increase the effectiveness of business processes and infrastructures supporting applications in areas such as transport, health, and energy. Derive innovative business models that strengthen the competitive position of European industry in sectors such as telecommunication, mobile devices, software and services, and content provision and media.

PROGRAM APPROACH: The Future Internet Public-Private Partnership Program follows an industry-driven, holistic approach encompassing R&D on network and communication infrastructures, devices, software, service, and media technologies. It promotes their experimentation and validation in real application contexts, bringing together demand and supply and involving users early in the research lifecycle. The new platform will thus be used by range actors, in particular SMEs and Public Administrations, to validate the technologies in the context of smart applications and their ability to support user-driven innovation schemes.

2.3 The Global Environment for Network Innovations (GENI)

The Global Environment for Network Innovations [6] is a project sponsored by the USA National Science Foundation. It is open and broadly inclusive, providing collaborative and exploratory environments for academia, industry, and the public to catalyze groundbreaking discoveries and innovation in emerging global networks. GENI is a virtual laboratory at the frontiers of network science and engineering for exploring future internets at scale. GENI creates major opportunities to understand, innovate, and transform global networks and their interactions with society.

GENI, a virtual laboratory for exploring future internets at scale, creates major opportunities to understand, innovate, and transform global networks and their interactions with society. Dynamic and adaptive, GENI opens up new areas of research at the frontiers of network science and engineering, and increases the opportunity for significant socio-economic impact. GENI:

- supports at-scale experimentation on shared, heterogeneous, highly instrumented infrastructure;
- enables deep programmability throughout the network, promoting innovations in network science, security, technologies, services and applications;

- provides collaborative and exploratory environments for academia, industry and the public to catalyze groundbreaking discoveries and innovation.

2.4 Future Internet Research and Experimentation Initiative

The Future Internet Research and Experimentation Initiative (FIRE) [7] creates a multidisciplinary research environment for investigating and experimentally validating highly innovative and revolutionary ideas for new networking and service paradigms. FIRE is promoting the concept of experimentally-driven research, combining visionary academic research with the wide-scale testing and experimentation that is required for industry. FIRE also works to create a dynamic, sustainable, large scale European Experimental Facility, which is constructed by gradually connecting and federating existing and upcoming test beds for Future Internet technologies. Ultimately, FIRE aims to provide a framework in which European research on Future Internet can flourish and establish Europe as a key player in defining Future Internet concepts globally.

2.5 WiMAX

As the first 4G wireless technology, WiMAX [8] combines the performance of WiFi with the range and quality of service (QOS) of a carrier-grade cellular technology. WiMAX can provide broadband wireless access (BWA) up to 30 miles (50 km) for fixed stations, and 3–10 miles (5–15 km) for mobile stations. In contrast, the WiFi/802.11 wireless local area network standard is limited in most cases to only 100–300 ft (30–100 m).

In emerging markets and rural areas, WiMAX is being deployed as a fixed wireless technology to provide basic internet connectivity to residential and business users, without the cost and difficulty of deploying fiber or DSL. In this fixed capacity, the technology can provide backhaul connectivity for Wi-Fi hotspots and other IP enabled devices such as VoIP phones and video surveillance cameras. In more developed markets, WiMAX is being used as a mobile wireless technology by large carriers and operators. The GENI WiMAX projects are creating open, programmable, GENI enabled "cellular-like" infrastructure on university campuses. The WiMAX base station provides network researchers with wide-area coverage and the ability to support both mobile and fixed end users.

3   Next-Generation Internet Techniques and Protocols

3.1 Software-Defined Networking

The key architectural principle of the Internet is based on the TCP/IP protocol. Potential Internet roadblocks are: IP networks are based on Autonomous Systems (AS). An autonomous system is a contiguous set of networks and routers under control of one "administrative authority." The basic IP forwarding paradigm is that all traffic from a given source to a given destination always follows the same path. The forwarding table in a router only contains one entry for a given destination.

Software-defined networking (SDN) is an approach to building computer networking equipment and software that allows network administrators to have programmable central control of network traffic without requiring physical access to the network's hardware devices. Conceptually, a router or switch is divided into two parts: Control Plane: performs configuration and control and Data Plane: handles packet processing. Vendors tightly couple these two planes. SDN avoid using embedded routing protocols and specify how to handle specific critical cases.

In Fig. 1 from [9] shows the old and the new architecture of the networks devices.

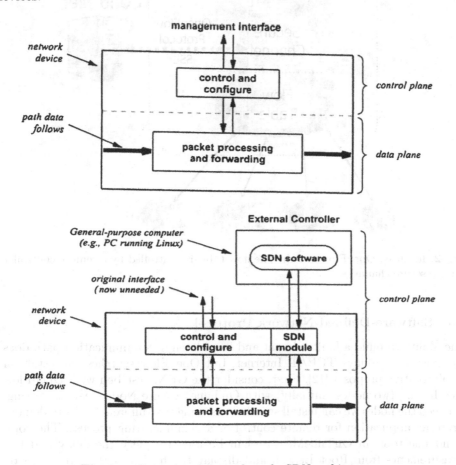

Fig. 1. a. Control and data plane, b. SDN architecture

3.2   OpenFlow Protocol

OpenFlow is a new protocol that is an instantiation of SDN. OpenFlow [10] enables networks to evolve by giving a remote controller the power to modify the behavior of network devices through a well-defined "forwarding instruction set". The growing OpenFlow ecosystem now includes routers, switches, virtual switches, and access points from a range of vendors. OpenFlow is based on an Ethernet switch, with an internal flow-table, and a standardized interface to add and remove flow entries. In Fig. 2 an example from [11] is given.

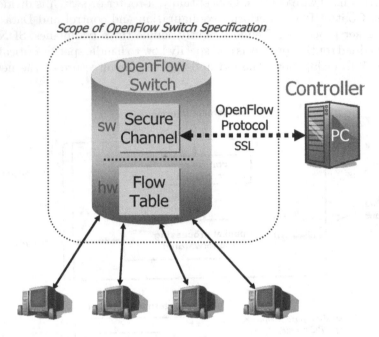

Fig. 2. Idealized OpenFlow switch. The flow table is controlled by a remote controller via the secure channel

3.3 Software-Defined Network Protocol

The ability to build a fault-tolerant and fixed latency communication path does not exist with todays TCP/IP Internet, but OpenFlow switches can establish an alternative proposal [12]. First, consider the GENI test bed with OpenFlow slice having two nodes (initially) with OnTimeMeasure Node Beacons setting. Between the nodes we can install switch flows using two different paths to demonstrate an application for remote control of a manufacturing process. The command line tools in OnTimeMeasure can be used to query the delay and loss measurements from Root Beacon and display the health of the two paths on a web page. Second, a preliminary version of the protocol running in a virtual environment on computers in a single lab is coded. Protocol experiments were carried out by team members at: Purdue, Ohio State, Kettering University, University of Missouri, and the Lit San Leandro gigabit metropolitan fiber project. With redundant paths, communication using the proposed protocol can tolerate faults increased transmission delay due to congestion of to hardware and, thus, reliable communication is achieved. In the event of congestion or failure of a path, the remaining path(s) support(s) communication between the access points while a replacement path is built to reestablish the desired level of fault tolerance.

4 New Internet Applications

Examples of these applications are: monitoring of medical procedures, control of Hazmat robots, monitoring and control of manufacturing robots, and remote unmanned vehicle operation. Remote Process Control - Reliable networks can provide new ways for controlling processes from afar, allowing engineers, artists, and experimenters to remotely control surgery robots, or advanced manufacturing processes, such as 3D printing, regardless of how close they are to the means of production. Such communication is needed for tasks including remote medical procedures, the control of robots used to clean up after a chemical or nuclear disaster, and the control of manufacturing processes.

4.1 Internet of Things

In 2010, Hans Vestberg, CEO of Ericsson group, predicted that by 2020, 50 billion devices would be connected to the internet. The Internet of Things (IoT) [13] is an emerging network superstructure that will connect physical resources and actual users. It will support an ecosystem of smart applications and services bringing hyper-connectivity to our society by using augmented and rich interfaces. Whereas in the beginning IoT referred to the advent of barcodes and Radio Frequency Identification (RFID), which helped to automate inventory, tracking and basic identification, today IoT is characterized by a dynamic trend toward connecting smart sensors, objects, devices, data, and applications. The main domains of IoT applications include the following.

- **Transport/Logistics.** By creating a connection between the two flows, an Internet of Things (IoT) for transport logistics may be created in which the logistics objects or "things" are capable of processing information, communicating with each other, and making their own decisions [14].
- **Smart Home:** The three main aspect of the smart home includes: (1) resource usage (water conservation and energy consumption), (2) security and (3) home comfort. Today, many of these products are still involved in small pilot programs. The security issues - complex security systems for detecting theft, fire or unauthorized entry. Do-it-yourself (DIY) kits are creating competition across previously separated home automation sectors. Most of them are cloud-based services. According to [15], custom-designed smart home systems will grow at only a 7-percent rate, compounded annually, to $2.2 billion in 2017. DIY kits will grow much faster but still only reach $200 million in annual sales by then. In contrast, connected home systems will explode from a $300-million base to $1.5 billion in 2017.

- **Smart Cities**: The term smart city is still not well defined. This is an urban area providing better quality of life. Mr. Rudolf Giffinger, professor at the Department of Spatial Planning, Centre of Regional Science, Vienna University of Technology et al.'s model [16] elucidates the characteristics of a smart city, encompassing economy, people, governance, mobility, environment and living. A smart city should not only be understood as an "intelligent" city but above all as an informed and needs-based city oriented towards the future. It will include smart metering, infrastructure for charging electric vehicles, and remote patient monitoring.
- **Smart Factory**: The industrial value chain, including product design, production planning, production engineering, production execution, and services were implemented separately. Today, new technologies are bringing these worlds together in exciting ways. Maintenance of machinery will be facilitated by connected sensors, allowing for real-time monitoring. Workers will be replaced by complex robots [17].
- **Retail**: Retailers realize both customer needs and business needs. M2M (Machine to Machine) Technology Building Blocks is described in [18]. Having information in real time helps enterprises to improve their business and to satisfy customer needs [19].
- **Smart Energy/Smart Grid**: The key issue is to detect ways to save energy and it is based on smart metering.

4.2   Applications in Education, Libraries and Museums

Holograms, lenticulars, and 3D television systems are the latest additions to high tech museum displays enabling the viewer to see museum artifacts in a whole new light. To bring these apps to life, developers leverage the unique capabilities of next-generation networks, including: High speed, multiple bi-directional streams of uncompressed video; Software-defined networks (e.g. OpenFlow), promising dramatically-improved control over network routing and optimization; Networks with capabilities such as virtual network slices matched to application requirements and distributed programmable resources throughout the network; Integrated wireless networks to facilitate, for example, sensor networks and continuous remote monitoring.

Some new-generation internet applications in education, libraries and museums are [20]:

- Cuyahoga County Public Library is partnering with Case Western Reserve University and One Community to bring a one-gigabit broadband connection to the new Warrensville Height branch, serving 20,000 residents from an economically disadvantaged community;
- Rutgers University Libraries is a lead partner on the Video Mosaic Collaborative (VMC), to create a portal to enable teachers and researchers to analyze and use over 20 years of classroom videos to transform mathematics research, teaching, and learning;

- San Francisco Public Library is developing a Teen Media Learning Lab in partnership with local education, museum, technology, and media organizations, to create a free, seven-day-per-week, interactive digital media learning space for youth;
- Graduate School of Library and Information Science at the University of Illinois Urbana-Champaign hold a series of four continuing education forums to enhance understanding of how libraries can adopt and use next-generation internet networks to address social inclusion through the organization US Ignite;
- High Quality Open Source Web Conferencing: Big Blue Button System will gives remote students multiple HD camera angles, high-quality audio and synchronized slides;
- Engage3D Conferencing: This in-browser application can bring the Aquarium's educational offerings live, interactive, and in 3D into these classrooms at no additional cost to the school. Unlike a simple video, this application allows students to interact with the content moving, seeing, exploring;
- Lynx Laboratories, Real-Time 3D Modeling Cameras. 3D Creation A camera for all your 3D creations that is as easy to use as a point-and-shoot (http://www.engadget.com/2013/02/02/lynx-a/);
- Hyperaudio Pad It is a transcript for audio and video based media, making them more accessible, searchable, navigable and index able. Hyperaudio Pad allows people to assemble and remix media as easily as they would a document (http://www.youtube.com/watch?v=Y-hZk4GI6a0);
- LITE Virtual Reality Workforce Development - Workforce development is based on virtual learning and interactive digital media technology. The use of advanced networking allows interactive training with several sites simultaneously;
- Luminosity - Easily creates interactive scientific visualizations in a web application focusing on simplicity. This web application enables a wider audience to participate in scientific research, data exploration and discovery.

5 Conclusions and Future Work

A vision of a new kind of global virtual museum of the future starts with exhibits anywhere, anytime. Example of such museums is the Vatican Museum. It provides visitors with the opportunity to take virtual tour of some of the dozen of museums and galleries from the Vatican collection. The visitors can view a three dimension video of the Sistine Chapel. We are trying to experiment some of the new techniques in the museum of Parzardjik. Historical Museum in Pazardjik is one of the most popular museum in Bulgaria. It is one of the first museums in the country. It is divided into several sections: Archaeology, History of Bulgaria XV-XIX century Ethnography, Modern History, Contemporary History, Foundations and scientific records. Archaeology will meet the moral culture of the region reflected in the pottery of the Stone Age, Middle and Western coins, weapons, ornaments, and has one of the most complete collection of the Thracian artifacts.

References

1. Digital Agenda for Europa. http://ec.europa.eu/digital-agenda/en/science-and-technology/future-internet
2. Geske, J., Stanchev, P.: The future of the next generation internet and possible applications into education and culture heritage. In: INESCO Digital Presentation and Preservation of Cultural and Scientific Heritage Conference, Sofia, issue 3, pp. 17–24 (2013)
3. Geske, J., Stanchev, P.: Next-generation internet. In: 17th International Conference on Distributed Computer and Communication Networks (DCCN-2013)-Control, Computation, Communications, 7–10 October 2013, Moscow, pp. 8–10 (2013)
4. US Ignite. http://us-ignite.org/
5. Future Internet Public-Private Partnership Programme. http://www.future-internet.eu/home/future-internet-ppp.html
6. GENI. http://www.geni.net/
7. The Future Internet Research and Experimentation Initiative (FIRE). http://www.ict-fire.eu/home.html
8. WiMAX. http://www.wimax.com/
9. Comer, D.: Internetworking with TCP/IP, vol. 1, 6th edn. Addison-Wesley, Boston (2014). ISBN-13: 9780136085300
10. OpenFlow. http://www.openflow.org/
11. McKeown, N., Anderson, T., Balakrishnan, H., Parulkar, G., Peterson, L., Rexford, J., Shenker, S., Turner, J.: OpenFlow: Enabling Innovation in Campus Networks. http://www.openflow.org/documents/openflow-wp-latest. https://mozillaignite.org/apps/418/pdf (2008)
12. Dr. Adams III, G.B.: Remote Process Control Using Reliable Communication Protocol, Mozilla Ignite, 3 April 2013. https://mozillaignite.org/apps/418/
13. Bassi, A., Bauer, M., Fiedler, M., Kramp, T., Kranenburg, R. (eds.): Enabling Things to Talk: Designing IoT solutions with the IoT Architectural Reference Model. Springer, Heidelberg (2013). ISBN-13: 9783642404023
14. Hribernik, K.A., Warden, T., Thoben, K.-D., Herzog, O.: An internet of things for transport logistics. In: An Approach to Connecting the Information and Material Flows in Autonomous Cooperating Logistics Processes, MITIP 2010, Aalborg University, Denmark. http://www.sfb637.uni-bremen.de/pubdb/repository/SFB637-C2-10-004-IC.pdf
15. Wolf, M., Foster, C.: Forecast: Smart Homes and the Internet of Things, 28 August 2013. http://research.gigaom.com/report/forecast-smart-homes-and-the-internet-of-things/
16. Giffinger, R., Haindlmaier, G.: Smart cities ranking: an effective instrument for the positioning of the cities? ACE: Architect. City Environ. 4(12), 7–26 (2010)
17. Hessman, T.: The Dawn of the Smart Factory, Industry Week, 14 February 2013. http://www.industryweek.com/technology/dawn-smart-factory
18. Retail and the Internet of Things: M2M Technology Building Blocks. Multi-Service Gateways and M2M Integration Platform, by Eurotech on 22 May 2013. http://www.slideshare.net/Eurotechchannel/eth-io-tmsgretail20130519
19. RFID Adoption into the Container Supply Chain: Proposing a framework more by Lisa Seymour. In: Proceedings of the 6th Annual ISOnEworld Conference, 11–13 April 2007, Las Vegas. www.isoneworld.org (2007)
20. The official blog of the institute of museum and library services. http://blog.imls.gov/?p=1463

Design and Scheduling in 5G Stationary and Mobile Communication Systems Based on Wireless Millimeter-Wave Mesh Networks

Vladimir Vishnevsky, Andrey Larionov[✉], and Sergey Frolov

V.A. Trapeznikov Institute of Control Sciences of Russian Academy of Sciences,
65 Profsoyuznaya Street, Moscow 117997, Russia
vishn@inbox.ru, larioandr@gmail.com, sergey@frolov.ru
http://www.ipu.ru/en

Abstract. The paper presents the concept of local and metropolitan wireless mesh networks operating in millimeter-wave band. These self-organizing networks may be used as the high-throughput backbones for emerging 5G communication systems, as well as standalone networks providing extensive QoS, mobility, reliability and ultra-high throughputs for the connected users. The paper observes the current state and the prospects of the development of the millimeter-wave mesh networks. The paper outlines the network logical structure including MAC level design, routing and resource allocation. The problem of delay-optimal scheduling within Spatial TDMA is formulated, and a simple fast algorithm for scheduling is proposed. The experimental results provided prove the algorithm allows to utilize the network resources at very high rates while building scheduling sufficient for delay-critical applications.

Keywords: Wireless mesh networks · STDMA · Millimeter-wave networks · Optimal scheduling · 5G networks

1 Introduction

Centimeter wave broadband wireless networks are currently one of the main trends in the telecommunication industry. Wireless networks including UMTS, cdma-200, WiFi (IEEE 802.11), WiMax (IEEE 802.16) are widely spread, have many advantages including rapid installation, mobility, price, often becoming the only cost-efficient solution.

However the high occupation of the centimeter band imposes strict restrictions on the allowed bandwidths and, consequently, limits the information transfere rate. In particular, WiFi network equipment that operates in the centimetric band 2.4–6.4 GHz, provides a nominal speed of 54 Mbit/s with a broadband of 20 MHz and 108 Mbit/s turbo at 40 MHz. The equipment that implements a new standard IEEE 802.11-2012 [2] and using MIMO technology provides the maximum data transfer speeds up to 600 Mbit/s. However, even these rates implemented on the basis of traditional technologies are not sufficient to satisfy the

V. Vishnevsky et al. (Eds.): DCCN 2013, CCIS 279, pp. 11–27, 2014.
DOI: 10.1007/978-3-319-05209-0_2, © Springer International Publishing Switzerland 2014

quickly and continuously growing volume of multimedia information. To this end, lots of research centers all over the world are working hard to dramaticaly increase the performance in the wireless networks.

One of the main trends in the development of the ultra-high capacity (over 1 Gbps) wireless networks is the transition from the traditional centimeter-wave band to the millimeter-wave band (60–100 GHz). This transition is characterized as a new wave of innovation in the area of wireless communications, comparable to the advent of the cellular standards and the invention of WiFi [4–6].

Although the millimeter-wave has been attracting the attention of the networks developers for a long time, but its practical usage in the telecommunication systems was limited by the 40 GHz until recently. Following the adoption of licensing regulations in 2005 by the Federal Communications Commission, a number of the first millimeter wave radio equipment appeared. In 2006, the European Institute of Telecommunications Standards (ETSI) has published technical rules on equipment operating in the E -band frequencies (71–76 and 81–86 GHz). These regulations conform to EU requirements and allow commercial use of the wireless E-band equipment in Europe. By today, many countries are adopting the E-band for wireless point-to-point communication systems, working in the short-wave part of the millimeter range.

The mm-wave electronic components with acceptable characteristics and price appearance, the increased centimeter-wave band occupancy and the novel multimedia applications development lead to the practical development of the E-band communication systems.

The low millimeter-wave band occupancy, the wide bandwidth allocation possibility (up to 5 GHz), the simplified spectrum regulations make the mm-wave band a unique choice for the implementation of personal, local, regional or metropolitan area networks, as well as for the point-to-point communication systems. Other advantages of the mm-wave band include but not limited to:

- ultra-high data transfere rate up to 10 Gbps;
- the possibility of creating miniature phased array antenna systems. It is even possible to create chipset-integrated antenna systems, as achieving a narrow radioation pattern for greater antenna gain require smaller antenna size;
- the heavy signal attenuation in the mm-wave band is an advantage as it limits the signal propagation distance and simplifies the frequency planning task;
- it is possible to implement different scrambling schemes, error-correction coding, utilize simple modulation schemes and multiple access methods;
- the narrow-band communications secrecy and integrity, i.e. the resistance to interference and attempts for unauthorized connections.

The nowadays existing mm-wave band telecommunication solutions may be broadly divided into two parts: point-to-point communication systems and personal area networks (PAN). The point-to-point communication systems are used rather often by LTE or WiMax networks operators to interconnect the base stations, or to build a wireless connection between buildings. These systems provide very fast connections with up to 10 Gbps speed, but are very expensive, require manual configuration and normally can not be used to build a cost-efficient network, covering a relatively large area, or to build a local network comprising many nodes with the up to 100 m distances. The PAN standard IEEE 802.15.3 amendement IEEE 802.15.3c "Millemeter-Wave-based Alternative Physical Layer Extension" was published in 2009, October. In November, 2009 the European standard ECMA-387 "High Rate 60 GHz PHY, MAC and HDMI PAL" was published [3]. These two standards cover the personal area networks operating in 60 GHz. One of the main applications of these technologies is to provide multimedia system components with wireless connections, replacing HDMI cables with wireless links. One should also mention the Wireless Gigabit Alliance (WiGig), found on May, 2009 by a number of industry leaders including Atheros, Broadcom, Dell, Intel, LG Electronics, Marvell, Microsoft, NEC, Nokia, Panasonic, Samsung Electronics and others. NXP, Realtek, STMicroelectronics, Tensorcom, Cisco, Texas Instruments and many other companies also entered the alliance later. By now, the specification produced under the WiGig alliance is included in the amendment IEEE 802.11ad, which final version was recently published. The standard enables to use 60 GHz spectrum in line-of-sight communications, at rather close distances to significantly boost the communications speed. Nevertheless, there are no existing solutions for building local, regional or metropolitan area networks with mm-wave equipment.

The outlined advantages of the mm-wave band for communications systems development coupled with the weaknesses of the existing centimeter-wave solutions, the increased throughput requirements introduced by the modern applications and the lack of scalable cost-effective mm-wave wireless solutions for implementing LANs and MANs, raise the actuality of the mm-wave multihop wireless mesh networks design and development.

The mm-wave mesh network will be overviewed in the next station. Afterwards, the base station protocol stack design will be presented. Finally, the paper will focus on time resources scheduling, which is one of the most significant part of the TDMA-based channel access method used in the network.

2 Millimeter-Wave Mesh Network Design

The self-organizing wireless millimeter-wave mesh network is formed by a number of base stations. The base stations functions include networks and neighborhood discovery, associations establishment, links management, user traffic transmission and bridging with existing wired and wireless networks like Ethernet, WiFi, WiMax, LTE and others. Each base station includes a frequency-duplexing transceiver, allowing to send and receive data simultaneously using two separate frequencies, which significantly increases the base station throughput. Figure 1 gives an example of a small mesh network.

The protocols forming the base station stack were designed with scalability in mind so the network is able to provide sufficient delays and throughputs characteristics when consisting of more then 100 base stations. Taking into account a typical distance of 500 m between neighbour stations the network may cover an area of 5 × 5 kilometers or even more.

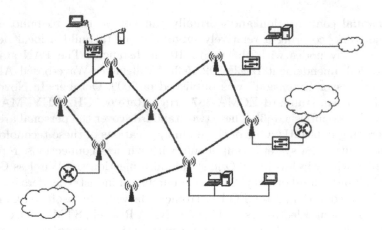

Fig. 1. Wireless mm-wave backbone mesh network example.

Each base station automatically selects the frequency pair when joining the network. The frequency pair choice made by the base station determine its possible connections. Say, if station A choosed frequency F_1 for transmitting and F_2 for receiving then it will be able to establish the connections with the stations using F_2 for transmitting and F_1 for receiving. The frequency assignment may be changed afterwards as a result of topology change, station movement or any other reason.

The network throughput is further increased due to the very low interference. To achieve high antenna gain the base stations transceivers use antenna arrays forming narrow beams which also allows avoiding collisions between densely spaced transmissions. The mm-wave signal heavy attenuation further reduces the collisions probability. These factors make it possible to ignore the interference in most cases however one must pay the attention to the transmissions coordination and synchronization both over the time and over the space.

Each base station comprises the only transceiver. Consequently, the base station is able to perform a single upload and a single download simultaneously. Because of this the maximum number of neighbours limit was forced (four stations, by default) as otherwise in some scenarios the link throughput may become arbitrary small.

The channel access method used in the network is Spatial TDMA [7]. Time scheduling is performed both localy by the stations, and centraly by the *root station*. For instance, the neighbour stations may decide to temporary boost their connection by allocating an additional time interval for it while the root station is in charge of more complicated scheduling functions, including time intervals allocation for multihop transmissions and spanning trees. The multihop transmission that was scheduled by the root station is called a *virtual channel*. The root allocates the same time value for all links transmitting a given virtual channel, so neither channel may have a bottleneck. Virtual channels and spanning trees are being used for extensive QoS support to allow transmitting the most critical or preprovisioned user data being sent with minimal delay and maximal priority. The STDMA implementation and scheduling will be discovered in more details in further sections. Time synchronization which is critical for any TDMA-based access is achieved using GPS/GLONASS adapters.

To bridge with miscellaneous local networks base stations implement IEEE 802.1D [14] bridge standard. This allows connecting the mesh network with existing networks including Ethernet, WiFi, WiMax and others.

3 MAC Layer Architecture

The base station MAC layer is in charge of links management, channel access and STDMA scheduling, network discovery, association and disassociation, mobility managment, frames routing, external networks discovery, root station election, and other essential functions. The layer structure is presented in the Fig. 2.

The primary MAC layer services include the following:

- *Traffic classification* and *queueing management* services for QoS management.
- A proactive routing protocol *Mesh Link State Routing (MLSR)* which is used as a default routing facility. The protocol is based on OSPF [15] and OLSR being adjusted for the mm-wave mesh network characteristics.
- Network discovery, association and disassociation, links management, local scheduling and other essential management functions are implemented in *Mesh Management (MMAN)* service.
- *Mesh Resource Allocation Protocol (MRAP)* is in charge of transmitting virtual channels creation, modification or deletion requests and schedules propagation.

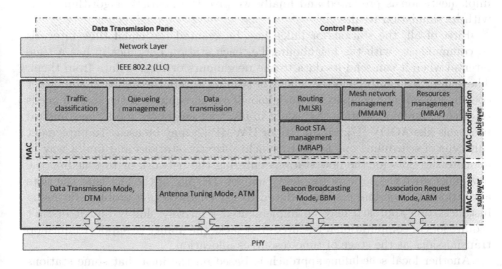

Fig. 2. MAC layer design.

- The root base station election, backup root stations selection, configuration and synchronization along with other root-related management functions are provided by the *Mesh Root Management Protocol*.

The base station uses several PHY-layer access modes:

- *Data transmission mode (DTM)* is used to transmit user, control and management frames through the link.
- Neighbour stations may adjust their antennas configurations to decrease the error rate using *Antenna Tuning Mode (ATM)*.
- Each associated base station periodically transmit a special Beacon-frame simulating broadcast in *Beacon Broadcasting Mode (BBM)*.
- *Association Request Mode (ARM)* is being used by the station that would like to associate with the network or to create a link with a new neighbour.

4   Scheduling in Spatial TDMA

The efficiency of any Spatial TDMA implementation depends heavily on the time scheduling algorithms. The quality of schedules being built affects the end-to-end delays for multi-hop transmissions, the stations and network throughputs and even the load balancing. These characteristics depend on the scheduling algorithm type, the criteria that was utilized and, of course, on the algorithm design and implementation. Many of the scheduling problems are NP-complete, so usually there is no straight way to achieve the optimal characteristics.

There are several ways to implement scheduling within Spatial TDMA, and some of them are used in the mm-wave mesh network MAC layer. These methods are observed briefly below, then a detailed review of the Spatial TDMA implementation is presented and finally we give the heuristic algorithm along with its simulation results.

First of all, the stations or links may be granted the equal time intervals to communicate with the neighbours, i.e. each station periodically has a time interval when it can send its data to the neighbours or receive data from them. These intervals can be selected when the station enters the network, so neither coordinator nor any complex network knowledge is required. If the station needs to discover a remote station that is not in its strict neighbourhood, the standard protocols like AODV [17], OLSR [16] or HWMP [2] may be used. To implement this type of scheduling one needs to synchronize the stations and find a way to allocate the slots for the new stations entering the network or to schedule the new links when the network topology changes. Despite of the fact that the slots allocation for each link may be reduced to the edge-coloring problem, in many instances this approach may be implemented relatively simply and it doesn't require lots of computational resources but lacks the ability to support multihop transmissions at the stage of time-resources allocation.

Another local scheduling approach is based on the idea that some stations act as access points to the others. These access points may broadcast beacons to indicate the superframe start, send out timing and resource allocations para-meters, and perform the scheduling [1]. Whenever a station A that is connected to the access point B needs a time slot to transmit its data, it sends a request to the access point. The access point schedules the requested transmission and informs the client about the slot starting time and its duration. This approach is widely used in many wireless networks including IEEE 802.11, IEEE 802.16 and IEEE 802.15.4.

On the other hand, the schedule may be built for a large group of stations. For example, some dedicated server station may accumulate all bandwidth requests from all stations, build the schedule for the whole network at once and propa-gate the schedule over the network. The slots in the schedule may be allocated to provide the optimal access for separate stations depending on their current load, as well as to provide the time resources for multihop transmissions based on the information of the running applications QoS requirements, including bandwidth and timing limitations. The latter approach allows uniform scheduling the mul-tihop transmissions over all their paths at once thus avoiding the bottlenecks: if the station A sends some data to the remote station D through their common neighbour C, the schedule will contain the equal number of slots for both A-C and C-D transmissions. The same task may be solved faster and in more reli-able way by some distributed resource allocation protocol without the dedicated central station, though using the server can lead to better schedules. The server-based approach for multihop transmissions scheduling is discussed below more detailing as it is used in the mm-wave mesh network base station MAC level.

Prior to build the scheduling algorithm one needs to determine the criterias to be optimized in the produced schedules. Lots of prior work, see [8–13], was focused on building the minimal-length schedules, as well as on maximizing the absolute and relative throughputs or minimizing the schedules delay. The widely used approach for building the algorithms maximizing the throughputs or minimizing the multihop transmissions delays is to decompose the problem into two parts: routing and time slots allocation. This decomposition allows simplifying the original task: the first part may be solved by the well-studied methods including Dijkstra, Bellman-Ford or Floyd-Warshall algorithms [18]. Unfortunately, the optimal time slots allocation problems for minimizing delay or maximizing the throughput are NP-complete.

4.1 Spatial TDMA Implementation

According to Spatial TDMA scheme the links should be granted time inter-vals. Considering the wireless networks, TDMA-based channel access methods usually have to account interference. To avoid the collisions while transmitting over interfering links these links must be scheduled in different time intervals. However, since the heavy attenuation in the mm-wave band and the narrow beams used by the base stations, the interference in the mm-wave mesh net-work may be neglected. This allows assuming just two possible types of con-flicts to be different transmissions simultaneous receptions by the same station and the simultaneous transmissions by the same station to different neighbours.

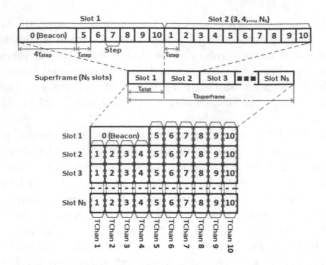

Fig. 3. STDMA time intervals and superframe structure.

The drawback of the narrow beams is that any time the station may send its data to only one neighbour, so it is impossible to use the broadcast nature of the wireless transmission.

MAC layer uses the three-layered time gradation, see Fig. 3. Time is divided into **steps**. The step is the shortest time interval that can be granted to any link. The step acts like a slot in the described above scheduling methods. A typical step duration is 100 us. Each ten consequent steps form a **slot**, which is 1 ms by default. The slot is a periodic structure used in the basic access: every link has one associated step to transmit in both directions during each slot. The step is assigned when the connection is established but can be changed afterwards. By default, first four steps in each slot are used for such assignment. Consequently, each station can have four connections at most by default. Other steps (six by default) are scheduled dynamically either by the negotiation between the neighbours, or by the root station. Several consequent slots forms a **superframe**. Superframes are being used by the root station for scheduling spanning trees and multihop transmissions. The superframe duration may vary depending on actual network size and load. A typical superframe duration for the network consisting of 100 stations operating under the heavy load is 50 ms. It should not be too large as this will increase the overall delay in the network while too small superframe duration values may decrease the quality of multihop transmissions serving.

The mm-wave mesh network implements several scheduling methods utilizing different scheduling approaches. The basic access method states that each station is being granted one step in each slot for each neighbour it has. Because of the full-duplex nature of the base stations, the network topology may be represented with a bipartite graph. The task of assigning each link to one step in each slot may be reduced to the edge coloring problem which is easily solved for bipartite graphs: the number of colors is always equal to the maximal node degree. Since the maximum number of neighbours is limited by four by default, the number of different steps needed to schedule all the links in the network is also four (if there exists a station with four neighbours).

The steps not distributed uniformly according to the first approach may be used in three different ways. First of all, these steps can be temporary captured by the neighbours to boost their connection in the case of increased traffic load. The neighbours may negotiate a step that is available for both stations and use it in the same slot to temporary increase their connection capacity.

Secondly, the root station provides the network with a schedule for transmitting data over the spanning tree. The trees transmitting schedule comprises two parts: data aggregation and data propagation. The data aggregation part is used to forward data from all stations to the root while the data propagation part following the aggregation is being used to propagate the data from the root to all other stations in the network. The whole tree schedule fits into the superframe, so any data fitting into the steps allocated for the tree will be delivered to the destination during the time interval, limited by the superframe length.

Finally, the central station builds the schedule minimizing the multihop transmissions delay and balancing the network load. If station A needs to create a virtual channel to station B with specific capacity and delay requirements, it sends the request to the root. The root collects the virtual channels creation, modification and deletion requests, rebuilds the schedule for all collected requests and propagates this schedule over the network. Each virtual channel may consist of several different paths, the only requirement is that each link that belongs to the path obtains the same number of steps in the schedule. The station that originated the virtual channel creation request is obliged to primarily use the created channel for the traffic it requested the channel for.

The root station builds spanning tree and multihop transmissions schedules for the duration of the superframe. For example, if the superframe consists of 50 consequent slots, and each slot uses its first 4 steps for the basic access (uniform distribution of the steps between the links), then the schedule will occupy 300 steps. Unless the schedule being updated by the root, all the stations use the same schedule. This means that if station A was scheduled to transmit data of the virtual channel h to station B during the step s then station A will transmit another portion of channel h in the same step s of the current superframe, the next superframe and so on. Because the first steps used in the basic access method are never utilized by the schedules, they will be further ignored to simplify the algorithms and computations. For the example given above, it is supposed that the superframe consists of the 300 sequential steps.

To figure out the nature of the delay in the multihop path, consider the schedule for path $A \rightarrow B \rightarrow C$. Let s_{AB} be the step number assigned to the link $A \rightarrow B$, and s_{BC} be the step number assigned to the link $B \rightarrow C$. If $s_{AB} < s_{BC}$, the link $A \rightarrow B$ is scheduled prior to the link $B \rightarrow C$ and the delay is $s_{BC} - s_{AB}$. Since all the steps are scheduled within a single superframe, the path delay will be limited by the superframe length d. But if the link $B \rightarrow C$ is scheduled prior to the link $A \to B$ then the single message can not be transmitted during one superframe: at the time a message arrives at station B, the station B order to transmit to station C had passed and station B has to wait for the next superframe to send the message over the link $B \to C$. This means that in case when $s_{AB} \geq s_{BC}$ the path delay will be calculated as $s_{BC} + d - s_{AB}$ where d is the superframe length. The occasion when some link is scheduled prior to the previous link will be called **inversion**. If path π has n inversions its delay $\delta \in [nd, (n+1)d]$.

Let the bipartide digraph $G = < V, E >$ where $|V| = N, |E| = M$ and $e_1 = (v, u) \in E \Leftrightarrow e_2 = (u, v) \in E \Leftrightarrow$ [stations v, u are connected] represent the network. Let $C = \{c_{ij}\} \in \{0, 1\}^{N \times N}$ be the vertex incidence matrix.

The superframe consists of T steps, E is the step duration and B is the link capacity. Each step can transmit $E \times B$ at most. The virtual channel request is divided into $k = \lceil r/(E \times B) \rceil$ subrequests. If the virtual channel requested capacity $r > E \times B$ it comprises more then one subrequest. Each subrequest capacity is assumed to be equal to B so there is no need to hold the capacity in the subrequest description. The subrequest can be represented with a couple $< u, v >$ where $u \in V$ is the channel source and $v \in V$ is the channel sink. Let $S = \{< u_1, v_1 >, < u_2, v_2 > \ldots, < u_R, v_R >\}$ be the set of all subrequests and $|S| = R$. In the following it will be assumed that all vertices, edges and subrequests are enumerated. Numeric values $1, 2, \ldots, N$ will be used to address vertices, $1, 2, \ldots, M$ to address edges and $1, 2, \ldots, R$ to address subrequests. If the request r was divided into subrequests $s_1, s_2 \ldots, s_c$, and the sets of steps $T_1, T_2 \ldots T_c$ were assigned for these subrequests the virtual channel will be formed by the union $\bigcup_{i=1}^{c} T_i$ of steps. In the following the same index r may be used to either address the subrequest, or the virtual channel that was created for the request the subrequest r was built for.

Finally, let the matrix $H = \{h_{fe}\} \in \{0, 1, \ldots, T\}^{R \times M}$ be the schedule: if h_{fe} is non-zero, the virtual channel f will be transmitted over the link $e = (i, j)$ in the time step h_{fe}. Otherwise, the link e is not used to transmit the virtual channel.

Let the virtual channel of the subrequest r being transmitted through the path $\phi_r = v_0, v_1, \ldots, v_{N_r} = D_r$ and $e_{m-1} = (k, i), e_m = (i, j)$. Then the delay on the m-th hop may be found as

$$l_{r,m} = \begin{cases} h_{r,e_m} - h_{r,e_{m-1}} & , h_{r,e_m} > h_{r,e_{m-1}} \\ T + h_{r,e_m} - h_{r,e_{m-1}} & , h_{r,e_m} < h_{r,e_{m-1}} \end{cases}$$

The total path delay L_r can be found as a sum $L_r = \sum_{k=1}^{N_r} l_{r,k}$. The problem of minimal delay scheduling can be formulated as a problem of finding the matrix $H : \max_{r \in \{1, 2, \ldots, R\}} L_r \to min$.

4.2 Problem Formulation

The problem of building the multipath delay-optimal schedule will be formulated as an integer linear program (ILP). The program is built with the following variables:

- $x_{i,j}^r$ is a binary variable that is assigned 1 iff the virtual channel r is being transmitted over the link (i,j) where $i,j = \overline{1,N}, r = \overline{1,R}$;
- y_i^r is a binary variable that is assigned 1 iff the node i receives the data of the virtual channel r where $i = \overline{1,N}, r = \overline{1,R}$;
- $u_{i,j}^{r,t}$ is the binary variable that is assigned 1, iff the virtual channel r is being scheduled to transmit over the link (i,j) in the step t where $i,j = \overline{1,N}, r = \overline{1,R}, t = \overline{1,T}$;
- $b_{i,j}^r = \overline{1,T}$ is an integer variable holding the step number in which the link i,j is scheduled to transmit the virtual channel r where $i,j = \overline{1,N}, r = \overline{1,R}$;
- p_i^r is a binary variable that is assigned 1 if the node v is scheduled to transmit the virtual channel r prior to receive it, i.e. the inversion takes place at the node v. Here $i = \overline{1,N}, r = \overline{1,R}$;
- q_i^r is a binary variable that is assigned 1 iff the node v is scheduled to transmit the virtual channel r prior to receive it, i.e. the inversion takes place at the node v. If the node v doesn't transmit the virtual channel r, this variable will be assigned 0. Here $i = \overline{1,N}, r = \overline{1,R}$;
- w^{-r} is the step that is scheduled for the virtual channel r source to transmit it where $r = \overline{1,R}$;
- w^{+r} is the step that is scheduled for the virtual channel r sink to receive it where $r = \overline{1,R}$;
- a^r is an integer variable that is assigned the number of iversions occuring on the virtual channel path. It is obvious that no inversion may take place at the source or at the sink. Here $r = \overline{1,R}$;
- l_r is an integer variable holding the accumulated delay of the virtual channel r where $r = \overline{1,R}$;
- L - the delay upper bound.

Paths unique and existence constraints:

P1: $\forall r = \overline{1,R}\ \forall i = \overline{1,N},\ i \neq S_r, D_r : \sum_{j=1}^{N} x_{i,j}^r = \sum_{k=1}^{N} x_{k,i}^r$ is the flow conservation constraint;

P2: $\forall r = \overline{1,R}\ \forall i = \overline{1,N} : \sum_{j=1}^{N} x_{i,j}^r \leq 1$ is the outgoing flow value constraint. Taking into account **M1** the same constraint will appear to the input flow value.

P3: $\forall r = \overline{1,R} : \sum_{j=1}^{N} x_{S_r,j}^r > \sum_{k=1}^{N} x_{k,S_r}^r$ is the source node flow value, must be positive.

P4: $\forall r = \overline{1,R} : \sum_{k=1}^{N} x_{k,D_r}^r > \sum_{j=1}^{N} x_{D_r,j}^r$ is the sink node flow value, must be negative.

P5: $\forall r = \overline{1, R} \, \forall i, j = \overline{1, N} : x_{i,j}^r \leq c_{ij}$ is a pair of nodes may be scheduled iff the nodes are connected.

P6: $\forall r = \overline{1, R}, \forall i = \overline{1, N} : y_i^r = \sum_{k=1}^{N} x_{k,i}^r$ is the variable is assigned 1 iff any incoming edge of the given node in the given slot is scheduled for transmission.

Steps assignment constraints:

A1: $\forall r = \overline{1, R}, \forall i, j = \overline{1, N} : \sum_{t=1}^{T} u_{i,j}^{r,t} = x_{i,j}^r$ is any link and any request is being scheduled only if the given link was selected to transmit the flow for the given request.

A2: $\forall t = \overline{1, T}, \forall i, j = \overline{1, N} : \sum_{r=1}^{R} u_{i,j}^{r,t} \leq 1$ is any link at any step may transmit one flow at most.

A3: $\forall t = \overline{1, T}, \forall i = \overline{1, N} : \sum_{r=1}^{R} \sum_{j=1}^{N} u_{i,j}^{r,t} \leq 1$ is any node at any step may transmit via one link at most.

A4: $\forall t = \overline{1, T}, \forall i = \overline{1, N} : \sum_{r=1}^{R} \sum_{k=1}^{N} u_{k,i}^{r,t} \leq 1$ is any node at any step may receive the data via one link at most.

Delays computation constraints:

D1: $\forall r = \overline{1, R}, \forall i, j = \overline{1, N} : \sum_{t=1}^{T} t u_{i,j}^{r,t} = b_{i,j}^r$ is the step number in the schedule. May take values $1, 2 \ldots T$.

D2: $\forall r = \overline{1, R}, \forall i = \overline{1, N} : \sum_{k=1}^{N} b_{k,i}^r \leq T p_i^r + \sum_{j=1}^{N} b_{i,j}^r$. If an inversion during transmitting the flow r in the intermediate node i takes place, the variable p_i^r will be assigned 1.

D3: $\forall r = \overline{1, R}, \forall i = \overline{1, N} : q_i^r = y_i^r p_i^r$. The variable is assigned 1 if an inversion at the node i takes place and the flow is being transmitted through that node.

D4: $\forall r = \overline{1, R} : a^r = \sum_{i=1}^{N} q_i^r$ is the upper bound on the number of inversions appearing during transmitting the flow r.

D5: $\forall r = \overline{1, R} : w^{-r} = \sum_{j=1}^{N} b_{S_r, j}^r$ is the step number when the flow source sends it.

D6: $\forall r = \overline{1, R} : w^{+r} = \sum_{k=1}^{N} b_{k, D_r}^r + T a^r$ is the upper bound on the step number when the flow sink receives it.

D7: $\forall r = \overline{1, R} : l^r = w^{+r} - w^{-r}$ is the upper bound on the accumulated delay through the path of the flow r.

D8: $\forall r = \overline{1, R} : L \geq l^r$ for computing L as the upper bound of the set $\{l^r : r = \overline{1, R}\}$.

The constraint **D3** uses the binary variables multiplication. To linearize the constraint it may be equally transformed into the two linear constraints:

D3.1: $\forall r = \overline{1,R}, \forall i = \overline{1,N} : y_i^r + p_i^r - q_i^r \le 1$
D3.2: $\forall r = \overline{1,R}, \forall i = \overline{1,N} : -y_i^r - p_i^r + 2q_i^r \le 0$
Finally, the integer linear minimization problem can be formulated:

$$L_0 = \min_{\substack{P1-P6 \\ A1-A4 \\ D1-D8}} L$$

4.3 Heuristic Scheduling Algorithm

To solve the minimum-delay scheduling problem in an efficient way, the heuristic algorithm was developed. The algorithm works fast enough and is easy to implement. The algorithm is based on Dijkstra routing [18] utilizing a very specific way to find the edges weights.

Let $IsFree(n : Integer, u : Node, v : Node) : Boolean$ be the function returning $True$ if and only if the step n is not scheduled yet for the link (u, v).

The function $NearestStep(s : Integer, u : Node, v : Node)$ returns the step number n that is the closest one to the step s in the future and $IsFree(n, u, v) = True$. If there is no free steps for the given link the function returns NIL. To simplify the computations, all functions using step numbers are assumed to operate with $(\mathbf{mod}\,T)$: step numbers s, $s + T$, $s + 2T$, ... address the same step. Using this assumption, the function $NearestSteps(s, u, v)$ is always expected to return either NIL, or the value greater then its first argument value s.

Finally, let w be the array of weights (this type will be denoted as $Weights$ below), as expected by Dijkstra algorithm. In the developed metric weight has the semantic of the step number at which the node receives the flow: if $w[v] < \infty$ the node v is scheduled to transmit the flow that is currently routed at the step $w[v](\mathbf{mod}\,T)$.

The relaxation $DelayRelax(linkSrc : Node, linkDst : Node, w : Weights)$ function used in the Dijkstra routing algorithm works as follows:

Input: $linkSrc$, $linkDst$, w
Result: the nearest empty slot offset or NIL
$n \leftarrow nearestStep(w[linkSrc], linkSrc, linkDst)$;
if $n \ne NIL$ **then**
| return n
else
| return NIL
end

To describe the algorithm we also need to define several additional functions:

- Let $DijkstraDelayRoute(s : Node, d : Node, w : Weights) : Path$ be the Dijkstra algorithm implementation using $DelayRelax()$ function instead of default $Relax()$, and returning the path with the minimal possible delay.
- The function $Schedule(path : Path, w : Weight, subreq : Subrequest)$ writes into the current schedule that each link $l = (u, v) \in path$ is being set to transmit the flow for the subrequest $subreq$ in the step $w[l.v](\mathbf{mod}\,T)$.

– The function $InitializeWeights(netw : NetworkGraph, source : Node, w : Weights)$ initializes the weights array by writing $w[v] \leftarrow \infty$ for all $v \neq source$, and $w[source] \leftarrow 0$.

To this end, the scheduling algorithm $DijkstraDelaySchedule(R : Requests, G : Network)$ can be defined:

Input: $reqs : Set\ of\ Requests, netw : NetworkGraph$
Data: $subs : Array\ of\ Subrequests$
Data: $path : Path\ in\ the\ network\ graph$
Data: $w : Weights\ array$
$subs \leftarrow BuildSubrequests(r);$
$Sort(subs);$
foreach $subreq \in subs$ **do**

> $InitializeWieghts(netw, subreq.source, w);$
> $path \leftarrow DijkstraDelayRoute(subreq.source, subreq.destination, w);$
> **if** $path \neq NULL$ **then**
> > $Schedule(path, w, subreq);$
>
> **end**

end

The subrequests array created on the first step is being sorted at the step two via function $Sort(s)$ call. The subrequests may be ordered with different strategies, but the default one is to put the subrequests built from the requests with a bigger capacity prior to the subrequests of the requests with a smaller capacity.

The algorithm is simple and achieves good delay results, which was proved by the simulation. A useful side effect is the increased load balancing: a large flow may be scheduled over several distinct paths, so when the network load increases the paths become longer, covering the nodes those were not used before. The simulation also proves that the under very heavy load the station service ratio may even be very close to 1.

4.4   Simulation

To analyze the performance of the $DijkstraDelaySchedule()$ algorithm the simulation experiment was built. The simulation was performed over the network sizes ranging from 10 to 1000 nodes with randomly generated virtual channels requests uniformly distributed over the whole network. During the experiment the data rate was limited by 1666 Mbps. As the first four steps in each 10-step slot were used for the uniform neighbours access (as described above), multi-hop scheduling algorithm was able to use 60 % of the time and consequently ≈1000 Mbps.

The number of virtual channels requests is varied from 1 to 600 for the network consisting of 100 stations, each of them required 100 Mbps capacity (consequently the requested network throughput varied from 100 Mbps to 60 Gbps). The average path length is 3.5 hops in all the experiments. One may observe that under these conditions the total required network capacity is varied up to

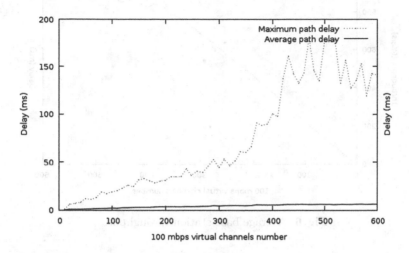

Fig. 4. Average and maximum path delay in ms against the number of 100 mbit 3-hops virtual channels requests.

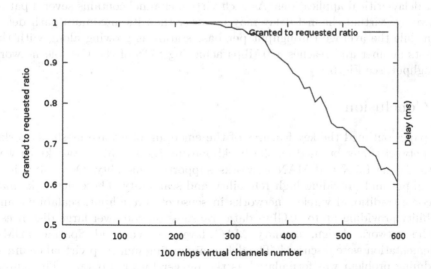

Fig. 5. Average granted capacity to the requested ratio.

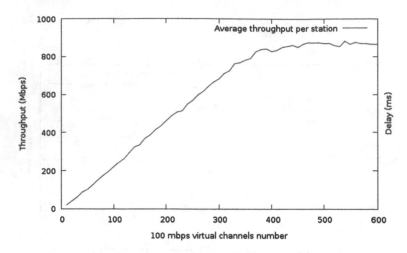

Fig. 6. Average base station throughput.

$2100 = 600 \times 3.5$ Mbps per station which is more then 1000 Mbps (the capacity the station may provide for virtual channels transmission). It is possible to schedule most of the virtual channels till the number of requests exceeds 300. Afterwards the ratio of the granted capacity to the requested one downgrades, see Fig. 5. The average and maximum delays are illustrated in Fig. 4. It can be noted that while the maximum delay of a single path may become rather large, the average path delay is still small enough to provide sufficient QoS for VoIP and other delay-critical applications. As each virtual channel contains several paths, the average virtual channel delay may be lower then its maximum path delay. Meanwhile the average throughput per base station is growing along with the requests number and reaches 870 Mbps achieving 87 % of the possible network throughput, see Fig. 6.

5   Conclusion

The paper outlined the key features of the emerging mm-wave mesh networks. The networks may be used as the backbones in 4G and 5G networks as well as standalone LAN and MAN networks supporting mobility, QoS, ultra-high throughput and providing high reliability and scalability. These networks may overcome traditional wireless networks in sense of throughput, scalability and reliability providing up to 10 Gbps data transmission rate over large distances.

The network design, primary MAC layer services and Spatial TDMA implementation were discussed. The delay minimizing multihop virtual channels scheduling problem was formulated as the integer linear program. The simple heuristic scheduling method based on Dijkstra algorithm allowing to route each virtual channel via multiple paths and minimizing the delays was presented. Finally, the simulation results proving the algorithm allows providing sufficient delays under very heavy load in large networks were given.

References

1. Vishnevsky, V., Semenova, O.: Polling Systems: Theory and Applications for Broadband Wireless Networks, p. 317. LAMBERT Academic Publishing, London (2012)
2. Part 11: Wireless LAN Medium Access Control (MAC) and Physical Layer (PHY) specifications. IEEE P802.11-2012. IEEE (2012)
3. Ecma/TC48/2010/025. ECMA-387 2nd Edition: High Rate 60GHz PHY, MAC and HDMI PAL. Whitepaper. Ecma International (2010)
4. Pitsiladis, G.T., Panagopoulos, A.D., Constantinou, P.: Improving connectivity in indoor millimeter wave wireless networks using diversity reception. In: 6th European Conference on Antennas and Propagation (EUCAP), pp. 510–514, Prague (2012)
5. Jabbar, J.P., Rohrer, V.S., Frost, J.P.G.: Sterbenz.: Survivable millimeter-wave mesh networks. Comput. Commun. 34(16), 1942–1955 (2011). ISSN 0140–3664
6. Xiao, S.-Q., et al.: Millimeter Wave Technology in Wireless PAN, LAN, and MAN. CRC Press, Boca Raton (2008)
7. Nelson, R., Kleinrock, L.: Spatial TDMA: a collision-free multihop channel access protocol. IEEE Trans. Commun. 33, 934–944 (1985)
8. Ronasi, K., Wong, V.W.S., Gopalakrishnan, S.: Distributed scheduling in multihop wireless networks with maxmin fairness provisioning. IEEE Trans. Wireless Commun. 11(5), 1753–1763 (2012)
9. Chaporkar, P., Kar, K., Luo, X., Sarkar, S.: Throughput and fairness guarantees through maximal scheduling in wireless networks. IEEE Trans. Inf. Theory 54(2), 572–594 (2008)
10. Djukic, P., Valaee, S.: Link scheduling for minimum delay in spatial re-use TDMA. In: 26th IEEE International Conference on Computer Communications (INFOCOM'2007), pp. 28–36 (2007)
11. Djukic, P., Valaee, S.: Delay aware link scheduling for multi-hop tdma wireless networks. IEEE/ACM Trans. Networking 17(3), 870–883 (2009)
12. Vishnevsky, V., Larionov, A.: Novel approach for scheduling in STDMA for high-throughput backbone wireless mesh networks operating within 60-80 GHz. In: International Conference on Advances in Mesh Networks (MESH'2010), Venice/Mestre, Italy (2010)
13. Vishnevsky, V., Larionov, A.: GWM-MAC protocol for high-throughput backbone wireless mesh networks operating within 60-80 GHz. In: IEEE International Congress on Ultra Modern Telecommunications and Control Systems (ICUMT'2010), Moscow (2010)
14. IEEE Standard for Local and metropolian area networks. Media Access Control (MAC) Bridges, IEEE 802.1D-2004, IEEE Std. (2004)
15. Moy, J.: OSPF Version 2, IETF RFC 2328 (1998)
16. Clausen, T., Jacquet, P.: Optimized Link State Routing Protocol (OLSR), IETF RFC 3626 (2003)
17. Perkins, C., Belding-Royer, E., Das, S.: Ad hoc On-Demand Distance Vector (AODV) Routing. RFC 3561, IETF (2003)
18. Cormen, T.H., Leiserson, C.E., Rivest, R.L., Stein, C.: Introduction to Algorithms, 3rd edn. MIT Press, Cambridge (2009)

A Survey of Congestion Control Mechanisms in Linux TCP

Christian Callegari[1], Stefano Giordano[1], Michele Pagano[1(✉)],
and Teresa Pepe[2]

[1] University of Pisa, Pisa, Italy
[2] Ericsson Research Italy, Pisa, Italy
{c.callegari,s.giordano,m.pagano}@iet.unipi.it, teresa.pepe@ericsson.com

Abstract. The Transmission Control Protocol (TCP) is used by the vast majority of Internet applications. Since its introduction in the 1970s, a lot of variants have been proposed to cope with the different network conditions we can have (e.g., wired networks, wireless networks, satellite links) and nowadays Linux OS includes 13 different TCP variants. The aim of this paper is to provide a complete survey of the different congestion control mechanisms used by the variants of the TCP implemented in the Linux Kernel 2.6.x.

1 Introduction

The Transmission Control Protocol (TCP) is a connection-oriented transport protocol that provides a reliable byte-stream data transfer service between pairs of processes, currently used by the vast majority of Internet applications.

The basic idea of TCP congestion control is that the rate of ACKs returned by the receiver determines the rate at which the sender can transmit data. The TCP uses several algorithms for congestion control and each of them controls the sending rate by manipulating a congestion window ($cwnd$) that limits the number of outstanding unacknowledged bytes that are allowed at any time. The $cwnd$ size is increased or decreased depending on the perceived congestion level. The sender can transmit up to the transmission window, which is the minimum of the $cwnd$ (flow control imposed by the sender) and the *advertised window* (flow control imposed by the receiver). All the modern implementations of the TCP contain at least four intertwined algorithms: Slow Start, Congestion Avoidance, Fast Retransmit, and Fast Recovery [1]. Moreover, the more recent implementations also include some other mechanisms and algorithms.

Although some experimental works on Linux TCP performance have been published (see, for instance [2] and references therein), to the best of our knowledge, a detailed survey of the different congestion control mechanisms in Linux TCP is not reported anywhere in the literature. The rest of the paper presents a brief description of all the TCP variants implemented in the Linux kernel 2.6.x, highlighting how they react to the *perceived congestion*, assuming that the reader is familiar with the standard congestion control in TCP [1].

V. Vishnevsky et al. (Eds.): DCCN 2013, CCIS 279, pp. 28–42, 2014.
DOI: 10.1007/978-3-319-05209-0_3, © Springer International Publishing Switzerland 2014

2   TCP Variants

2.1   TCP Reno

Reno version of the TCP [3] is nowadays considered the "standard" TCP and referred to as "TCP" in the following. Indeed, it basically implements the four *classical* congestion control mechanisms of TCP [1].

In a nutshell it implements Slow Start and Congestion Avoidance as two different phases during transmission. During the Slow Start phase the *cwnd* is initialized to one MSS and since there are no packet losses, TCP Reno continues to increase the window size by one at every incoming ACK, resulting in an exponential growth of the *cwnd* per RTT (to be noted that, when not differently specified, the *cwnd* increment is measured in MSS). If the *cwnd* exceeds the value of *ssthresh*, TCP Reno enters the Congestion Avoidance phase, which corresponds to an Additive Increase Multiplicative Decrease (AIMD) procedure. Roughly speaking, the *cwnd* is increased by one MSS per RTT resulting in a linear increasing. When the TCP transfer encounters congestion (packet losses are detected), the window is decreased. TCP Reno distinguishes between two levels of congestion: heavy and mild. If a loss is detected because the timeout has expired, it is assumed that there is an heavy congestion. In this case Reno restarts from Slow Start. Otherwise, if a loss is detected by receiving three duplicate ACKs, the congestion is considered mild and the Fast Retransmit/Fast Recovery algorithm is performed.

An example of the behavior of the *cwnd* is depicted in Fig. 1(a) (all the Figures in the section are related to experimental results, obtained by simulating a TCP connection lasting 300 s over a network described by a one-way delay of 150 ms and a loss probability of 0.0005 %). To be noted that, from the figure it is not possible to analyze the Slow Start phase. This is simply due to the time-scale and to the connection parameters.

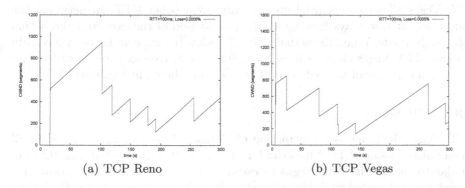

(a) TCP Reno   (b) TCP Vegas

Fig. 1. *Cwnd* behavior

2.2   TCP Vegas

TCP Vegas was proposed in 1994 by Brakmo, O'Malley, and Peterson in [4]. It is based on modifications of TCP Reno. Unlike TCP Reno, it is not only able to detect congestion when it occurs, but also allows the sender to avoid congestion using a more sophisticated bandwidth estimation scheme. TCP Vegas monitors changes in the flow rate (and RTT) to predict congestion before losses occur. The idea is as follows: when the network is not congested, the actual flow rate will be close to the expected flow rate; otherwise, it will be smaller than the expected flow rate. Thus, using this difference in flow rates, TCP Vegas can estimate the congestion level in the network and update the window size accordingly. In more detail, the algorithm evaluates the expected rate

$$Expected = cwnd/RTT_{base}, \tag{1}$$

where RTT_{base} is the minimum experimented RTT, and the actual rate

$$Actual = cwnd/RTT. \tag{2}$$

Based on the difference of the rates

$$Diff = Expected - Actual, \tag{3}$$

the source updates its window size as follows:

$$cwnd = \begin{cases} cwnd + 1 & \text{if } Diff < \alpha \\ cwnd & \text{if } \alpha < Diff < \beta \\ cwnd - 1 & \text{if } Diff > \beta \end{cases} . \tag{4}$$

Typical values for α and β are 1 or 2 and 3 or 4 respectively. A modified version of the previously described congestion detection algorithm is incorporated into the Slow Start phase. To be able to avoid congestion during Slow Start, TCP Vegas allows exponential growth only every other RTT. In between, the congestion window stays fixed so a valid comparison of the expected and actual rate can be made. When the actual rate falls below the expected rate by a certain amount, TCP Vegas changes from Slow Start mode to linear increase/decrease mode. An example of the behavior of the *cwnd* is shown in Fig. 1(b).

2.3   TCP Veno

TCP Veno (the name is a contraction of Vegas + Reno) was introduced in [5] as an enhancement of TCP Reno and TCP Vegas. It makes use of a mechanism similar to the one in TCP Vegas to estimate the state of the connection. If a packet loss is detected while the connection is in the congestion state, TCP Veno assumes that the loss is due to congestion; otherwise, it assumes that the loss is "random".

TCP Veno only refines the additive increase, multiplicative decrease of TCP Reno; all other parts remain intact. In more detail, during the additive increase phase the sender measures the expected and actual rates and $Diff$ as in TCP Vegas. When $RTT > RTT_{base}$, there is a bottleneck link where the packets of the connection accumulate. To evaluate the network status, TCP Veno uses the queue backlog N (measured in packets) and defined by the following equation:

$$RTT = RTT_{base} + N/Actual. \tag{5}$$

If $N < \beta$ (available bandwidth not fully utilized), where β is a suitable threshold, at each new received ACK, we set $cwnd = cwnd + 1/cwnd$; if $N \geq \beta$ (available bandwidth fully utilized), at every other received ACK, we set $cwnd = cwnd + 1/cwnd$. The first part is the same as the algorithm in TCP Reno. In the second part, the backlogged packets N in the buffer exceed β and TCP Veno increases the $cwnd$ by one every two RTT, so that the connection can stay in this operating region longer.

Regarding the multiplicative decrease mechanism, TCP Veno only modifies the way the $ssthresh$ is updated when a loss is detected by means of three duplicate ACKs (Fast Recovery case). In more detail, when a packet is lost, if $N < \beta$ TCP Veno assumes that it is a random loss and fixes $ssthresh = cwnd \cdot (\frac{4}{5})$. Otherwise, if $N \geq \beta$ TCP Veno assumes that there is congestion and the $cwnd$ is modified using TCP Reno rule $(cwnd = cwnd/2)$. An example of the behavior of the $cwnd$ is shown in Fig. 2(a).

2.4   TCP Westwood

TCP Westwood [6] is a sender-side modification of the TCP intended to better handle large bandwidth-delay product paths (large pipes), with potential packet losses due to transmission or other errors (leaky pipes), and with dynamic load (dynamic pipes). The innovative idea is to continuously measure, at the TCP sender side, the bandwidth (BW) used by the connection via monitoring the rate of returning ACKs. An "Eligible Rate" is estimated and used by the sender to update $ssthresh$ and $cwnd$ after a congestion episode, that is after three

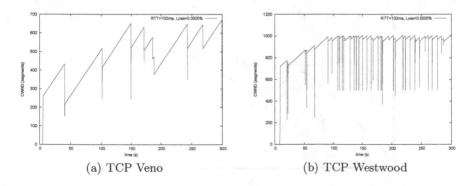

(a) TCP Veno   (b) TCP Westwood

Fig. 2. *Cwnd* behavior

duplicate ACKs or after a timeout. If a loss is detected because the timeout has expired

$$ssthresh = \frac{BW \cdot RTT_{base}}{Segment\_size} \quad cwnd = 1 \tag{6}$$

where $Segment\_size$ is the length of the current TCP segment measured in bits. Otherwise, if a loss is detected by receiving three duplicate ACKs

$$ssthresh = \frac{BW \cdot RTT_{base}}{Segment\_size} \quad cwnd = \min(cwnd, ssthresh) \tag{7}$$

The proposed mechanism is particularly effective over wireless links, where sporadic losses due to radio channel problems are often misinterpreted as a symptom of congestion by "classical" TCP schemes and thus lead to an unnecessary window reduction. It is worth noticing that all the other mechanisms are the same as for TCP Reno. The behavior of the $cwnd$ is depicted in Fig. 2(b).

2.5 TCP BIC

Binary Increase Congestion Control (BIC) [7] is an implementation of the TCP with an optimized congestion control algorithm for high speed networks with high latency (long fat networks). TCP BIC algorithm consists of three parts:

- Additive increase phase
- Binary search increase phase
- Maximum probing phase

In [7], the following parameters are defined (see Fig. 3):

- Minimum Window Size (mWS): $cwnd$ size, which ensures no losses
- Maximum Window Size (MWS): maximum permitted $cwnd$ size
- Target Window Size (TWS): midpoint between mWS and MWS
- S_{max} and S_{min}: two given constants

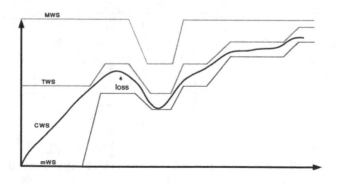

Fig. 3. TCP BIC parameters

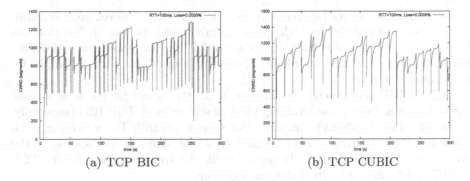

(a) TCP BIC   (b) TCP CUBIC

Fig. 4. *Cwnd* behavior

Initially TCP BIC increments the *cwnd* according to the Slow Start algorithm as in TCP Reno. Packet losses are taken as an indication for congestion. After a loss event, TCP BIC reduces its window by a multiplicative factor, β (usually $\beta = 0.8$). The window size just before and after the reduction are set to MWS and mWS respectively. While the distance between TWS and mWS is larger than S_{max}, TCP BIC increments the *cwnd* linearly by a quantity S_{max} (*additive increase*). This non-aggressive phase ensures faster convergence and RTT-fairness. If TCP BIC does not get packet losses at the updated window size, the latter becomes the new mWS; otherwise, it becomes the new MWS.

Then, the algorithm enters the *binary search increase* phase. The *cwnd* setting is viewed as a searching problem, which employs a form of binary search algorithm: the algorithm repeatedly computes TWS and sets the *cwnd* to the midpoint. When a packet loss occurs, the midpoint is taken as the new MWS, otherwise the midpoint becomes the new mWS. So the growth function after a window reduction will be most likely a linear one followed by a logarithmic one.

If the window grows over the maximum (*cwnd* reaches MWS), a new maximum must be found. TCP BIC enters a new phase called *max probing*, characterized by a window growth function exactly symmetric to those used in additive increase and binary search. In more detail, at first it uses the inverse of binary search (it is logarithmic so its reciprocal will be exponential) and then additive increase. During max probing, the window grows slowly initially to find the new maximum nearby, and after some time of slow growth, if it does not find the new maximum, it switches to a faster increase, which consists of an additive increase phase. During this phase *cwnd* is incremented by a large fixed increment. The good performance of TCP BIC come from the slow increase around MWS and linear increase during additive increase and max probing. Figure 4(a) presents the overall behavior of the *cwnd*.

2.6   TCP CUBIC

In 2005 Rhee and Xu proposed TCP CUBIC [8], an enhanced version of their previous TCP BIC variant. It simplifies the TCP BIC window control and improves

TCP-friendliness and RTT-fairness. Although TCP BIC achieves good scalability, fairness, and stability in high speed environments, its growth function can still be too aggressive, especially under short RTT or low speed networks. Furthermore, the several different phases of window control add a lot of complexity in analyzing the protocol.

TCP CUBIC acts exactly the same as TCP BIC, apart for the new window growth function that, while retaining most of strengths of TCP BIC (especially, its stability and scalability), simplifies the window control. The window growth function of TCP CUBIC is a cubic function, whose shape is very similar to the growth function of TCP BIC. In more detail, the congestion window of TCP CUBIC is determined by the following function:

$$cwnd = C(t - K)^3 + MWS, \tag{8}$$

where C is a scaling factor, t is the elapsed time from the last window reduction, MWS is the window size just before the last window reduction, and $K = \sqrt[3]{\dfrac{MWS \cdot \beta}{C}}$, where β is a multiplication decrease factor applied for window reduction at the time of loss event (i.e., the window reduces to $\beta \cdot MWS$).

Similarly to the TCP BIC case, the window grows very fast upon a window reduction, but as it gets closer to MWS, its growth slows down. Around MWS, the window increment becomes almost zero. Above that, TCP CUBIC starts probing for more bandwidth and the window grows slowly initially, accelerating its growth as it moves away from MWS. This slow growth around MWS enhances the stability of the protocol, and increases the utilization of the network while the fast growth away from MWS ensures the scalability of the protocol. Moreover, to enhance the fairness and stability, the window increment per second is limited to S_{max}. Finally, to assure that the $cwnd$ growth is not less than that offered by the standard TCP (i.e., TCP Reno), the protocol also computes the simulated window size of the standard TCP as:

$$cwnd_{TCP} = MWS \cdot \beta + 3\frac{1 - \beta}{1 + \beta}\frac{t}{RTT}, \tag{9}$$

where t is the elapsed time from the last window reduction. If $cwnd_{TCP}$ is larger than $cwnd$, TCP CUBIC sets $cwnd = cwnd_{TCP}$. In Fig. 4(b) we show the overall behavior of the congestion window.

2.7   HSTCP

High-Speed TCP (HSTCP) has been developed by S. Floyd in [9] to overcome the limitations of the standard TCP congestion control in networks with a large bandwidth delay product.

HSTCP sets $cwnd$ using the rules reported below. It uses two new increasing and decreasing functions, $a(cwnd)$ and $b(cwnd)$, respectively. When an ACK is received (in Congestion Avoidance), the window is increased as follows:

$$cwnd = cwnd + \frac{a(cwnd)}{cwnd}. \tag{10}$$

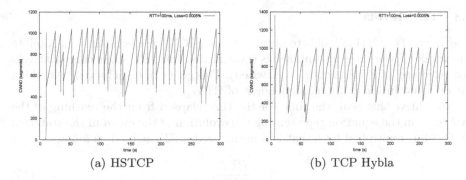

(a) HSTCP   (b) TCP Hybla

Fig. 5. $Cwnd$ behavior

When a loss is detected through triple duplicate ACKs, the window is decreased using the following function:

$$cwnd = cwnd(1 - b(cwnd)). \tag{11}$$

When the congestion window is small, HSTCP behaves exactly like the standard TCP so $a(cwnd)$ is 1 and $b(cwnd)$ is 0.5. When the TCP congestion window is beyond a given threshold, $a(cwnd)$ and $b(cwnd)$ become functions of the current window size. The idea is to modify the response function of the TCP. In more detail, for a given large $cwnd W1$ and a given loss probability HP, the following response function (packets per round) has to be respected:

$$W1 = \frac{\sqrt{\frac{a(cwnd)(2-b(cwnd))}{2b(cwnd)}}}{\sqrt{HP}}. \tag{12}$$

Thus, the two functions are defined by:

$$a(cwnd) = \frac{W1^2 \cdot HP \cdot 2b(cwnd)}{2 - b(cwnd)} \tag{13}$$

$$b(cwnd) = \frac{(HD - 0.5)(log(cwnd) - log(W))}{log(W1) - log(W)} + 0.5 \tag{14}$$

where HD is such that $b(W1) = HD$. In this region, as the congestion window increases, the value of $a(cwnd)$ increases and the value of $b(cwnd)$ decreases. This means that HSTCP window will grow faster than the standard TCP and also recover from losses more quickly. This behavior allows HSTCP to quickly utilize available bandwidth in networks with large bandwidth delay products.

Moreover, to better reach this goal, HSTCP also makes use of a modified Slow Start phase: at first the $cwnd$ increases as in the standard Slow Start, but when it exceeds a given threshold (max_thresh) the $cwnd$ is set equal to $cwnd + MSS/K$ at each received ACK, where $K = 2 \cdot cwnd/max\_thresh$. Figure 5(a) depicts an example of the resulting behavior of the $cwnd$.

2.8 TCP Hybla

The aim of TCP Hybla [10] is to obtain, for long RTT connections (e.g., satellite and wireless), the same instantaneous transmission rate, $B(t)$, of a comparatively fast reference TCP connection (e.g., wired). This goal can be achieved by making the $cwnd$ evolution in time independent of RTT.

To achieve this goal, the time (or the time elapsed from the reaching of the $ssthresh$) in the equation representing the evolution of the $cwnd$ in the standard TCP [10] is multiplied by ρ that is a normalized RTT, defined as follows:

$$\rho = \frac{RTT}{RTT_0}, \tag{15}$$

where RTT_0 is the RTT of the reference connection to which we aim to equalize TCP Hybla performance. So the $cwnd$ is evaluated as follows:

$$cwnd = \begin{cases} \rho 2^{\frac{\rho t}{RTT}} & 0 \leq t < t_{\gamma,0} \quad \text{Slow Start} \\ \rho \left[\rho \dfrac{t - t_{\gamma,0}}{RTT} + \gamma \right] & t \geq t_{\gamma,0} \quad \text{Congestion Avoidance} \end{cases} \tag{16}$$

where γ is the $ssthresh$ and $t_{\gamma,0}$, is the switching time, defined as the time at which the $cwnd$ reaches the value $\rho\gamma$, and it is the same for every RTT, being $t_{\gamma,0} = RTT_0 \cdot log_2\gamma$. From (16) the segment transmission rate is given by

$$B = \begin{cases} \dfrac{2^{\frac{t}{RTT_0}}}{RTT_0} & 0 \leq t < t_{\gamma,0} \quad \text{Slow Start} \\ \dfrac{1}{RTT_0} \left[\dfrac{t - t_{\gamma,0}}{RTT_0} + \gamma \right] & t \geq t_{\gamma,0} \quad \text{Congestion Avoidance} \end{cases} \tag{17}$$

which is clearly independent of RTT and equal to the segment transmission rate of the reference TCP connection. Figure 5(b) reports the behavior of the $cwnd$ in TCP Hybla.

2.9 Scalable TCP

Scalable TCP [11] is a modified version of the traditional TCP that improves the TCP performance in high-speed wide area networks. Scalable TCP is based on simple modifications of the algorithm used to update the $cwnd$, during the Congestion Avoidance phase only. In more detail the algorithm works as follows:

- for each ACK received not in loss recovery $cwnd = cwnd + 0.01$
- when congestion is detected (on each loss event) $cwnd = 0.875 \cdot cwnd$

To be noted that Scalable TCP algorithm is only used for windows above a certain threshold, obtained by studying the intersection of the TCP Reno response curve and the Scalable TCP response curve. The behavior of the $cwnd$ is reported in Fig. 6(a).

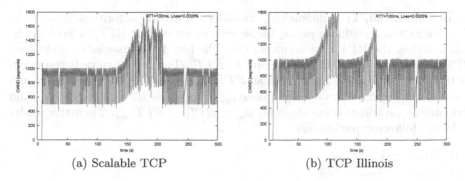

(a) Scalable TCP   (b) TCP Illinois

Fig. 6. *Cwnd* behavior

2.10   TCP Illinois

TCP Illinois is a variant of the TCP congestion control particularly indicated for high-speed long-distance networks. It was introduced by S. Liu et al. in [12] to achieve high throughput and allocate network resources fairly. It is a loss-delay based algorithm, which uses packet loss as the primary congestion signal to determine the direction of window size change, and uses queuing delay as the secondary congestion signal to adjust the pace of window size change. Similarly to the standard TCP, TCP Illinois increases the *cwnd* for each received ACK using a decreasing function α and decreases the *cwnd* using an increasing function β.

The key idea is the following: when the average queueing delay d_a is small, the sender assumes that the congestion is not imminent and sets a large α and small β; when d_a is large, the sender assumes that the congestion is imminent and sets a small α and large β.

A possible choice, suggested in [12], for α and β is:

$$\alpha = f_1(d_a) = \begin{cases} \alpha_{max} & \text{if } d_a \leq d_1 \\ \dfrac{k_1}{k_2 + d_a} & \text{otherwise} \end{cases} \tag{18}$$

$$\beta = f_2(d_a) = \begin{cases} \beta_{min} & \text{if } d_a \leq d_2 \\ k_3 + k_4 d_a & \text{if } d_2 < d_a < d_3 \\ \beta_{max} & \text{otherwise} \end{cases} \tag{19}$$

$f_1(\cdot)$ and $f_2(\cdot)$ are continuous functions and thus $\alpha_{max} = \frac{k_1}{k_2+d_1}$, $\beta_{min} = k_3 + k_4$, and $\beta_{max} = k_3 + k_4 d_3$. Suppose d_m is the maximum average queueing delay and let $\alpha_{min} = f_1(d_m)$, then we also have $\alpha_{min} = \frac{k_1}{k_2+d_m}$. From these conditions we have

$$\begin{array}{ll} k_1 = \dfrac{(d_m - d_1)\alpha_{min}\alpha_{max}}{\alpha_{max} - \alpha_{min}} & k_2 = \dfrac{(d_m - d_1)\alpha_{min}}{\alpha_{max} - \alpha_{min}} - d_1 \\ k_3 = \dfrac{\beta_{min}d_3 - \beta_{max}d_2}{d_3 - d_2} & k_4 = \dfrac{\beta_{max} - \beta_{min}}{d_3 - d_2} \end{array} \tag{20}$$

In more detail, TCP Illinois only modifies the AIMD algorithm. That is, in the Congestion Avoidance phase, the sender measures the RTT for each ACK, and averages the RTT measurements over the last W acknowledgements (one RTT interval) to derive the average RTT, RTT_a. The sender records the maximum and minimum RTT ever seen as RTT_{max} and RTT_{base}, respectively, and computes the maximum (average) queueing delay $d_m = RTT_{max} - RTT_{base}$ and the current (average) queueing delay $d_a = RTT_a - RTT_{base}$. Then the sender picks the following parameters:

- $0 < \alpha_{min} \le 1 \le \alpha_{max}$
- $0 < \beta_{min} \le \beta_{max} \le 1/2$
- $cwnd_{thresh} > 0$
- $0 \le \eta_1 < 1$
- $0 \le \eta_2 \le \eta_3 \le 1$

The sender sets $d_i = \eta_i \cdot d_m$ $(i = 1, 2, 3)$, computes k_i $(i = 1, 2, 3, 4)$, α, and β. Thus, if $cwnd < cwnd_{thresh}$ TCP Illinois sets $\alpha = 1$ and $\beta = 1/2$. Otherwise the k_i values are updated if RTT_{max} or RTT_{base} is updated, while the α and β values are updated once per RTT and $cwnd = cwnd + \alpha/cwnd$ for each ACK or $cwnd = cwnd - \beta cwnd$, if in the last RTT there is packet loss detected through triple duplicate ACK. Once there is a timeout, the sender sets the $ssthresh$ to $cwnd/2$, enters the Slow Start phase, and resets $\alpha = 1$ and $\beta = 1/2$; then, α and β values are unchanged until one RTT after the Slow Start phase ends.

As a results, TCP Illinois increases the throughput much more quickly than the TCP, when congestion is far, and increases the throughput very slowly, when congestion is imminent. As a result, the window curve is concave and the average throughput achieved is much larger than the standard TCP. An example of the resulting $cwnd$ behavior is shown in Fig. 6(b).

2.11   TCP YeAH

TCP YeAH (the name stands for Yet Another Highspeed TCP) is a new version of the TCP introduced in [13]. It allows an efficient use of available bandwidth without inducing "network stress".

TCP YeAH can operate into two different modes: "fast" and "slow" mode. During the "fast" mode it increments the $cwnd$ according to an aggressive rule (typically STPC rule, since it is very simple to implement). Instead, in the "slow" mode, it acts as TCP Reno. The working state is decided according to the estimated number of packets in the bottleneck queue. Let RTT_{base} be the minimum RTT measured by the sender (i.e., an estimate of the propagation delay) and RTT_{min} the minimum RTT estimated in the current data window of $cwnd$ packets. The current estimated queuing delay is $RTT_{queue} = RTT_{min} - RTT_{base}$. From RTT_{queue} it is possible to infer the number of packets enqueued by the flow as:

$$Q = RTT_{queue} \cdot G = RTT_{queue} \cdot \left(\frac{cwnd}{RTT_{min}} \right), \tag{21}$$

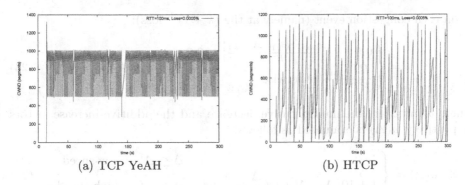

(a) TCP YeAH   (b) HTCP

Fig. 7. *Cwnd* behavior

where G is the goodput. It is also possible to evaluate the ratio between the queuing and the propagation delays, $L = \dfrac{RTT_{queue}}{RTT_{base}}$ that indicates the network congestion level. If $Q < Q_{max}$ and $L < \dfrac{1}{\varphi}$, the algorithm is in the "fast" mode, otherwise it is in the "slow" mode. Q_{max} and φ are two tunable parameters; Q_{max} is the maximum number of packets a single flow is allowed to keep into the buffers and $\dfrac{1}{\varphi}$ is the maximum level of buffer congestion with respect to the bandwidth-delay product. During the "slow" mode, a precautionary decongestion algorithm is implemented: whenever $Q > Q_{max}$, the *cwnd* is diminished by Q and *ssthresh* set to *cwnd*/2. Since RTT_{min} is computed once per RTT, the decongestion granularity is one RTT. Finally it is important to highlight that the use of the SACK option is mandatory in TCP YeAH. We report the behavior of the *cwnd* in Fig. 7(a).

2.12 H-TCP

H-TCP was presented in [14] as an alternative protocol, suitable for deployment in high-speed and long distance networks. In [14] H-TCP is shown to be fair when deployed in homogeneous networks, to be friendly with conventional TCP sources, to be able to utilize network resources efficiently, and to be rapid to respond to changes in available bandwidth. Moreover, when deployed in conventional networks, H-TCP behaves as a conventional TCP variant.

The innovative idea of this approach is that the increase rule is designed to maintain symmetry in the manner in which competing flows acquire available bandwidth. This is achieved by using an increasing function that is a function of the time elapsed since the last packet drop experienced.

In more detail, the *cwnd* is evaluated as follows:

– on each ACK:

$$a = 2(1 - b)a(\Delta) \qquad cwnd = cwnd + a/cwnd \qquad (22)$$

– on each congestion event (namely at the $(k+1)^{\text{th}}$ loss):

$$b = \begin{cases} 0.5 & \left| \frac{B^-(k+1) - B^-(k)}{B^-(k)} \right| > 0.2 \\ \dfrac{RTT_{min}}{RTT_{max}} & \text{otherwise} \end{cases} \qquad cwnd = b \cdot cwnd \qquad (23)$$

where b and a are the multiplicative decrease and the additive increase factors respectively. $a(\Delta)$ is evaluated as follows:

$$a(\Delta) = \begin{cases} 1 & \Delta \leq \Delta^L \quad \text{low-speed} \\ 1 + 10(\Delta \quad \Delta^L) \mid (\dfrac{\Delta - \Delta^L}{2})^2 & \Delta > \Delta^L \quad \text{high-speed} \end{cases} \qquad (24)$$

where Δ is the time in seconds since the last congestion event, $\Delta^L = 1\,\text{s}$ (typical value), $\dfrac{RTT_{min}}{RTT_{max}}$ is the ratio of minimum and maximum RTT experienced by the flow, and $B^-(k)$ is the throughput achieved immediately before the kth loss. Optionally, by scaling a with RTT, the increase rate can became effectively invariant with RTT, in which case convergence time is also invariant with RTT. RTT unfairness between competing flows is also mitigated when RTT scaling is employed. The behavior of the *cwnd* is depicted in Fig. 7(b).

2.13   TCP-LP

In [15] a new variant of the TCP, TCP Low-Priority (TCP-LP), is proposed to provide low priority service in the presence of TCP traffic. To achieve this aim it is necessary for TCP-LP to infer congestion earlier than the TCP.

TCP-LP evaluates one-way packet delay and employs a simple delay threshold-based method for early inference of congestion.

Denote d_i as the one-way delay for a packet with sequence number i, and d_{min} and d_{max} as the minimum and maximum one-way packet delays experienced through the connection lifetime. The algorithm computes the smoothed one-way delay sd_i as follows:

$$sd_i = (1 - \gamma)sd_{i-1} + \gamma d_i \,, \qquad (25)$$

where γ denotes the delay smoothing parameter.

Comparing this value with a threshold, within the range of the minimum and maximum delay, TCP-LP is able to early detect congestion. More specifically the congestion indication condition is

$$sd_i > d_{min} + (d_{max} - d_{min})\delta \,, \qquad (26)$$

where $0 < \delta < 1$ denotes the threshold parameter.

When the congestion is detected, TCP-LP halves the *cwnd* and goes into the "inference detected" state by starting a *inference timeout timer*. During this period, TCP-LP only observes responses from the network, without increasing its *cwnd*. If before the timer expires another congestion alarm arrives, the *cwnd* is reduced to MSS. Otherwise TCP-LP goes back into the Congestion Avoidance phase and the *cwnd* is increased of one MSS for each RTT. In order to prevent TCP-LP from overreacting to burst of congestion indicating packets, TCP-LP ignores succeeding congestion indications if the source has reacted to a previous delay-based congestion indication or to a dropped packet in the last RTT. Finally, the minimum *cwnd* for TCP-LP flows in the *inference phase* is set to 1. In this way, TCP-LP flows conservatively ensure that an excess bandwidth of at least one packet per RTT is available before probing for additional bandwidth.

It is worth noticing that the Slow Start phase is the same as in TCP Reno. In Fig. 8 we show an example of the behavior of the *cwnd* for TCP-LP.

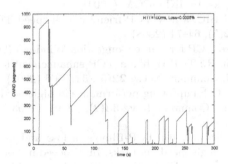

Fig. 8. *Cwnd* behavior (TCP-LP)

3 Conclusions

The TCP is for sure the most used transport layer protocol in the Internet. In the years several variants have been proposed to overcome the limitations that the standard TCP has in particular scenarios, such as wireless or satellite networks. As a results, the current kernel of the Linux OS (version 2.6.x) includes 13 different versions, going from the standard TCP (TCP Reno) and its improved version (TCP Vegas), to the variants for wireless networks (TCP Veno and TCP Westwood), high-speed networks (TCP BIC, TCP CUBIC, HSTCP, TCP Hybla, TCP Illinois, Scalable TCP, TCP YeAH, and HTCP), and also a low-priority version (TCP-LP).

References

1. Allman, M., Paxson, V., Blanton, E.: TCP Congestion Control, RFC 5681 (Draft Standard). http://www.ietf.org/rfc/rfc5681.txt (September 2009)
2. Callegari, C., Giordano, S., Pagano, M., Pepe, T.: Behavior analysis of TCP Linux variants. Comput. Netw. **56**, 462–476 (2012)

3. Floyd, S., Henderson, T., Gurtov, A.: The NewReno Modification to TCP's Fast Recovery Algorithm, RFC 3782 (Proposed Standard). http://www.ietf.org/rfc/rfc3782.txt (April 2004)
4. Brakmo, L.S., O'Malley, S.W., Peterson, L.L.: TCP vegas: new techniques for congestion detection and avoidance. In: SIGCOMM, pp. 24–35 (1994)
5. Fu, C.P., Liew, S.: TCP veno: TCP enhancement for transmission over wireless access networks. IEEE J. Sel. Areas Commun. **21**(2), 216–228 (2003)
6. Mascolo, S., Casetti, C., Gerla, M., Sanadidi, M.Y., Wang, R.: TCP westwood: bandwidth estimation for enhanced transport over wireless links. In: Proceedings of the 7th Annual International Conference on Mobile Computing and Networking (MobiCom '01), pp. 287–297. ACM, New York (2001). doi:10.1145/381677.381704
7. Xu, L., Harfoush, K., Rhee, I.: Binary increase congestion control (BIC) for fast long-distance networks. In: INFOCOM (2004)
8. Rhee, I., Xu, L.: CUBIC: a new TCP-friendly high-speed TCP variant. SIGOPS Oper. Syst. Rev. **42**(5), 64–74 (2008)
9. Floyd, S.: HighSpeed TCP for Large Congestion Windows (2002)
10. Caini, C., Firrincieli, R.: TCP Hybla: a TCP enhancement for heterogeneous networks. Int. J. Satell. Commun. Netw. **22**(5), 547–566 (2004)
11. Kelly, T.: Scalable TCP: improving performance in highspeed wide area networks. SIGCOMM Comput. Commun. Rev. **33**(2), 83–91 (2003). doi:10.1145/956981.956989
12. Liu, S., Başar, T., Srikant, R.: TCP-Illinois: a loss and delay-based congestion control algorithm for high-speed networks. In: Proceedings of the 1st International Conference on Performance Evaluation Methodolgies and Tools (Valuetools '06), p. 55. ACM, New York (2006). doi:10.1145/1190095.1190166
13. Baiocchi, A. Castellani, A.P., Vacirca, F.: YeAH-TCP: yet another highspeed TCP. In: Proceedings of PFLDnet (2007)
14. Shorten, R., Leith, D.: H-TCP: TCP for high-speed and long-distance networks. In: Proceedings of the PFLDnet (2004)
15. Kuzmanovic, A., Knightly E.W.: TCP-LP: a distributed algorithm for low priority data transfer. In: Proceedings of INFOCOM (2003)

Analysis of the Resource Distribution Schemes in LTE-Advanced Relay-Enhanced Networks

Tatiana Efimushkina$^{(\boxtimes)}$ and Konstantin Samuylov

Telecommunication Systems Department, Peoples' Friendship University of Russia,
Ordzhonikidze 3, Moscow 115419, Russia
tvefimushkina@gmail.com, ksam@sci.pfu.edu.ru
http://www.rudn.ru

Abstract. Nowadays with the introduction of smart phones along with the development of audio and video applications increase people's demands for the ubiquitous high-data rate coverage. One of the cost-effective solutions adopted in the Long-Term Evolution-Advanced (LTE-Advanced) standard that extends coverage and enhances throughput is the placement of heterogeneous nodes: low-power Relay Nodes (RNs) are deployed to assist transmissions between an evolved Node B (eNB) and multiple User Equipments (UEs). In this framework resource allocation becomes an important and crucial problem that directly influences the potential capacity and coverage improvements. In this paper we overview the concepts of basic relaying and resource allocation in a relay-enhanced LTE-Advanced network. We investigate the performance of an analytical model for a relay-enhanced LTE-Advanced downlink channel, which is presented in detail. Various resource allocation schemes are analyzed, among which, according to the experimental analysis, the proportional one with constraints achieves the best performance.

Keywords: LTE-Advanced · Relay node · Resource allocation · Analytical model · Performance measures

1 Introduction

The Long Term Evolution (LTE) is a radio platform technology that was designed by the 3rd Generation Partnership Program (3GPP) to provide high peak throughputs and low latencies ubiquitously, and to achieve flexible utilization of the available frequency spectrum [1]. The given requirements can be supported by exploiting efficient transmission schemes that vary in appliance with the traffic direction:

Downlink. LTE exploits the Orthogonal Frequency Division Multiple Access (OFDMA) as the multiple access technique for downlink (from evolved Node B (eNB) to User Equipment (UE)) to improve the spectrum efficiency and facilitate flexible user resource allocation [2]. OFDMA is a multi-user version of a digital modulation Orthogonal Frequency Division Multiplexing (OFDM), in

V. Vishnevsky et al. (Eds.): DCCN 2013, CCIS 279, pp. 43–57, 2014.
DOI: 10.1007/978-3-319-05209-0_4, © Springer International Publishing Switzerland 2014

which the available bandwidth is divided into a large number of independent closely-spaced orthogonal to each other subcarriers that are used to carry the data. With multiple subcarriers transmitting in parallel, longer symbol duration is used, and therefore, OFDMA is robust to time delays caused by multipath fading or frequency selectivity of the radio channel. Moreover, the use of a cyclic prefix minimizes the Inter-Symbol Interference (ISI).

Uplink. Although OFDMA is considered to be the optimum multiple access technique for downlink transmission, it comes along with the high Peak-to-Average Power Ratio (PAPR) that makes it less favorable for the uplink transmission (from UE to eNB). Instead, the Single-Carrier Frequency Division Multiple Access (SC-FDMA) is used, which is quite similar to the OFDMA but with an extra Discrete Fourier Transform (DFT) processing block before the subcarrier mapping routine. The extended transformations make each information bit spread over all the subcarriers, which results in significantly smaller variations in the instantaneous power of the transmitted signal. Moreover, the performance of SC-FDMA system is highly affected by the type of the exploited mapping scheme: localized or distributed. In a localized scheme each UE considers a set of adjacent subcarriers to transmit its data, whereas in a distributed scheme the subcarriers allocated to a UE are spread over the entire bandwidth.

Although LTE substantially outperforms the performance of the 3rd generation systems, such as High Speed Packet Access (HSPA), etc., it could not be used for future data traffic requirements that have been addressed within International Mobile Telecommunications-Advanced (IMT-Advanced) specifications [3]. So LTE-Advanced, which implies a natural evolution of LTE, has been introduced by 3GPP for fulfilling the IMT-Advanced requirement. Note that one of the technological novelties of the LTE-Advanced is a heterogeneous networks deployment.

In this paper we investigate the problem of resource allocation in relay-enhanced LTE-Advanced networks. In the next section the basic relaying concept and resource allocation in a relay-enhanced LTE networks are discussed. Section 3 presents in detail an efficient analytical model for a relay-enhanced LTE downlink channel. Here, the balance equations and the main performance measures are demonstrated. Moreover, various resource allocation schemes are discussed. In Sect. 4 the extensive experimental analysis is conducted, according to which the extended proportional resource allocation scheme achieves the best performance. At last, conclusion is presented.

2   Relay-Enhanced LTE Networks

Heterogeneous LTE networks are characterized by the placement of eNBs that transmit at high power levels (5 W–40 W), overlaid with the Pico/Femto/Relay Nodes (RNs) that transmit at substantially lower power levels (100 mW–2 W). The utilization of the heterogeneous base stations is considered to be a promising solution for LTE communications due to a number of reasons. Firstly, the deployment of the low-power nodes improves the system capacity and coverage by offering alternative paths to users located in shadow areas [4]. Secondly, the deployment costs of the heterogeneous cellular systems decrease. Thirdly, the exibility in Pico/Femto/Relay positioning allows a faster enhanced network construction. Lastly, RNs offer additional flexibility as they do not need a wired backbone access.

2.1 Relay Nodes

Relay Nodes (RNs) are relatively small devices that come in various types according to the relay technology adopted:

1. **The layer 1 relay node.** It is an Amplifier and Forward (AF) type of relay technology, a.k.a. repeater that simply amplifies and transmits the received signal in both downlink and uplink directions [5]. It comes with low cost and short processing delay, however, deteriorates the received Signal to Interference plus Noise Ratio (SINR) by amplifying the intercell interference and noise together with the desired signal.
2. **The layer 2 relay node.** It is a Decode and Forward (DF) type of relay technology that includes the demodulation/decoding processing to verify the correctness of the received data, and then encodes and modulates the data for further transmission. So the noise is eliminated, however, along with the processing delay new radio-control functions, such as mobility control, retransmission control and user-data concatenation/segmentation, are to be added between the eNB and UE transparently.
3. **The layer 3 relay node.** It is very similar to layer 2 relay technology but incorporates several similar functions as eNB that leads to a small impact on standard specifications [5]. Principally, the layer 3 RN can almost be considered a wireless eNB, which include functionalities such as radio resource management, scheduling and Hybrid Automatic Repeat Request (HARQ) retransmissions [6].

The backhaul link can either use an additional frequency band (out-of-band RNs) or operate in the same spectrum as communication from/to UEs (in-band RNs). The in-band RNs are universally deployable since they do not require additional frequency licenses. Note that the RNs fundamentals can be exploited in the Self-Organizing Networks (SON) [7].

2.2 Resource Allocation

In relay-enhanced LTE networks communication can be performed directly with the eNB in a single-hop, or via a RN in multi-hops. Note that two-hop relaying has been proven to give the highest system throughput, and when the number of hops is larger than three, the system overhead for exchanging control messages uses a great amount of resources [8]. In order to gain the potential capacity and coverage improvements of two-hop relaying, resource allocation on different hops should be cooperated to avoid data shortage or overflow in RNs.

There are two types of resource allocation architectures for relay-enhanced networks named centralized and semi-distributed architectures [9,10]. In centralized allocation, eNB is responsible for allocating the available resources to all links. So to perform efficiently, eNB needs to be aware of the Channel State Information (CSI) of each link and the queue length on every RN. The centralized allocation can reduce the complexity of RNs, but it consumes more resources for control message exchange.

In semi-distributed allocation, eNB assigns each RN a number of resources, which are allocated to the UEs by using RN's scheduler. Note that the proper allocation of the channel resources can cooperatively reduce the wastes of radio resources, and thus, increase network efficiency.

3   Analytical Modeling

The problem of modeling of the relay-enhanced LTE networks in conjunction with the resource allocation has recently gained much attention in the literature (refer to [11–16] and their bibliographies). We further would like to study the model of the relay-enhanced LTE network, with two types of nodes: eNB and RNs, that is formulated for downlink channel. Note that the following model represents an efficient way to analyze various resource allocation algorithms based on the derived performance measures.

3.1   Description of the Model

We consider a centralized resource allocation scenario of a cellular network with one eNB and K RNs, $K < \infty$. The downlink subframe is divided into S channels (Chs), which occupy the smallest unity of channel bandwidth both in frequency and time bands. All the S Chs are distributed between the eNB and K RNs to transmit the packets in the direction of UEs.

In the model description we frequently refer to the request, which has a physical meaning of a packet. Let us suppose that there are $K + 1$ types of requests in the cell. Here, k-request corresponds to the k-type, which is to be transmitted to the UE in the coverage area of the eNB in case $k = 0$, or in the coverage area of the k-RN (RN_k), $k = \overline{1, K}$. The arriving requests at the eNB and RNs are stored in the buffers, the capacity of which for eNB is $r_0, r_0 < \infty$ and for RN_k corresponds to $r_k, r_k < \infty, k = \overline{1, K}$. Moreover, we presume that the arrived requests at the system with fully occupied buffers are lost and do not influence its functioning. We will consider the system as a queuing system.

The system functions in discrete time measured in time slots with the constant length $h = 1$, which corresponds to the duration of the downlink data subframe in LTE network. Moreover, the scheduling of the resources are conducted every Time Transmission Interval (TTI) that is equal to $1\,\text{ms}$. Assume that all the changes in the system occur at time moments $nh, n = 1, 2, \dots$. Therefore, during the time slot n, or the time interval $[nh, (n+1)h)$, the following sequence of events can take place:

1. The requests are serviced by the Chs of RN_k and the space they took in the buffers of RN_k is released;
2. The requests are serviced by the Chs of eNB, and the space 0-requests took in the buffer of eNB is released;
3. The serviced requests arrive to the buffers of the corresponding RN_k, while the space they took in the buffers of eNB is released;
4. New requests arrive to the buffer of eNB;
5. All the S Chs are reallocated between eNB and RNs;
6. The state is fixed.

Let η_n, $\eta_n \in \{0,1\}$ is a number of group arrival of requests during the time slot n, taking into account that all of the $\eta_n, n \geq 0$, - are independent identically distributed random variables (RVs) with a generation function (GF)

$$A(z) = Mz^{\eta_n} = 1 - a + az, |z| \leq 1, a = P\{\eta_n = 1\}, 0 < a < 1, n \geq 0. \quad (1)$$

The number of requests χ_n in the arrived group is an independent to n RV with the GF

$$G(z) = Mz^{\chi_n} = \sum_{i \geq 1} g_i z^i, |z| \leq 1, G(1) = 1, g_i = P\{\chi_n = i\}, n \geq 0. \quad (2)$$

Thus, the arrival requests follow a group geometric distribution Geom^G as the time duration between the group arrivals conforms a geometric distribution with the mean $1/a$ and is characterized by the GF

$$A(G(z)) = 1 - a + aG(z) = \sum_{i \geq 1} a_i z^i, |z| \leq 1, a_0 = \bar{a} = 1 - a, a_i = ag_i, i \geq 1. \quad (3)$$

Assume that each request from the arrival group belongs to type k (k-request) with the probability $c_k, k = \overline{0, K}, c_\bullet = 1$. Here and further, dot in place of index means a full sum of the variable. All in all, the arrival rate corresponds to the $(K+1)-$ dimensional group geometrical distribution. In order to simplify the modeling, the service time is considered to follow the deterministic law with the duration of the request's service equal to one slot. Hence, every request is serviced during one slot, and releases the occupied buffer space. The described system can be denoted in the following way: $\mathrm{Geom}_{K+1}^G \mid D = 1 \mid dif(s_\bullet) = S \mid \bar{r}$. The notation $dif(s_\bullet) = S$ indicates that the number of servers (Chs) varies from slot to slot and constitutes to the overall S Chs, which allows to investigate different resource allocation schemes. The structure of the designed system is presented in Fig. 1. We can suppose that the requests in the buffer of the eNB do not differentiate in types, whereas the type selection is conducted using the polynomial scheme with the probabilities $c_0, ..., c_K$. The functioning of the network system is described by the homogeneous Markov chain ξ_n at time moments $nh + 0, n \geq 0$, with the state space

$$X = \{\boldsymbol{x} = (x_0, x_1, ..., x_K)^{\mathrm{T}} : x_k = \overline{0, r_k}, k = \overline{0, K}\}, |X| = \prod_{k=0}^{K} (r_k + 1), \quad (4)$$

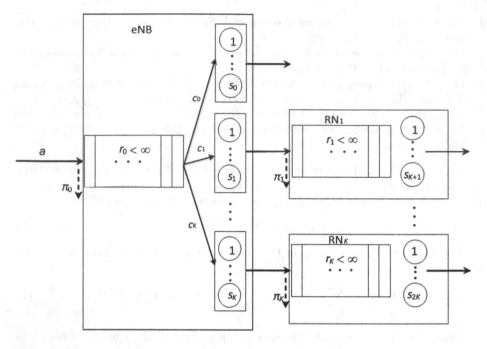

Fig. 1. The structure of the queuing model with overall S Chs, which are distributed between eNB ans RNs, and $K + 1$ types of incoming requests.

where x_k- is a number of k-requests in the buffer of the corresponding eNB or RN_k. If $0 < a < 1$ the Markov chain $\xi_n, n \geq 0$ is aperiodic, and there exists a stationary probability distribution $[x], x \in X$, which is found from the balance equations

$$a[0] = \overline{a} \sum_{\Omega_0} c_0^{x_0} [x], \tag{5}$$

$$1 - \sum_{\Omega_1} c_0^{s_0^{\min}} \prod_{k=1}^{K} c_k^{s_k^{\min} + \delta(x_k, r_k)(1 + \ldots + n_k)} a'[x] =$$

$$\sum_{\Omega_2} c_0^{s_{0,q}^{\min}} \prod_{k=1}^{K} c_k^{s_k^{\min} - q_k + \delta(x_k, r_k)(1 + \ldots + n_{k,q})} a_q'[x + \sum_{k=0}^{K} q_k e_k], \tag{6}$$

and normalizing equation

$$\sum_{x \in X} [x] = 1, \tag{7}$$

where $\delta(a, b) = \begin{cases} 0, a \neq b; \\ 1, a = b. \end{cases}$ Let us now define the notations used in (5)-(6):

1. $s_k^{\min} = \min(x_k, s_{K+k}) -$ a number of the serviced requests.

2. $s_\bullet^{\min} = \sum_{k=1}^{K} s_k^{\min} -$ an overall number of the serviced requests by the Chs of RN_k.

3. $s_0^{\min} = \min(s_0, x_0 - s_\bullet^{\min}) -$ a number of the serviced 0-requests at eNB.

4. $r_0' = x_0 - s_\bullet^{\min} - s_0^{\min} -$ a number of requests in the buffer of eNB after 0-requests are serviced, taking into account that the requests serviced by Chs of RN_k arrive at the same quantity to their buffers from the corresponding Chs of eNB.

5. $n_k = \min(s_k - s_k^{\min}, r_0' - \sum_{i=1}^{k-1} \delta(x_i, r_i) n_i) -$ a number of requests serviced at eNB and headed for the RN_k, but which do not have any free space in the buffers of RN_k. This particular case is considered when the number of k-requests coincide with the buffer's capacity.

6. $s_{0,q}^{\min} = \min(s_0, x_0 - \sum_{k=1}^{K}(s_k^{\min} - q_k)) -$ a number of the serviced 0-requests at eNB.

7. $r_{0,q}' = x_0 - \sum_{k=1}^{K}(s^{\min}{}_k - q_k) - s_{0,q}^{\min} -$ a number of requests in the buffer of eNB after 0-requests are serviced, taking into account that the requests serviced by the Chs of RN_k arrive at the quantity q_k to their buffers from the corresponding Chs of eNB.

8. $n_{k,q} = \min(s_k - s_k^{\min} + q_k, r_{0,q}' - \sum_{i=1}^{k-1} \delta(x_i, r_i) n_{i,q}), k = \overline{1, K} -$ a number of requests serviced at eNB and headed for RN_k, but which do not have any free space in the buffers of the RN_k. This particular case is considered when the number of k-requests coincide with the buffer's capacity.

9. $\Omega_0 = \{x : x_k \le s_{K+k}, k = \overline{1, K}\}$ for $\forall x \in X \backslash 0 -$ a state space, which satisfies the condition: the number of requests in the buffer of RN_k does not exceed the number of allocated Chs. Note that the Chs at eNB are not taken into account.

10. $\Omega_1 = \{x : x_0 \ge s_\bullet^{\min}, s_k \ge s_k^{\min} k = \overline{1, K}\}$ for $\forall x \in X \backslash 0 -$ a state space, which satisfies the condition: the number of allocated Chs at eNB for the requests that head to RN_k exceeds the number of serviced requests at RN_k. The following condition guarantees that the number of serviced requests at RN_k will be able to occupy the released buffers.

11. $\Omega_2 = \{x : x_0 \ge \sum_{k=1}^{K}(s_k^{\min} - q_k) \ge 0\}, q_k = \overline{-x_k, r_k - x_k}, k = \overline{1, K}$ for $\forall x \in X \backslash 0 -$ a state space, which satisfies the condition: the number of allocated Chs at eNB for the requests that head to the RN_k exceeds the number of serviced requests at RN_k, taking into account the variable q_k.

12. $a' = \begin{cases} \sum_{i=1}^{\infty} a_{s_0^{\min} + s_\bullet^{\min} + \sum_{k=1}^{K} \delta(x_k, r_k) n_k + i}, & \text{if } \delta(x_0, r_0) = 1 \\ a_{s_0^{\min} + s_\bullet^{\min} + \sum_{k=1}^{K} \delta(x_k, r_k) n_k}, & \text{otherwise} \end{cases} -$ the probability of the arrival of the number of requests to the buffers of eNB during the slot to preserve exactly the same state of the system.

13. $a_q' = \begin{cases} \sum_{i=1}^{\infty} a_{s_{0,q}^{\min} + \sum_{k=1}^{K}(s_k^{\min} - q_k) - q_0 + \sum_{k=1}^{K} \delta(x_k, r_k) n_{k,q} + i}, & \text{if } \delta(x_0, r_0) = 1 \\ a_{s_{0,q}^{\min} + \sum_{k=1}^{K}(s_k^{\min} - q_k) - q_0 + \sum_{k=1}^{K} \delta(x_k, r_k) n_{k,q}}, & \text{otherwise} \end{cases}$

$-$ the probability of arrival of the number of requests to the buffers of eNB to preserve the state of the system, taking into account the change q_k.

3.2   Resource Allocation Schemes

In order to take into consideration the resource distribution at slot n, we introduce the following vector

$$s^n = (s_0^n, s_1^n, ..., s_{2K}^n)^{\mathrm{T}} = (f_0^n(x), f_1^n(x), ..., f_{2K}^n(x)) = f^n(x), \qquad (8)$$

where $f^n(x)-$ is a function that defines the strategy of the resource allocation. The function $f^n(x)$ may indicate one of the following resource allocation methods (assume that $f^n(x) = f(x), n \geq 0$).

1. Method M1 (deterministic, does not depend on x) [15]

$$s_k = \lfloor \frac{S}{2K+1} \rfloor, k = \overline{1, 2K}, s_0 = S - \sum_{k=1}^{2K} s_k. \qquad (9)$$

2. Method M2 (deterministic, does not depend on x) [15]

$$s_{K+k} = \lfloor \frac{S - \lfloor \frac{S}{2} \rfloor}{K+1} \rfloor, k = \overline{1, K}, S' = S - \sum_{k=1}^{K} s_{K+k}, s_k = \lfloor \frac{S'}{K+1} \rfloor, k = \overline{1, K},$$

$$s_0 = S' - \sum_{k=1}^{K} s_k. \qquad (10)$$

Note that resource allocation schemes M1, M2 are deterministic and do not adapt to the varying traffic volumes. The difference between M1 and M2 is that in method M1 less number of the time-frequency channels are allocated to eNB, and accordingly, more Chs are provided to RNs.

3. Method M3 (proportional) [15]

$$s_{K+k} = \lfloor \frac{x_k S}{x_\bullet} \rfloor, S' = S - \sum_{k=1}^{K} s_{K+k}, s_k = \lfloor \frac{\lfloor x_0 c_i \rfloor S'}{x_0} \rfloor, k = \overline{1, K},$$

$$s_0 = S' - \sum_{k=1}^{K} s_k. \qquad (11)$$

4. Method M4 (proportional with constraints) [16]. Method M4 is an enhanced proportional scheme with constraints that is fully based on the network load state. It gives the priority in terms of resources to the RNs, and limits the number of Chs allocated to eNB when the network load has a tendency to rise. Let us describe the algorithm, which is divided into two parts: distribution of the Chs among the RNs and resource allocation at eNB.

Part I. Resource allocation at RNs. If the number of requests in the buffers of the $RN_k, k = \overline{1, K}$ exceeds the overall number of S Chs, i.e. $\sum_{i=1}^{K} x_i > S$, then the proportional division of the Chs among RNs is conducted

$$s_{K+k} = \lfloor \frac{x_k S}{\sum_{i=1}^{K} x_i} \rfloor. \qquad (12)$$

In the opposite case when $\sum_{i=1}^{K} x_i \leq S$, then the following distribution of the resources without allocating the Chs to eNB for transmission the k-requests is considered

$$s_{K+k} = \min(x_k, S'_k), S'_k = S - \sum_{i=0}^{k-1} s_{K+i}, s_k = 0, k = \overline{1, K}. \tag{13}$$

The remaining Chs $S'' = S - \sum_{i=1}^{K} s_{K+i}$ are used for resource allocation at eNB.

Part II. Resource allocation at eNB. The main advantage of the method M4 is an ability to constrain the allocation of the Chs to the eNB for k-requests transmission, when there is no free space in the buffers of the $RN_k, k = \overline{1, K}$. Therefore, resource distribution at eNB is based on the following mathematical recursive algorithm

Step 1. Input: $n = 0 : S''_{(n)} = S'', c_k^{(n)} = c_k, s_k^{(n)} = 0, k = \overline{1, K}$.
Step 2. If $S''_{(n)} > n$ move to Step 3, otherwise, terminate the algorithm.
Step 3.

$$n = n+1, s_k^{(n)} = \min(\sum_{m=0}^{n-1} \lfloor \frac{\lfloor x_0 c_k^{(m)} \rfloor S''_{(m)}}{x_0} \rfloor, r_k - (x_k - s_{K+k})), k = \overline{1, K}$$

$$s_0^{(n)} = \sum_{m=0}^{n-1} \lfloor \frac{\lfloor x_0 c_0^{(m)} \rfloor S''_{(m)}}{x_0} \rfloor, S''_{(n)} = S''_{(n-1)} - \sum_{k=0}^{K}(s_k^{(n)} - s_k^{(n-1)}),$$

$$c_k^{(n)} = \frac{c_k^{(n-1)}(1 - \delta(s_k^{(n)}, r_k - (x_k - s_{K+k})))}{c_\bullet^{(n)}}, \tag{14}$$

where $c_\bullet^{(n)} = \sum_{k=1}^{K}(c_k^{(n-1)}(1 - \delta(s_k^{(n)}, r_k - (x_k - s_{K+k})))) + c_0^{(n-1)}$.
Step 4. If $S''_{(n)} \neq S''_{(n-1)}$ move to Step 2, otherwise, the remaining resources are allocated to transmit 0-requests

$$s_0^{(n)} = s_0^{(n)} + S''_{(n)}, \tag{15}$$

after which the algorithm is terminated.

3.3   Performance Measures

The blocking probability as well as other performance characteristics can be obtained from the steady-state probability distribution, and are defined below.

1. The blocking probability L_k of k-requests, $k = \overline{0, K}$:
   $L_0 = \sum_{g=r_0-r'_0+1}^{\infty} a_g \sum_X [\boldsymbol{x}]$,
   $L_k = \sum_{s'_k=r_k-x_k+s_k^{\min}}^{n_k} c_k^{s'_k} \sum_{s:s_k>r_k-x_k+s_k^{\min}} [\boldsymbol{x}], k = \overline{1, K}$.
2. The overall blocking probability L of the requests: $L = 1 - \prod_{k=0}^{K} L_k$.

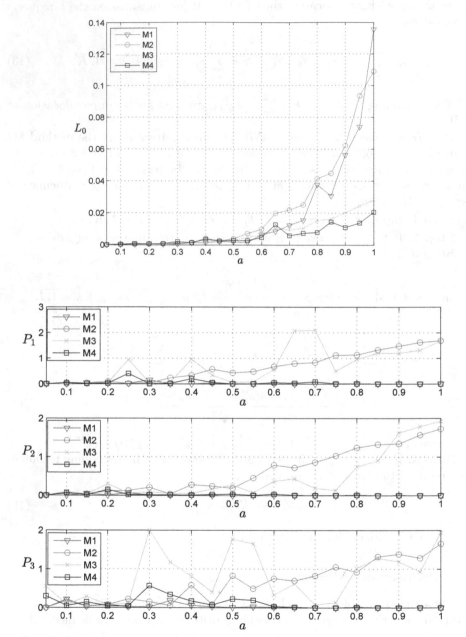

Fig. 2. The blocking probability of the requests at the eNB, and the mean number of lost k-requests at RN_k from the first to the second row, respectively.

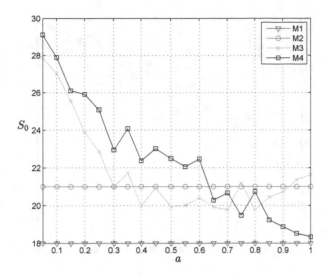

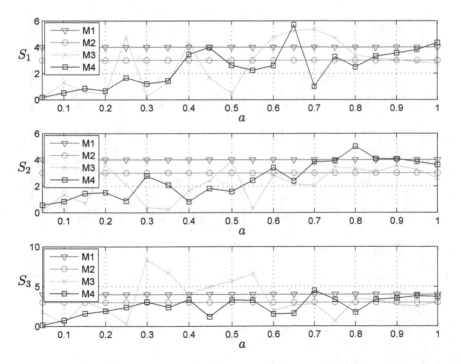

Fig. 3. The mean number of Chs at eNB, and the mean number of Chs at RN_k from the first to the second row, respectively.

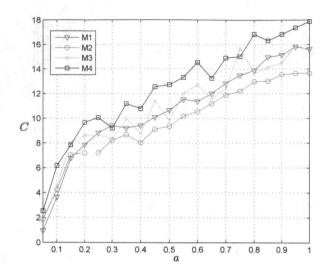

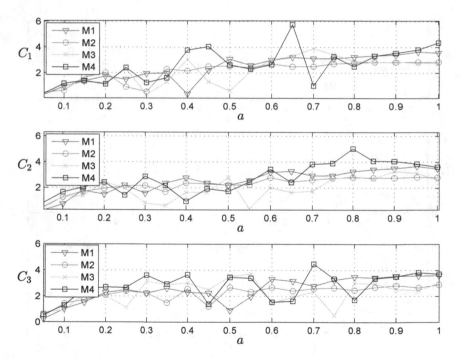

Fig. 4. The mean number of serviced requests in the system, and the mean number of serviced k-requests at RN_k from the first to the second row, respectively.

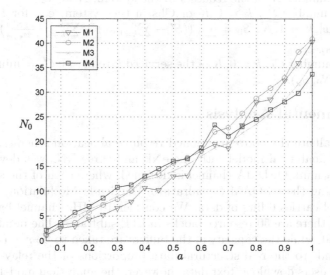

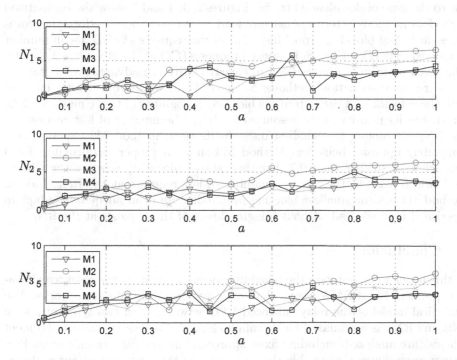

Fig. 5. The mean number of requests at eNB, and the mean number of requests at RN_k from the first to the second row, respectively.

3. The mean number N_k of k-requests, $k = \overline{0, K}$: $N_k = \sum_X x_k[\boldsymbol{x}]$.
4. The mean number N of the requests in the system: $N = N_\bullet$.
5. The mean number S_k, $k = \overline{0, K}$ of Chs in the system, e.g. for M3: $S_k = \sum_X \lfloor \frac{x_k S}{x_\bullet} \rfloor [\boldsymbol{x}], k = \overline{1, K}$, $S_0 = \sum_X ((S' - \sum_{k=1}^{K} s_k) + \sum_{k=1}^{K} \lfloor \frac{x_k S'}{x_\bullet} \rfloor)[\boldsymbol{x}]$, $S' = S - \sum_{k=1}^{K} s_{K+k}$.
6. The mean number C_k, $k = \overline{0, K}$ of the serviced k-requests: $C_k = \min(N_k, S_k)$, $C_0 = \min(N_0, S_0)$.

4   Experimental Analysis

In order to evaluate the above mentioned resource allocation schemes, the simulation of the analytical model with one eNB and three RNs was developed in C++. The downlink OFDMA channel is designed, where each of the subcarrier is modeled using the 64-QAM (Quadrature Amplitude Modulation), so every OFDM symbol carries 6 bits of data. We consider 10 MHz channel bandwidth, and therefore, there are 50 resource blocks in LTE subframe. The number of the Chs is set to 30, while the capacity of the buffers are taken to be $r = (45, 7, 7, 7)$ that corresponds to the real structural data proportions of the relay-enhanced LTE cell. The users download text data, however, the analytical model is invariant to the type of downloaded traffic. Figures 2, 3, 4 and 5 show the comparison of the four methods in terms of network load (a) versus various performance measures, including blocking probability (L_0) of the requests at eNB, mean number of lost requests (P_k), Chs (S_k), serviced requests (C_k) at the RN$_k$, $k = \overline{1, K}$, etc. The given performance measures can clearly clarify the advantages/limitations of the resource allocation methods. Method M1 shows quite a drastic increase with the network load rise in terms of blocking probability of the requests at eNB, but due to its priority of the resources for RN$_k$, the number of lost requests at the RN$_k$ are among the smallest ones. In its turn, method M2 experiences a completely opposite behavior. Method M3 shows a proper performance for a network load smaller than 0.7, but then starts allocating most of the Chs to eNB, which leads to the rise in a number of lost requests at the RN$_k$. Indeed, method M4 has the smallest blocking probability, and can adequately adapt to the rise of network load without bringing losses of the requests at the RN$_k$.

5   Conclusion

In this paper we investigate the relaying concept along with the resource allocation problem in OFDMA relay-enhanced heterogeneous cellular networks. The analytical model of the relay-enhanced LTE network, with two types of nodes: eNB and RNs, is formulated for downlink channel. Various resource assignment schemes are analyzed, including fixed approach and several dynamic ones. Performance measures that enable the evaluation of the proposed resource allocation algorithms are discussed. According to the numerical analysis, the proposed improved proportional method with constraints allows adequately adapting to the demands of the relay-enhanced LTE network with the rise of the network load.

Acknowledgments. The reported study was partially supported by RFBR, research project No. 13-07-00953-a.

References

1. 3GPP TR 36.913 v8.0.0: Requirements for Further Advancements for E-UTRA (LTE-Advanced). Release 8 (2008)
2. 3GPP TS 36.201 v1.0.0: LTE Physical Layer General Description (2012)
3. ITU-R: Requirements Related to Technical Performance for IMT-Advanced Radio Interfaces. Rep. M.2134 (2008)
4. Beniero, T., Redana, S., Hmlinen, J., Raaf, B.: Effect of relaying on coverage in 3GPP LTE-advanced. In: 69th IEEE Vehicular Technology Conference, pp. 1-5. Barcelona (2009). doi:10.1109/VETECS.2009.5073520
5. Iwamura, M., Takahashi, H., Nagata, S.: Relay technology in LTE-advanced. J. NTT DOCOMO Tech. J. **12**(2), 29–36 (2010)
6. Saleh, A., Riihonen, T., Hmlinen, J., Redana, S., Raaf, B., Wichman, R.: Performance of amplify-and-forward and decode-and-forward relays in LTE-advanced. In: IEEE 70th Vehicular Technology Conference (VTC2009-Fall), pp. 1-5. Anchorage, AK (2009). doi:10.1109/VETECF.2009.5378824
7. Osterbo, O., Grondalen, O.: Benefits of self-organizing networks (SON) for mobile operators. J. Comput. Netw. Commun. 2012 (2012). doi:10.1155/2012/862527
8. Wang, L.: Resource allocation in OFDMA relay-enhanced cellular networks. In: SOKENDAI Publ. (2010)
9. Kwak, R., Cio, J.: Resource allocation for OFDMA multi-hop relaying downlink systems. In: IEEE Global Telecommunication Conference, pp. 3225–3229, Washington, DC (2007). doi:10.1109/GLOCOM.2007.611
10. Kaneko, M., Popovski, P.: Adaptive resource allocation in cellular OFDMA systems with multiple relay stations. In: 65th IEEE Vehicular Technology Conference, pp. 3026–3030, Dublin (2007). doi:10.1109/VETECS.2007.620
11. Ma, Z., Xiang, W., Long, H., Wang, W.: Proportional fair resource partition for LTE-Advanced networks with type I relay nodes. In: IEEE International Conference on Communications, pp. 1–5, Kyoto (2011). doi:10.1109/icc.2011.5963280
12. Huang, L., Rong, M., Wang L., Xue Y.: Resource allocation for OFDMA based relay enhanced cellular networks. In: 65th Vehicular Technology Conference, pp. 3160–3164, Dublin (2007). doi: 10.1109/VETECS.2007.647
13. Moraes, T., Gonzalez, A., Nisar, M., Seidel, E. QoS-aware resource allocation for in-band relaying in LTE-advanced. In: 8th International Conference on Wireless and Mobile Communications, pp. 195–201, Venice, Italy (2012)
14. Moraes, T., Nisar, M., Gonzalez, A., Seidel, E.: Resource allocation in relay enhanced LTE-advanced networks. J. EURASIP Wirel. Commun. Netw. (2012). doi:10.1186/1687-1499-2012-364
15. Efimushkina, T., Samuylov, K.: Load distribution schemes investigation in LTE heterogeneous networks. In: VI Industry Scientific Conference Technologies of the Information Society, Moscow, Russia (2012) (In Russian)
16. Efimushkina, T.: Performance measures investigation of the improved resources allocation scheme in heterogeneous LTE network. J. T-Comm **7**, 58–65 (2013). (In Russian)

Modelling Transient States in Queueing Models of Computer Networks: A Few Practical Issues

Tadeusz Czachórski[1(✉)], Monika Nycz[2], and Tomasz Nycz[2]

[1] Institute of Theoretical and Applied Informatics of Polish Academy of Sciences, Gliwice, Poland
tadek@iitis.pl
[2] Institute of Informatics, Silesian Technical University, Gliwice, Poland
{Monika.Nycz,Tomasz.Nycz}@polsl.pl

Abstract. The article summarises author's experience in two problems related to the use of queueing models in performance evaluation of computer networks: modelling transient states of queues and computations for queueing network models having large number of nodes. Both issues are not well represented in classical queueing theory, yet important to applications, because the observed traffic is time dependant and network topologies that should be considered become larger and larger. The article discusses two approaches: diffusion approximation and fluid-flow approximation that can cope with much larger models that are attainable with the use of Markov chains.

1 Introduction

Queueing models are frequently used in modelling and evaluation of computer networks. Queueing theory, introduced a century ago to model telephone exchanges was successfully adapted to the needs of computer science but new problems arise following the constant development of computer networks. The problems are mainly related to computational aspects of queueing models and more precisely to the need of taking into account very large topologies, corresponding to real ones encountered in computer networks and the necessity to analyse transient behaviour of queues, as the intensity of traffic flows generated by users, e.g. internet applications is permanently changing. The quality of transmission services depends on current load of links and not on its average value. Also modelling and understanding the performance of traffic control mechanisms, control stability and its impact on quality of service needs transient state analysis, e.g. [27].

The use of analytical models known in classical queueing theory is limited to single M/M/1 and M/M/1/N stations and even there the transient state solutions are quite complex. Moreover, the results refer to transient states but it is assumed that the model parameters, the input rate in particular, are constant. Therefore, they do not fit well the problem of modelling IP routers, where the incoming streams of packets are not Poisson and the size of packets is not

V. Vishnevsky et al. (Eds.): DCCN 2013, CCIS 279, pp. 58–72, 2014.
DOI: 10.1007/978-3-319-05209-0_5, © Springer International Publishing Switzerland 2014

exponentially distributed. We need models describing constantly changing non-Poisson flows and considering general distributions of service times. We need also the possibility to include in these models the description of self-similarity of flows. The models should also be scalable to meet very large topologies characteristic to the Internet.

In sections that follow we describe our experience and present simple numerical examples referring diffusion approximation, and fluid-flow approximation – the approaches we think are most suitable to this purposes.

2   Diffusion Approximation of a Single Queue

This approach is merging states of the considered queueing system and thus needs much less computations than the Markov models. We present here the principles of the method following [9] where steady-state solution of a single G/G/1/N model was given and then extended to the network of queues in [10]. We supplemented these results with semi-analytical, semi-numerical transient state solution [3] given for constant model parameters but it could be applied also in case of time-dependent parameters if we only make them constant within small intervals.

Let $A(x)$, $B(x)$ denote the interarrival and service time distributions at a service station and $a(x)$ and $b(x)$ be their density functions. The distributions are general but not specified, the method requires only the knowledge of their two first moments. The means are denoted as $E[A] = 1/\lambda$, $E[B] = 1/\mu$ and variances are $\text{Var}[A] = \sigma_A^2$, $\text{Var}[B] = \sigma_B^2$. Denote also squared coefficients of variation $C_A^2 = \sigma_A^2 \lambda^2$, $C_B^2 = \sigma_B^2 \mu^2$. $N(t)$ represents the number of customers present in the system at time t.

Diffusion approximation, replaces the process $N(t)$ by a continuous diffusion process $X(t)$, the incremental changes $dX(t) = X(t + dt) - X(t)$ of which are normally distributed with the mean βdt and variance αdt, where β, α are coefficients of the diffusion equation

$$\frac{\partial f(x,t;x_0)}{\partial t} = \frac{\alpha}{2} \frac{\partial^2 f(x,t;x_0)}{\partial x^2} - \beta \frac{\partial f(x,t;x_0)}{\partial x}. \tag{1}$$

This equation defines the conditional pdf of $X(t)$:

$$f(x,t;x_0)dx = P[x \leq X(t) < x + dx \mid X(0) = x_0].$$

The both processes $X(t)$ and $N(t)$ have normally distributed changes; the choice $\beta = \lambda - \mu$, $\alpha = \sigma_A^2 \lambda^3 + \sigma_B^2 \mu^3 = C_A^2 \lambda + C_B^2 \mu$ ensures that the parameters of these distributions increase at the same rate with the length of the observation period. In the case of G/G/1/N station, the process evolves between barriers placed at $x = 0$ and $x = N$ where barriers *with instantaneous jumps* are placed, [9]. When the diffusion process comes to $x = 0$, it remains there for a time exponentially distributed with a parameter λ_0 and then it returns to $x = 1$. The time when the process is at $x = 0$ corresponds to the idle time of the system.

When the process comes to the barrier at $x = N$, it stays there for a time which is exponentially distributed with a parameter μ_0 which corresponds to the time when the system is full and do not accept new customers (the completion time of current service from the moment when the queue becomes full). The assumption on exponential sojourn times in barriers will be dropped below where transient model is presented. Diffusion equation becomes and is supplemented by balance equations for probabilities $p_0(t)$ and $p_N(t)$ of being at the barriers

$$\frac{\partial f(x,t;x_0)}{\partial t} = \frac{\alpha}{2}\frac{\partial^2 f(x,t;x_0)}{\partial x^2} - \beta\frac{\partial f(x,t;x_0)}{\partial x} +$$
$$+ \lambda_0 p_0(t)\delta(x-1) + \lambda_N p_N(t)\delta(x-N+1) ,$$

$$\frac{dp_0(t)}{dt} = \lim_{x\to 0}\left[\frac{\alpha}{2}\frac{\partial f(x,t;x_0)}{\partial x} - \beta f(x,t;x_0)\right] - \lambda_0 p_0(t) ,$$

$$\frac{dp_N(t)}{dt} = \lim_{x\to N}\left[-\frac{\alpha}{2}\frac{\partial f(x,t;x_0)}{\partial x} + \beta f(x,t;x_0)\right] - \lambda_N p_N(t) , \qquad (2)$$

where $\delta(x)$ is Dirac delta function.

Our solution of these equations is based on the representation of the density function $f(x,t;x_0)$ of the diffusion process with barriers with jumps by a superposition of the density functions $\phi(x,t;x_0)$ of diffusion processes with absorbing barriers at $x = 0$ and $x = N$, which has the following form

$$\phi(x,t;x_0) = \begin{cases} \delta(x-x_0) & \text{for } t = 0 \\ \frac{1}{\sqrt{2\Pi\alpha t}}\sum_{n=-\infty}^{\infty}\left\{\exp\left[\frac{\beta x_n'}{\alpha} - \frac{(x-x_0-x_n'-\beta t)^2}{2\alpha t}\right] \right. \\ \left. - \exp\left[\frac{\beta x_n''}{\alpha} - \frac{(x-x_0-x_n''-\beta t)^2}{2\alpha t}\right]\right\} & \text{for } t > 0 , \end{cases} \qquad (3)$$

where $x_n' = 2nN$, $x_n'' = -2x_0 - x_n'$. If the initial condition is defined by a function $\psi(x)$, $x \in (0,N)$, $\lim_{x\to 0}\psi(x) = \lim_{x\to N}\psi(x) = 0$, then the pdf of the process has the form $\phi(x,t;\psi) = \int_0^N \phi(x,t;\xi)\psi(\xi)d\xi$.

The probability density function $f(x,t;\psi)$ of the diffusion process with elementary returns is composed of the function $\phi(x,t;\psi)$ which represents the influence of the initial conditions and of a spectrum of functions $\phi(x,t-\tau;1)$, $\phi(x,t-\tau;N-1)$ which are pd functions of diffusion processes with absorbing barriers at $x = 0$ and $x = N$, started at time $\tau < t$ at points $x = 1$ and $x = N-1$ with densities $g_1(\tau)$ and $g_{N-1}(\tau)$:

$$f(x,t;\psi) = \phi(x,t;\psi) + \int_0^t g_1(\tau)\phi(x,t-\tau;1)d\tau + \int_0^t g_{N-1}(\tau)\phi(x,t-\tau;N-1)d\tau .$$
$$(4)$$

Densities $\gamma_0(t)$, $\gamma_N(t)$ of probability that at time t the process enters to $x = 0$ or $x = N$ are

$$\gamma_0(t) = p_0(0)\delta(t) + [1 - p_0(0) - p_N(0)]\gamma_{\psi,0}(t) + \int_0^t g_1(\tau)\gamma_{1,0}(t - \tau)d\tau$$

$$+ \int_0^t g_{N-1}(\tau)\gamma_{N-1,0}(t - \tau)d\tau ,$$

$$\gamma_N(t) = p_N(0)\delta(t) + [1 - p_0(0) - p_N(0)]\gamma_{\psi,N}(t) + \int_0^t g_1(\tau)\gamma_{1,N}(t - \tau)d\tau$$

$$+ \int_0^t g_{N-1}(\tau)\gamma_{N-1,N}(t - \tau)d\tau , \tag{5}$$

where $\gamma_{1,0}(t)$, $\gamma_{1,N}(t)$, $\gamma_{N-1,0}(t)$, $\gamma_{N-1,N}(t)$ are densities of the first passage time between corresponding points, e.g.

$$\gamma_{1,0}(t) = \lim_{x \to 0} [\frac{\alpha}{2} \frac{\partial \phi(x,t;1)}{\partial x} - \beta\phi(x,t;1)] . \tag{6}$$

For absorbing barriers

$$\lim_{x \to 0} \phi(x,t;x_0) = \lim_{x \to N} \phi(x,t;x_0) = 0 ,$$

hence $\gamma_{1,0}(t) = \lim_{x \to 0} \frac{\alpha}{2} \frac{\partial \phi(x,t;1)}{\partial x}$. The functions $\gamma_{\psi,0}(t)$, $\gamma_{\psi,N}(t)$ denote densities of probabilities that the initial process, started at $t = 0$ at the point ξ with density $\psi(\xi)$ will end at time t by entering respectively $x = 0$ or $x = N$.

Finally, we may express $g_1(t)$ and $g_N(t)$ with the use of functions $\gamma_0(t)$ and $\gamma_N(t)$:

$$g_1(\tau) = \int_0^\tau \gamma_0(t)l_0(\tau - t)dt , \qquad g_{N-1}(\tau) = \int_0^\tau \gamma_N(t)l_N(\tau - t)dt , \tag{7}$$

where $l_0(x)$, $l_N(x)$ are the densities of sojourn times in $x = 0$ and $x = N$; the distributions of these times are not restricted to exponential ones as it is in Eq. (2).

The above equations are transformed by the Laplace transform, and the transform of $f(x,t,x_0)$ is obtained analytically and then its original is computed numerically using e.g. Stehfest algorithm [22].

In case of unlimited queue of $G/G/1$ type we just remove the barrier at $x = N$ and related to it terms and equations.

3   Open Network of G/G/1, G/G/1/N Queues, One Class, Steady State and Transient Solution

The steady-state open networks models of $G/G/1$ queues were studied in [10]. Let M be the number of stations and suppose at the beginning that there is one class of customers. The throughput of station i is, as usual, obtained from traffic equations

$$\lambda_i = \lambda_{0i} + \sum_{j=1}^M \lambda_j r_{ji} , \qquad i = 1, \ldots, M, \tag{8}$$

where r_{ji} is routing probability between station j and station i; λ_{0i} is external flow of customers coming from outside of network.

Second moment of interarrival time distribution is obtained from two systems of equations; the first defines C_{Di}^2 as a function of C_{Ai}^2 and C_{Bi}^2; the second defines C_{Aj}^2 as another function of $C_{D1}^2, \ldots, C_{DM}^2$:

(1) The formula (9) is exact for $M/G/1$, $M/G/1/N$ stations and is approximate in the case of non-Poisson input [1]

$$d_i(t) = \varrho_i b_i(t) + (1 - \varrho_i) a_i(t) * b_i(t) , \qquad i = 1, \ldots, M, \qquad (9)$$

where * denotes the convolution operation. From (9) we get

$$C_{Di}^2 = \varrho_i^2 C_{Bi}^2 + C_{Ai}^2 (1 - \varrho_i) + \varrho_i (1 - \varrho_i) . \qquad (10)$$

(2) Customers leaving station i according to the distribution $D_i(x)$ choose station j with probability r_{ij}: intervals between customers passing this way has pdf $d_{ij}(x)$

$$d_{ij}(x) = d_i(x) r_{ij} + d_i(x) * d_i(x)(1 - r_{ij}) r_{ij} + d_i(x) * d_i(x) * d_i(x)(1 - r_{ij})^2 r_{ij} + \cdots \qquad (11)$$

hence

$$E[D_{ij}] = \frac{1}{\lambda_i r_{ij}} , \qquad C_{Dij}^2 = r_{ij}(C_{Di}^2 - 1) + 1 . \qquad (12)$$

$E[D_{ij}]$, C_{Dij}^2 refer to interdeparture times; the number of customers passing from station i to j in a time interval t has approximately normal distribution with mean $\lambda_i r_{ij} t$ and variation $C_{Dij}^2 \lambda_i r_{ij} t$. The sum of streams entering station j has normal distribution with mean

$$\lambda_j t = [\sum_{i=1}^{M} \lambda_i r_{ij} + \lambda_{0j}] t \qquad \text{and variance} \qquad \sigma_{Aj}^2 t = \{\sum_{i=1}^{M} C_{Dij}^2 \lambda_i r_{ij} + C_{0j}^2 \lambda_{0j}\} t ,$$

hence

$$C_{Aj}^2 = \frac{1}{\lambda_j} \sum_{i=1}^{M} r_{ij} \lambda_i [(C_{Di}^2 - 1) r_{ij} + 1] + \frac{C_{0j}^2 \lambda_{0j}}{\lambda_j} . \qquad (13)$$

Parameters λ_{0j}, C_{0j}^2 represent the external stream of customers. For K classes od customers with routing probabilities $r_{ij}^{(k)}$ (let us assume for simplicity that the customers do not change their classes) we have

$$\lambda_i^{(k)} = \lambda_{0i}^{(k)} + \sum_{j=1}^{M} \lambda_j^{(k)} r_{ji}^{(k)} , \qquad i = 1, \ldots, M; \; k = 1, \ldots, K, \qquad (14)$$

and

$$C_{Di}^2 = \lambda_i \sum_{k=1}^{K} \frac{\lambda_i^{(k)}}{\mu_i^{(k)2}} [C_{Bi}^{(k)2} + 1] + 2\varrho_i(1 - \varrho_i) + (C_{Ai}^2 + 1)(1 - \varrho_i) - 1 . \qquad (15)$$

A customer in the stream leaving station i belongs to class k with probability $\lambda_i^{(k)}/\lambda_i$ and we can determine $C_{Di}^{(k)^2}$ in the similar way as it has been done in Eqs. (11–12), replacing r_{ij} by $\lambda_i^{(k)}/\lambda_i$:

$$C_{Di}^{(k)^2} = \frac{\lambda_i^{(k)}}{\lambda_i}(C_{Di}^2 - 1) + 1 \, ; \qquad (16)$$

then

$$C_{Aj}^2 = \frac{1}{\lambda_j}\sum_{l=1}^{K}\sum_{k=1}^{K} r_{ij}^{(k)}\lambda_i \left[\left(\frac{\lambda_i^{(k)}}{\lambda_i}(C_{Di}^2 - 1)\right)r_{ij}^{(k)} + 1\right] + \sum_{k=1}^{K} \frac{C_{0j}^{(k)^2}\lambda_{0j}^{(k)}}{\lambda_j} \, . \qquad (17)$$

Equations (10), (13) or (15), (17) form a linear system of equations and allow us to determine C_{Ai}^2 and, in consequence, parameters β_i, α_i for each station.

In our approach to transient analysis, the time axis is divided into small intervals (equal e.g. to the smallest mean service time) and at the beginning of each interval the Eqs. (8), (10), (13) are used to determine the input parameters of each station based on the values of $\varrho_i(t)$ obtained at the end of the precedent interval. As the values of parameters are changed at each interval, also external flows $\lambda_{0j}^{(k)}(t)$ may be modelled following any, possibly self-similar process.

Numerical example 1. The considered exemplary network topology is presented in Fig. 1. It was generated by aSHIIP generator [29] allowing generation of hierarchical networks, typical for Internet - this sample network consists of 6 levels. The same topology was used in an example referring to the fluid flow

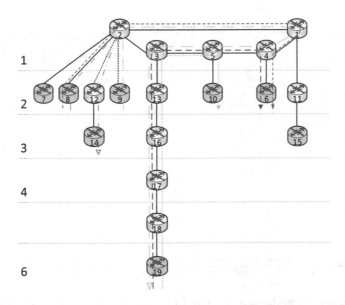

Fig. 1. Topology of the network considered in numerical examples.

Table 1. Generated flows λ_i, $i = 8, 9, 19$, as a function of time

time units	0–20	20–25	25–65	65–70	70–75	75–100
node 8	0.8	1.7	1.7	1.7	1.7	1.3
node 9	1.8	1.8	1.5	1.0	1.0	1.0
node 19	1.0	1.0	2.5	2.5	2.5	1. 5

approximation presented in the next section. We do not compare the results, as diffusion model does not incorporate the TCP congestion window mechanism (although it is possible) and the loads of networks are different in both examples. The examples demonstrate rather the possibilities of both approaches. Here, flows are generated by nodes 8, 9, 19 and their intensity is piecewise, given in Table 1, the routing of flows is indicated in the figure. All nodes have the same service intensity $\mu = 3$, the queue capacity is $N = 20$, and the squared coefficient of variation for all flaws and stations are: $C_A^2 = C_B^2 = 1$. The propagation time between nodes is null (it is easy to compute knowing the length o the links and the speed of light in nodes).

In Figs. 2, 3 the time evolution of mean queues in a few chosen nodes, predicted by diffusion approximation and simulation (we treat simulation results as almost exact) are compared, giving the idea of the errors of the approach. The model naturally gives not only these mean values but also the distributions of queues. Figure 4 compares the time-dependant flow intensities at certain nodes obtained via diffusion approximation and simulation. With our software we may we may generate and solve numerically much larger models, having hundreds of thousands or millions nodes and flows. It seems that our solver concerning transient diffusion models is the unique existing one.

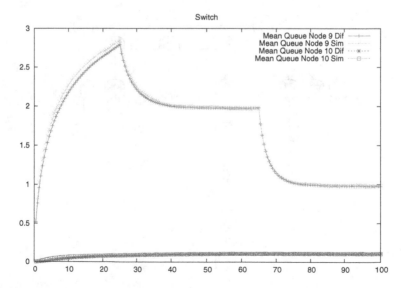

Fig. 2. Mean queue lengths at nodes 9 and 10, diffusion approximation and simulation results

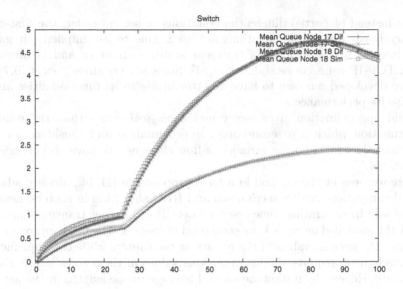

Fig. 3. Mean queue lengths at nodes 9 and 10, diffusion approximation and simulation results

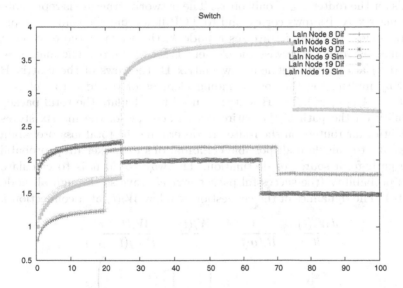

Fig. 4. Flow intensities at nodes 8, 9, and 19, diffusion approximation and simulation results

4 Fluid-Flow Approximation

Fluid-flow approximation is a well-known approach of modelling transient behaviour where only mean values of traffic intensity and service times are considered. Compared to the diffusion approximation, mathematical side of the model is simple: instead of partial differential equations of second order, the first-order ordinary linear differential equations are used. Due to its simplicity, it gained much interest in the analysis of transient states in Internet and in investigation of TCP-IP connections stability [4,31]. Some solvers already exist [6,7] but we have developed our own to have a better inside to its functionalities and to optimise its performance.

Fluid approximation gives larger *methodological errors* than the diffusion approximation which is a second-order approximation and considers not only the mean values but also the variance of flow changes. We have also observed it our tests [5].

Here we present the method in a form proposed in [11,14], already adapted to TCP congestion window mechanism and RED algorithm in routers, hence in an easy way incorporating some essential details of Internet transmissions.

Let the modelled network V be composed of routers. The fluid approximation computes the average values of the queues at the routers while the implemented RED mechanism requires the instantaneous values of them to estimate discard probability. Hence, the instantaneous and average queue lengths in the network are noted by the vectors $\mathbf{q}$ and $\mathbf{x}$. The values of the routers' discard probabilities depend on the instantaneous queue length and are recorded in vector $\mathbf{p}(\mathbf{x})$ which depends for the router $v \in V$ only on x_v. The network structure is represented by binary matrix $\mathbf{A}$. Its rows correspond to TCP flows and the columns represent network nodes. If a flow i traverses a node k, the value of the element a_{ik} is determined as "one", otherwise the element is set to "zero". The matrix $\mathbf{A}$ and vector $\mathbf{p}(\mathbf{x})$ are used to define a new matrix $\mathbf{B}$: the rows of the matrix $\mathbf{B}$ are formed by multiplying the corresponding element of $\mathbf{p}$ and a row of $\mathbf{A}$, such that $B_{ij} = A_{ij} \cdot p_j(x_j)$. The $\mathbf{B}$ matrix is used to calculate the total packet loss probability on the path of the entire flow. Each row of the matrix stores the probabilities for routers on the route. To determine the total loss probability, it is necessary to calculate all possible combinations of packet drop probabilities on the path from source to destination. The way to do it is to calculate the success probability (the successful packet arrival traverse through all nodes on the path). The dynamics of the congestion window $W_i(t)$ at a connection i is:

$$\frac{dW_i(t)}{dt} = \frac{1}{R_i(\boldsymbol{q}(t))} - \frac{W_i(t)}{2} \cdot \frac{W_i(t - \tau)}{R_i(\boldsymbol{q}(t - \tau))} \cdot$$
$$\cdot \left(1 - \prod_{j \in V}(1 - B_{ij})\right) . \tag{18}$$

The other equations concern the mean queue length q_v of router v at this instant, the average queue x_v and delays R_i.

The router's AQM packet discard probability $p(x(t))$, Eq. 20, included in the above formula is determined with the use of the weighted average queue length:

$$x_v(k\delta) = (1 - \gamma_v) \cdot x_v((k-1)\delta) + \gamma_v q_v(k\delta) \tag{19}$$

where:

δ – queue sampling parameter,

γ – weight parameter, specifying the percentage of current queue q taken in the moving average,

k – iteration step,

$$p_v(x_v) = \begin{cases} 0, & 0 \leqslant x_v < t_{min_v} \\ \frac{x_v - t_{min_v}}{t_{max_v} - t_{min_v}} p_{max_v}, & t_{min_v} \leqslant x_v < t_{max_v} \\ 1, & t_{max_v} \leqslant x_v \leqslant B_v \end{cases} \qquad (20)$$

$x_v(t)$, $q_v(t)$ are respectively the average and instantaneous queue lengths in router v. With the transmission capacity C_v the time change of q_v is

$$\frac{dq_v(t)}{dt} = \sum_{i=1}^{K} \frac{W_i(t)}{R_i(q(t))} - \mathbf{1}(q_v(t) > 0) \cdot C_v, \qquad (21)$$

C_v is transmission capacity of the router v. A router allows reception of traffic from K TCP flows ($K \leqslant N$). Each flow $i(i = 1, ..., N)$ is determined by time varying congestion window size W_i, measured in packets and constant propagation delay Tp_i throughout flow route. Total packet delay for flow i (Round Trip Time) consists of queue delay and propagation delay.

$$R_i(q(t)) = \sum_{j=1}^{M} \frac{q_j(t)}{C_j} + Tp_i. \qquad (22)$$

Numerical example 2. The topology of the considered network is the same as in Example 1 and generated by aSHIIP. The flows, propagation times, and buffer sizes at routers were also generated randomly. Table 2 illustrates the network parameters: B - maximum buffer capacity; C - service intensity; Q_0 - initial queue length; γ, t_{min}, t_{max}, p_{max} - weight, thresholds and maximum probability for RED algorithm. On the whole structure, we ran our algorithm for selecting the pair of border routers as a flow endpoints and searched for the shortest paths from the sender to the receiver using the Dijkstra algorithm. Table 3 shows the flow parameters: W_0 - starting window size; Tp - total propagation delay that is the sum of the link delays on the route.

Table 2. Router parameters

parameter \ node	1	2	3	4	5	6	7	8	9	10	11	12	13	14	15	16	17	18	19	
B	91	370	90	114	353	246	169	152	160	271	162	141	34	43	80	81	18	17	15	[pack]
C	0.15	0.225	0.225	0.225	0.225	0.15	0.15	0.15	0.075	0.075	0.15	0.075	0.225	0.075	0.15	0.225	0.225	0.225	0.225	[pack/t.u.]
Q_0	0																			[pack]
γ	0.05																			[-]
t_{min}	15	61	15	19	58	41	28	25	26	45	27	23	5	7	13	13	3	2	5	[pack]
t_{max}	45	185	45	57	176	123	84	76	80	135	81	70	17	21	40	40	9	8	10	[pack]
p_{max}	0.1																			[-]

Table 3. Flow parameters

parameter \ flow	1	2	3	4	5	
Tp	16.61	14.98	53.38	42.01	30.72	[t.u.]
W_0			1			[pack]

Few results are displayed, e.g. the RTT as a function of time for flow classes, Fig. 5. The value of RTT specifies the time needed to propagate an information through the network after which a sender may react on losses. The first overload occurs roughly at $t = 300$ time units (t.u.) and the RTT time of classes 3–5 at that moment ranges between [260, 325], therefore the expected moment when the sender reduces the transmitted traffic is after about 300 t.u. It is visible in Fig. 6 – the window sizes of class 3–5 are reduced at a time close to $t = 600$ t.u. For other classes the losses are so small that the continuous increase of window sizes is observed.

The reduction of window sizes contributed to an immediate decrease of queue length of congested nodes (Fig. 7). The queue in 17th and 18th node started to empty after 600th time unit because of exceeding the RED maximum threshold by average queue around 300th t.u. and rejecting new packets since then. However, the queues did not reach the maximum buffer sizes that were set respectively on 18 and 17 packets.

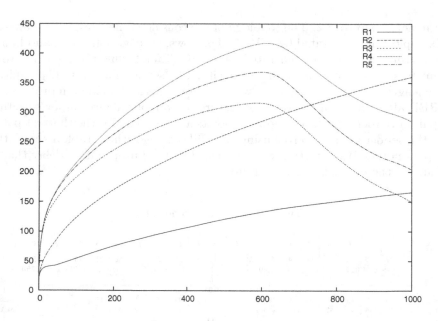

Fig. 5. The RTT time R_i in each flow class

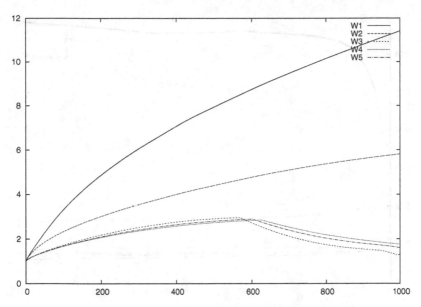

Fig. 6. The congestion window sizes W_i in each flow class

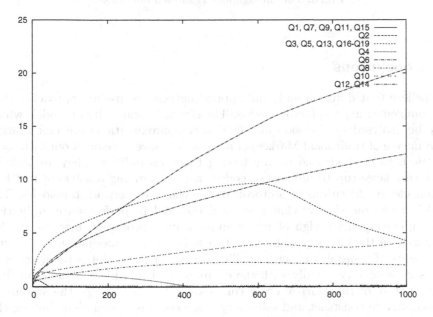

Fig. 7. The queue lengths Q_i in each node

Figure 8 displays the throughputs of flow classes. Each class has different characteristic, however the classes with a shared overloaded part of the route had similar behaviour. The flows that traversed through shared non-congested network fragments have different throughputs – as it is seen for class 1 and 2.

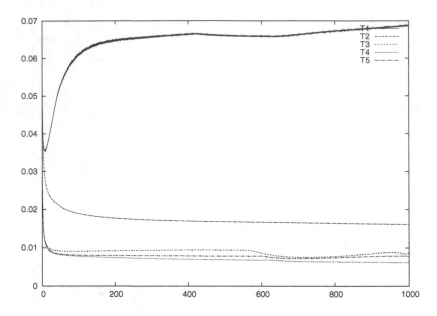

Fig. 8. The throughputs T_i in each flow class

5 Conclusions

We believe that diffusion and fluid approximations are useful approaches that are complementary to Markov models. The size and complexity of models which may be analysed by diffusion and fluid flow approximations are much larger than in case of traditional Markovian models. We have developed our own tools for the both methods and we are testing their possibilities. They are able to treat very large (up to millions of nodes) networks giving a software test bed to consider modifications of protocols or the choice of network topologies. The models based on Markov chains are still essential in performance evaluation and supporting the design of new communication protocols, mechanisms for regulation of the intensity of Internet transmissions and mechanisms to ensure the quality of transmission services. Their principal constraint is the number of states growing very rapidly with the complexity of an object being modelled; as each state of the Markov chain corresponds to one state of the system, it is necessary to construct and solve very large systems of equations linking the states probabilities. The existing solvers as e.g. XMARCA [28], PEPS [20], or PRISM [21] consider only steady state Markov chains.

We are developing our own Markovian package Olymp [19]. It is a library for generating transition matrices of continuous time Markov chains, and solve the resulting model. Olymp uses Java language to define network nodes and their interactions - that gives a lot of flexibility in defining a network structure and functions.

At the moment we are able to generate and solve Markov chains of the 150 million of states. The main method of solution is one of projection methods based on Krylov subspace with Arnoldi process, e.g. [23,26,28]. We plan to increase the size of tractable Markov chains by several orders through the use of a GPU-CPU (graphical processing unit) and a better design of computational algorithms for parallel computing and optimization of memory usage, [17].

An alternative to analytical models is discrete event simulation – also used here to evaluate diffusion approximation results. We have developed an extension of OMNET++ (a popular simulation tool written in C++, [18]) allowing simulation of transient state models. In this case a simulation run should be repeated a sufficient number of times (e.g. 500 thousands in our examples) and the results for a fixed time should be averaged. That makes transient simulation models time-consuming.

Acknowledgments. This work was supported by Polish project NCN nr 4796/B/T02/2011/40 *"Models for transmissions dynamics, congestion control and quality of service in Internet"* and the European Union from the European Social Fund (grant agreement number: UDA-POKL.04.01.01-00-106/09).

References

1. Burke, P.J.: The output of a queueing system. Oper. Res. **4**(6), 699–704 (1956)
2. Carrasco, J.A.: Transient analysis of some rewarded markov models using randomization with quasistationarity detection. IEEE Trans. Comput. **53**(9), 1106–1120 (2004)
3. Czachórski, T.: A method to solve diffusion equation with instantaneous return processes acting as boundary conditions. Bull. Pol. Acad. Sci. Tech. Sci. **41**(4) (1993)
4. Czachórski, T., Grochla, K., Pekergin, F.: Stability and dynamics of TCP-NCR(DCR) protocol in presence of UDP flows. In: García-Vidal, J., Cerdà-Alabern, L. (eds.) Euro-NGI 2007. LNCS, vol. 4396, pp. 241–254. Springer, Heidelberg (2007)
5. Czachórski, T., Pekergin, F.: Diffusion approximation as a modelling tool. In: Kouvatsos, D.D. (ed.) Next Generation Internet: Performance Evaluation and Applications. LNCS, vol. 5233, pp. 447–476. Springer, Heidelberg (2011)
6. Simulating large networks using Fluid Flow Models (FFM). http://www-net.cs.umass.edu/fluid/ffm.html
7. FSIM. http://www.ispl.jp/oosaki/software/fsim
8. Genin, D., Marbukh, V.: Bursty fluid approximation of TCP for modeling internet congestion at the flow level. In: Allerton'09: Proceedings of the 47th annual Allerton Conference on Communication, Control, and Computing, IEEE Press (2009)
9. Gelenbe, E.: On approximate computer systems models. J. ACM **22**(2), 261–263 (1975)
10. Gelenbe, E., Pujolle, G.: The behaviour of a single queue in a general queueing network. Acta Inf. **7**(2), 123–136 (1976)
11. Hollot, K., Liu, Y., Misra, V., Towsley, D., Gong, W.B.: Fluid methods for modeling large heterogeneous networks. Technical report AFRL-IF-RS-TR-2005-282 (2005)

12. Kleinrock, L.: Queueing Systems. Volume I: Theory. Wiley, New York (1975)
13. Kleinrock, L.: Queueing Systems. Volume II: Computer Applications. Wiley, New York (1976)
14. Liu, Y., Lo Presti, F., Misra, V., Gu, Y.: Fluid models and solutions for large-scale IP networks. In: ACM/SigMetrics (2003)
15. Liu, J.: Packet-level integration of fluid TCP models in real-time network simulation. In: Proceedings of the 38th Conference on Winter Simulation, Monterey, California, 03–06 December , pp. 2162 - 2169 (2006)
16. Misra, V., Gong, W., Towsley, D.: A fluid-based analysis of a network of AQM routers supporting TCP flows with an application to RED. In: Proceedings of the Conference on Applications, Technologies, Architectures and Protocols for Computer Communication (SIGCOMM 2000), pp. 151–160 (2000)
17. Numerical computation for Markov chains on GPU: building chains and bounds, algorithms and applications. Project POLONIUM 2012–2013, bilateral cooperation PRISM-Université de Versailles and IITiS PAN, Polish Academy of Sciences
18. OMNET++ Community Site. http://www.omnetpp.org
19. Pecka, P., Deorowicz, S., Nowak, M.: Efficient representation of transition matrix in the Markov process modeling of computer networks. In: Czachórski, T., Kozielski, S., Stańczyk, U. (eds.) Man-Machine Interactions 2. AISC, vol. 103, pp. 457–464. Springer, Heidelberg (2011)
20. PEPS. http://www-id.imag.fr/Logiciels/peps/userguide.html
21. PRISM - probabilistic model checker. http://www.prismmodelchecker.org/
22. Stehfest, H.: Algorithm 368: numeric inversion of Laplace transform. Comm. ACM 13(1), 47–49 (1970)
23. Saad, Y.: Analysis of some Krylov subspace approximations to the matrix exponential operator. SIAM J. Numer. Anal. 29(1), 208 (1992)
24. Sakumoto, Y., Asai, R., Ohsaki, H., Imase, M.: Design and implementation of flow-level simulator for performance evaluation of large scale networks. In: 15th Annual Meeting of the IEEE International Symposium on Modeling, Analysis, and Simulation of Computer and Telecommunication Systems (MASCOTS) (2007)
25. Sakumoto, Y., Ohsaki, H., Imase, M.: Accelerating flow-level network simulation with low-pass filtering of fluid models. In: SAINT 2012, Izmir (2012)
26. Scientifique, C., Philippe, B., Sidje, R.B.: Transient solutions of Markov processes by Krylov subspaces. In: 2nd International Workshop on the Numerical Solution of Markov Chains (1989)
27. Srikant, R.: The Mathemtics of Internet Congestion Control. Springer, Heidelberg (2004)
28. Stewart, W.: Introduction to the Numerical Solution of Markov Chains. Princeton University Press, Chichester (1994)
29. Tomasik, J., Weisser, M.A.: Internet topology on AS-level: model, generation methods and tool. In: 29th IEEE International Performance Computing and Communications Conference IPCCC 2010 (2010)
30. Weltzl, M.: Network Congestion Control: Managing Internet Traffic. Wiley, New York (2005)
31. Gu, Y., Liu, Y., Towsley, D.: On integrating fluid models with packet simulation. In: Proceedings of IEEE INFOCOM (2004)

New Generation Computer Networks Survivability Analysis and Optimization

Yurij Zaychenko[✉] and Helen Zaychenko

Institute for Applied System Analysis, NTTU "KPI", Kiev, Ukraine
baskervil@voliacable.com, syncmaster@bigmir.net

Abstract. The problem of survivability analysis of new generation computer networks is considered. Survivability index for computer networks is suggested and method for its estimation is described. The problem of computer networks structure optimization by survivability indices is also considered and method for its solution described. The experimental investigations of the suggested methods for computer networks survivability analysis and optimization are presented.

1 Introduction

New generation networks (NGN) are the networks which utilize the perspective networking technologies (e.g. multiprotocol label switching (MPLS) and Generalized MPLS) aimed at extending information services for users and ensuring the given QoS (Quality of Service) for different information types: audio, video and data.

The main QoS factors are Packets Transfer Delay (PTD), Packets Delay Variance (PDV) and Packets Loss Ratio (PLR) [1,2].

The main advantages of this network technology are the following: it provides unified technique for fast transmission of various information types — data, video and audio via high speed channels and integrates with upper level protocol — TCP/IP.

The important problem arising when analyzing existing NGN networks or their development is the problem of network survivability analysis. In order to perform such analysis it's necessary to introduce adequate survivability indices which would take into account the peculiarities of this technology and to develop an algorithm for its estimation. The main goals of this paper are the presentation of the suggested survivability analysis method, its investigations and structural network optimization by survivability.

2 Problem Statement and Model for Estimation Survivability

Like the work [3] we consider the system survivability as the ability to preserve its functioning and to ensure the fulfillment of its main functions (perhaps in

V. Vishnevsky et al. (Eds.): DCCN 2013, CCIS 279, pp. 73–81, 2014.
DOI: 10.1007/978-3-319-05209-0_6, © Springer International Publishing Switzerland 2014

the shortened amount) under given QoS. As the main MPLS network function is the transmission of different classes of packets flows so we'll estimate the survivability level as maximal flow value to be transmitted in a network under its channels and nodes failures under given values of QoS factors.

Let's consider the survivability analysis problem statement.

Let it be MPLS network which is defined by an oriented graph $G = \{X, E\}$ where $X = \{x_j\}$ is a set of network nodes, $E = (r, s)$ is a set of channels, μ_{rs} are channels capacities.

Assume that in network K classes of service (CoS) are transmitted due to so-called demand matrices $H(k) = ||h_{ij}(k)||$, $i = 1 \ldots N$, $j = 1 \ldots N$ (Mbits per sec). For each flow the corresponding QoS is introduced as a given value mean delay time $T_{cp,k}$ estimated by the following expression [3].

$$T_{cp,k} = \frac{1}{H_{\Sigma}^{(k)}} \sum_{(r,s) \in E} \frac{f_{rs}^{(k)} \sum_{j=1}^{k} f_{rs}^{(j)}}{\left(\mu_{rs} - \sum_{j=1}^{k-1} f_{rs}^{(j)}\right) \cdot \left(\mu_{rs} - \sum_{j=1}^{k} f_{rs}^{(j)}\right)}, \qquad (1)$$

where $H_{\Sigma}^{(k)} = \sum_{i=1}^{n} \sum_{j=1}^{n} h_{ij}^{(k)}$ is a total intensity of input flow of the k-th CoS, μ_{rs} is the channel (r, s) capacity, $f_{rs}^{(k)}$ is a k-th class of flow value in the channel (r, s).

It's necessary to determine the survivability indices for a given network.

In papers [3,4] the following complex factor was suggested for survivability analysis of MPLS networks

$$P\{H_{\Sigma}^{\Phi}(1) \geq r \,\% H_{\Sigma}^{0}(1)\}, P\{H_{\Sigma}^{\Phi}(2) \geq r \,\% H_{\Sigma}^{0}(2)\} \ldots P\{H_{\Sigma}^{\Phi}(k) \geq r \,\% H_{\Sigma}^{0}(k)\}, \qquad (2)$$

where $H_{\Sigma}^{0}(k)$ is k-th class flow value in the faultless state; $H_{\Sigma}^{\Phi}(k)$ is a real flow value of class k in case of failures, $r = (50 \div 100)$, $P\{H_{\Sigma}^{\Phi}(k) \geq r \,\% H_{\Sigma}^{0}(k)\}$, $k = 1 \ldots K$ is the probability that flow value of k-th class transmitted in a network would be not less than a fraction r of the nominal flow value in the faultless state $H_{\Sigma}^{0}(k)$.

As it is not known a priori the maximal flow value under failures, the hypotheses is introduced that general flow structure under failures is preserved, that is the ratio of flow values of different classes (CoS) should be preserved, namely

$$H_{\Sigma}^{\Phi}(1) : H_{\Sigma}^{\Phi}(i) : H_{\Sigma}^{\Phi}(k) = H_{\Sigma}^{0}(1) : H_{\Sigma}^{0}(i) : H_{\Sigma}^{0}(k), \qquad (3)$$

This assumption allows to suggest the following method of MPLS networks survivability analysis [3].

3   The Algorithm of MPLS Networks Survivability Analysis

Let MPLS network $G = (X, E)$ be considered consisting of n elements, channels and nodes exposed to influence of environment due to which they fail. It's assumed that reliability characteristics of network elements: readiness coefficients (probability of faultless state) of channels $k_{\Gamma_{r,s}}$ and nodes k_{Γ_i}, $(r, s) \in E$, $i = 1 \ldots n$ are known. Consider the following network failure states:

1. Failure of one channel: Z_1;
2. Failure of one node: Z_2;
3. Failure of two channels: Z_3;
4. Simultaneous failure of one channel and one node: Z_4;
5. Failure of three channels: Z_5.

Assuming failures of network elements to be statistically independent events we may determine the probability of each state $P(Z_i)$. For example, if Z_i is the channel (r_i, s_i) failure, then

$$P(Z_i) = (1 - K_{\Gamma_{r_i,s_i}}) \prod_{(r,s) \neq (r_i,s_i)} K_{\Gamma_{r,s}} \prod_{i=1}^{n} K_{\Gamma_i}, \qquad (4)$$

where $K_{\Gamma_{r_i,s_i}}$ is a probability of faultless state of the channel $(r, s) \neq (r_i, s_i)$; $1 - K_{\Gamma_{r_i,s_i}}$ is a probability of the channel (r_i, s_i) failure.

In [3] MPLS networks survivability estimation algorithm was suggested, which consists of the following steps:

1. Compute the total flow value in the faultless state (so-called nominal flow) for all classes of service (CoS) $H_\Sigma^{(0)}(1), H_\Sigma^{(0)}(2), \ldots, H_\Sigma^{(0)}(K)$.
2. Simulate network different failure states: Z_1, Z_2, Z_3, Z_4, Z_5. For each failure state calculate the corresponding probability $P(Z_i)$ according to (4).
3. Find the maximal flow value for all CoS in the state Z_j : $H_\Sigma^{\Phi}(k, z_j)$, $k = 1 \ldots K$. We use for it a specially developed algorithm of finding maximal flow [4].
4. Calculate the complex survivability index for each CoS: for class k

$$P\{H_\Sigma^{\Phi}(k) \geq r \% H_\Sigma^0(k)\} = \sum_{Z_i} P(Z_i), \qquad (5)$$

where summing in (5) is performed over all states Z_i such that $H_\Sigma^{\Phi}(k) \geq r \% H_\Sigma^0(k)$, H_Σ^0 is the nominal flow value of the class k in network faultless state; $H_\Sigma^{\Phi}(k)$ is a real flow value of the class k in case of failures, $r = (50 \div 100), k = 1 \ldots K$.

The found dependencies $P\{H_\Sigma^{\Phi}(1) \geq r \% H_\Sigma^0(1)\}, P\{H_\Sigma^{\Phi}(2) \geq r \% H_\Sigma^0(2)\}$ $\ldots P\{H_\Sigma^{\Phi}(k) \geq r \% H_\Sigma^0(k)\}$, are further presented as curves in coordinates - $P\{H_\Sigma^{\Phi}(k)\} - r \% H_\Sigma^0$. And by its change we may estimate the survivability of the corresponding network.

4 Accelerated Algorithm of Survivability Determination in MPLS Networks

While estimating networks survivability it's necessary to analyze all the failure states in a network including the states of multiple failures, namely the state sets: Z_3 – two channels failure; Z_4 – failure of one channel and one node and Z_5 – a failure of three channels, etc.

Let it be in a network m channels and n nodes, then the total number of states of the set Z_3 is equal to $|Z_3| = C_m^2 = \frac{m(m-1)}{2!}$, in the set Z_4 is equal to $|Z_4| = m \cdot n$, and in the set Z_5 is equal to $|Z_5| = C_m^3 = \frac{m(m-1)(m-2)}{3!}$.

As for survivability calculations it's necessary to solve the problem of maximal flow search for each state the problem of survivability estimation demands huge volume of calculations even for a small network. Therefore to obtain substantial cut in calculations in the problem of network survivability estimation approximate algorithm based on the method of statistical trials (Monte-Carlo) was developed. The algorithm runs in the following way.

Set the number of failure states to be accounted in % for each subset Z_i, e.g.: for subset

Z_1: $n_1 = 100\%$; Z_2: $n_2 = 100\%$; Z_3: $n_3 = 30\%$; Z_4: $n_4 = 20\%$; Z_5: $n_5 = 10\%$; These subsets are chosen randomly Z_3, Z_4, Z_5 and denote them correspondingly: $Z_{3,y}$, $Z_{4,y}$, $Z_{5,y}$. Then calculate the contribution in the total survivability function $P\{H_\Sigma(k) \geq r\%H_\Sigma^{(0)}\}$ by each subset as follows:

$P_3\{H_\Sigma(k) \geq r\%H_\Sigma^{(0)}\} = \Sigma P(z_j)$, $z_j : H_\Sigma(z_j) \geq r\%H_\Sigma^{(0)}$, $z_j \in Z_{3,y}$.

$P_4\{H_\Sigma(k) \geq r\%H_\Sigma^{(0)}\} = \Sigma P(z_j)$, $z_j : H_\Sigma(z_j) \geq r\%H_\Sigma^{(0)}$, $z_j \in Z_{4,y}$.

$P_5\{H_\Sigma(k) \geq r\%H_\Sigma^{(0)}\} = \Sigma P(z_j)$, $z_j : H_\Sigma(z_j) \geq r\%H_\Sigma^{(0)}$, $z_j \in Z_{5,y}$.

Then calculate the integral estimate for network survivability with the account of all failure states.

The experimental investigations were carried out for estimation of the suggested approximate algorithm depending on the values $n_i\%$, $i = 3, 4, 5$. As the results of these experiments for a network with parameters $n = 25$, $m = 39$, the mean percentage estimation error of survivability for values $n_1 = 100\%$, $n_2 = 100\%$, $n_3 = 30\%$, $n_4 = 20\%$, $n_5 = 10\%$ lies in the range $4 \div 5\%$.

5   The Network Survivability Optimization Problem Statement

In the process of network design after analysis of its survivability the problem arises to ensure the desired survivability level. Naturally, this problem may be solved by the way of reserving its channels and nodes and the structural optimization, which demands the additional expenses. Consider the corresponding problem statement of network structural optimization by survivability indices.

Let it be MPLS network which as earlier is defined by oriented graph $G = \{X, E\}$, where $X = \{x_j\}$ is a set of network nodes, $E = \{(r, s)\}$ is a set of channels; μ_{rs} are channels capacities.

Assume K classes of flows (CoS) are transmitted in the network according to demand matrices $H(k) = ||h_{ij}(k)||, i = 1 \dots N, j = 1 \dots N$ (Mbit/s). The reliability characteristics of channels and nodes are given, namely readiness coefficients for channels $K_{\Gamma_{rs}}$ and nodes K_{Γ_i} and corresponding failure probabilities $P_{omk_i} = K_{\Gamma_i}$. For each class k quality of service (QoS) is given as a mean delay time value $T_{cp,k}$. Let the following values of survivability indices be established for each flow class: $P_{0,predef}^{(k)}, P_{1,predef}^{(k)}, \dots, P_{5,predef}^{(k)}$.

It is demanded to determine such network structure for which the following requirements on survivability level will be ensured:

$$P\{H_\Sigma^\Phi(k) \geq r \,\% H_\Sigma^{(0)}(k)\} \geq P_{r,predef}^{(k)}, r = (50 \div 100), k = 1 \ldots K, \quad (6)$$

and the additional costs would be minimal:
$C_\Sigma = \Sigma C_{rs}^{res}(\mu_{rs}) \to min.$

The achievement of the desired survivability level we'll obtain by corresponding reservation of the most responsible channels and nodes.

For reservation efficiency estimation we propose to introduce the following index for channels:

$$\alpha_{r_i s_i} = -\frac{\Delta P(Z_i)}{C_{r_i s_i}}, \quad (7)$$

where Z_i is a state of failure of the channel (r_i, s_i);
$\Delta P(Z_i)$ is a probability change of the state Z_i in case of channel reservation,
$C_{r_i s_i}$ is a cost of this reservation.

The value $\Delta P(Z_i)$ is estimated by the following formula:

$$P_{res}(Z_i) - P(Z_i) =$$
$$P_{omkr_i s_i}^2 \cdot \prod_{(r,s) \neq (r_i,s_i)} K_{\Gamma_{r,s}} \prod_{i=1}^n K_{\Gamma_i} - P_{omkr_i s_i} \prod_{(r,s) \neq (r_i,s_i)} K_{\Gamma_{r,s}} \prod_{i=1}^n K_{\Gamma_i} =$$

$$= -(1 - P_{omkr_i s_i}) \cdot P_{omkr_i s_i} \prod_{(r,s) \neq (r_i,s_i)} K_{\Gamma_{r,s}} \cdot \prod_{i=1}^n K_{\Gamma_i} = -(1 - P_{omkr_i s_i}) \cdot P(Z_i),$$
$$(8)$$

where $P_{omkr_i s_i}$ is failure probability of the channel (r_i, s_i).

The similar expressions are used for estimation of nodes reservation.

The index $\alpha_{r_i s_i}$ is used for selection of the proper elements, nodes and channels to be reserved in the first turn. The following method of MPLS network optimization by survivability level is suggested.

6 Method of MPLS Network Optimization by Survivability

The algorithm consists of finite number of iterations. On each iteration the next element (node or channel) is reserved.

k-th iteration.

1. For each channel and node the index $\alpha_{r_i s_i}$ is computed by formula (7).
2. Select channel (r^*, s^*) such that $\alpha_{r^*, s^*} = max_{(r_i, s_i)} \alpha_{r_i, s_i}$.
3. Reserve channel (r^*, s^*) and recalculate survivability indices for all the flow classes after reservation using the following formula:

$$P^H\{H_\Sigma^\Phi(k) \geq r \,\% H_\Sigma^{(0)}\} = P\{H_\Sigma^\Phi(k) \geq r \,\% H_\Sigma^0\} + |\Delta P(Z_i^*)|, \quad (9)$$

where $\Delta P(Z_i^*)$ is a probability change of the state Z_i after the channel (r^*, s^*) reservation.

4. Check the fulfillment of condition (6):
$$P^H\{H_\Sigma^\Phi(k) \geq r\,\%H_\Sigma^{(0)}\} \geq P_{r,predef}^{(k)}, r = (50 \div 100), k = 1\ldots K.$$

If the conditions (6) are fulfilled for each r and all the classes K, then stop, otherwise, go to $(k+1)$ iteration.

The described iterations are repeated until the condition (6) would be true. As at each iteration the values of survivability indices increase and their upper bound are limited by 1 the algorithm converges after finite number of iterations not exceeding $m + n$, where m is a number of channels, n is a number of nodes.

7 The Experimental Investigations of the Survivability Analysis Algorithm

The suggested algorithms for NGN networks survivability analyses and optimization were implemented and corresponding software kit was developed. The experimental investigations of the suggested algorithms were carried out. The experiments were performed on the basis of global MPLS Ukrainian network, its structure is presented on Fig. 1.

In the first series of experiments the investigations of survivability algorithm were carried out. In the process of experiments dependence of survivability indices on variations of the set values of QoS-mean delay time (MDT) for different classes (CoS) was explored.

All the experiments were performed for readiness coefficients of channels distributed uniformly in the range (0.9–0.95) and readiness coefficients of nodes distributed uniformly in the interval (0.95–0.99).

In the first experiment the survivability sensibility to the variations of the constraint on MDT for the first class of service was explored. The corresponding results are presented in the Table 1 and on Fig. 2.

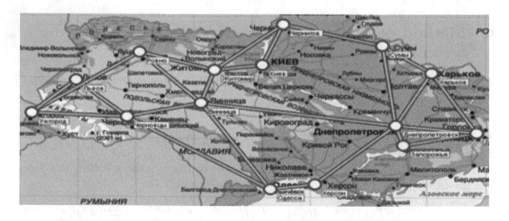

Fig. 1. The structure of Global MPLS network

Table 1. Survivability versus MDT (mean delay time)

	Tcp = 0.01	Tcp = 0.05	Tcp = 0.15	Tcp = 0.7
P (100%)	0.352779	0.389927	0.410044	0.432404
P (90%)	0.352779	0.389927	0.410044	0.432404
P (80%)	0.482439	0.520454	0.539737	0.565827
P (70%)	0.483985	0.521005	0.541752	0.565827
P (60%)	0.484394	0.521005	0.543298	0.566516
P (50%)	0.484394	0.521005	0.545879	0.566516

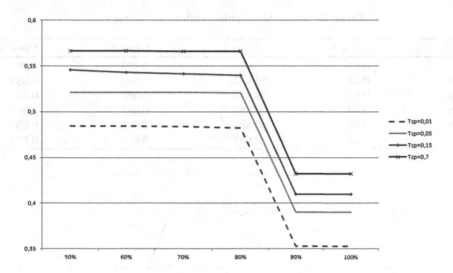

Fig. 2. Survivability sensibility versus MDT variation

As we may easily see from the curves survivability index is more sensitive to variations of T_{cp} in the interval 0.01–0.05 than in the interval 0.15–0.7. The similar results were obtained for other CoS.

In the next experiments the investigations of survivability optimization algorithm by reservation were carried out. In process of experiments the influence of demands matrices variations on survivability indices were explored for all CoS.

The demands matrix was varied using multiplying coefficient k which may have the following values 0.5; 0.75; 1; 1.25; 1.5; 1.75. $H_* = kH$.

The analysis was performed with the next parameters: channels readiness coefficient 0.99, nodes coefficient -0.94 and $k = 0.5 - 1.4$, $k = 2$.

In the Table 2 the survivability indices values for different classes (CoS) are presented before optimization and in the Table 3 after optimization by the suggested algorithm for $k = 0.5$.

The corresponding survivability indices after optimization for 4 classes are presented also on Fig. 3.

Comparing results in Tables 2 and 3 we may conclude the application of suggested algorithm has substantially improved network survivability.

Table 2. Survivability indices before optimization

H %	Class 1	Class 2	Class 3	Class 4
0.5	0.385	0.385	0.385	0.385
0.6	0.385	0.385	0.385	0.385
0.7	0.385	0.385	0.385	0.385
0.8	0.385	0.385	0.385	0.385
0.9	0.385	0.385	0.385	0.385
1	0.385	0.385	0.385	0.385

Table 3. Survivability indices after optimization

H %	Class 1	Class 2	Class 3	Class 4
0.5	0.82	0.685	0.685	0.685
0.6	0.82	0.685	0.685	0.685
0.7	0.82	0.685	0.685	0.66
0.8	0.82	0.685	0.685	0.66
0.9	0.82	0.685	0.685	0.66
1	0.685	0.685	0.685	0.66

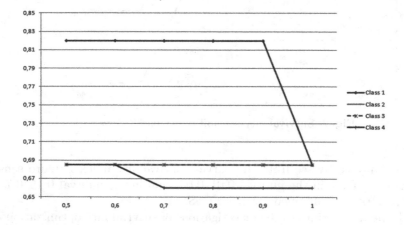

Fig. 3. The optimized survivability indices for different CoS

8   Conclusion

1. The problem of survivability analysis for new generation computer networks is stated. The complex survivability index is introduced and the algorithm of its estimation is developed.
2. The problem of NGN computer networks optimization by survivability indices is stated and the algorithm of its solution is suggested.
3. The experimental investigations of the suggested algorithms were carried out and their efficiency was estimated.

References

1. Vivek, O.: Structure and realization of modern MPLS technology. Wiliams, p. 480 (2004)
2. Goldshtein, A.B., Goldshtein, B.S.: MPLS Technology and Protocols, p. 304. St. Petersburg (2005)
3. Zaychenko, H.Y., Zaychenko, Y.P.: Networks on MPLS technology: modeling, analysis, optimization. Kyiv: NTTU KPI, p. 240 (2008)
4. Zaychenko, H.Y., Zaychenko, Y.P.: Searching the max flow in networks with asynchronous information transfer mode. Inf. Select. Process. **17**(93), 52–64 (2002)

Optimisation of Extended Generalised Fat Tree Topologies

Adamantini Peratikou[✉] and Mo Adda

School of Computing, University of Portsmouth, Portsmouth, PO1 3HE, UK
{Adamantini.Peratikou,Mo.Adda}@port.ac.uk

Abstract. Extended generalised fat tree (XGFT) are interconnection networks with bidirectional multistage properties, (BMIN) which can be extended or scaled to accommodate different system sizes and requirements. However, these extended topologies do not address power consumption and traffic constraints. In this paper, we extract a sub-set of the generalised fat tree topologies that are power consumption and performance aware. We called this sub-set optimised OXGFT. The cost which is proportional to the relative power is the objective function that is minimised based on the traffic constraints to maintain a lower delay and a higher throughput. The simulation results show that the extracted OXGFT topologies perform well under various load conditions.

Keywords: Fat-tree · Extended generalised fat tree · Optimisation · Interconnections · High performance architectures

1 Introduction

Fat tree networks [1] were proposed as binary tree based topologies. The only difference is that the processors of the fat tree are located to the leaves of a binary tree (Fig. 1), and the fact that moving upwards to the root of the tree the communication links increase and therefore the communication bandwidth increases as-well. K-ary n tree architectures were later proposed with the difference that the upward links are quicker by a factor k than the downward links in order to achieve a non-changing bisection bandwidth. Fat trees have a constraint that when implementing them the switch port rates become too high near the root of the tree, thus the use of switches with the same radix and port speed is inevitable.

The most suitable candidate for the fat tree topologies is k-ary n-trees as it allows the switches to be configured at all levels in a similarly way as illustrated in Fig. 1(c). In fat trees the system size depends on the degree of the switch such as $p = 2^n$, where n is the height of the level, while in K-ary n-trees the descendants of the node are at index positions therefore $p = k^n$.

Topologies based on fat tree that do not have full bisectional bandwidth are call extended fat trees such as m-ary trees [2].

V. Vishnevsky et al. (Eds.): DCCN 2013, CCIS 279, pp. 82–90, 2014.
DOI: 10.1007/978-3-319-05209-0_7, © Springer International Publishing Switzerland 2014

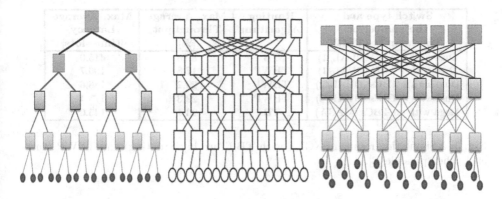

Fig. 1. a 4-level fat tree, b binary 4 tree [2], c K-ary n-tree with $k = 3$ and $n = 3$

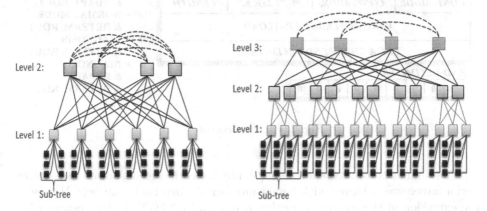

Fig. 2. a XGFT configuration of (2, 6, 6, 4, 0) with 36 processors and b XGFT (3, 4, 3, 5, 2, 2, 2) with 60 processors [3]

While fat tree class topologies include some interesting characteristics they also experience some known issues such as the bottleneck caused from the limited availability of paths as in some cases only a single path exists. Extended generalised fat tree or XGFT was proposed by [2] as an optimisation of the standard fat tree topologies. Extended generalised fat tree or XGFT [3] on the other hand, unlike k-ary n-tree, are interconnections that can be extended, or scaled to accommodate different system sizes and requirements. Switches in various stages of the network have different number of bi-directional ports. Like k-ary n-tree, extended generalised fat trees can be regenerated recursively to accommodate a larger system, and the connectivity along with the links used, depends on the configuration requirements. Figure 2 illustrates two examples of XGFT each with different number of routing switches and number of leaf nodes. Each different stage of routing switches from top switch to the bottom switch is considered to have different levels of switches, with each level consisting different sub-trees.

Switch type and the number of TB-channels (TBC)	Routing Algorithm	Max. Average Throughput [%]	Max. Average Latency [time slots]
Dual with 1 TBC (DUAL/1)	TB	8.18	415.0
Dual with 1 TBC (DUAL/1)	TBWP	17.8	196.7
Dual with 2 TBCs (DUAL/2)	TBWP	20.2	186.6
Dual with 3 TBCs (DUAL/3)	TBWP	22.3	175.4
Mega with no TBCs (MEGA)	TB	23.1	145.6

Fig. 3. a Simulation results of XGFT with TB (Turn back routing) and TBWP (turn back when possible [3].

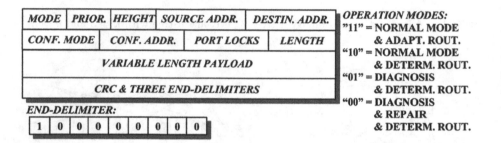

Fig. 4. Packet structure [3]

The simulation results reported in [3], illustrated in Fig. 3, showed that better performance was achieved with higher number of Turn back channels. However, the extension of the routing algorithm proposed in XFGT [4] does not provide any performance enhancements, thus the added complexity that is introduced in the configuration of XGFT is unnecessary. This can be proved by our future work where we propose a different generalisation.

The addressing used is based on the space encoding for TB (Turn Back) and TBWP (Turn back when possible) routing algorithms [3]. The addresses are of integer vectors which specify the routing path, both the source address and the destination address are attached into the packet header to ensure correct shortest path calculations (Fig. 4). Two routing options exist, the up routing and the down routing. In the up routing, a packet is routed in the upwards direction until it finds the common routing switch ancestor or reaches the root switch of the destination node. The common ancestor switch can be found, by comparing the destination address with the source address carried along with the packet at each switch stage of the network. Once the ancestor switch is reached, then the down routing checks the destination address to determine the proper port to route the packet through to reach its destination. This routing is entirely deterministic, whereas the up routing is adaptive.

2   Optimal Configuration

To find the optimal configuration among the space of all the endless XGFTs topologies, a simulator was developed that uses the power consumption as the objective functions with traffic constraints. That simulator takes the number of processors and runs a set of constrain in order to find the optimum configuration of both sub-trees and routing switches to be set to achieve the higher performance possible with the lower cost.

2.1   Objective Function

The objective function is proportional to the power consumption. By minimising the cost of the architecture, one can minimise the power associated with it. The connectivity cost for each level depends on the number of ports, the number of sub-trees, and the number of routing switches. Overall the cost of the XGFT of level n can be expressed as

$$Cost = \sum_{i=1}^{n} R_i^T \times L_i^2 \tag{1}$$

where R_i^T is the total number of routing switches at level i and L_i is the total number of ports per routing switch at level i. The total number of routing switches and ports per routing switch at level i, which determines the complexity of the level and hence the total complexity of the network, can be defined by the following two equations:

$$R_i^T = R_i \prod_{j=i+1}^{n} S_j \tag{2}$$

$$L_i = S_i + \frac{R_{i+1}}{R_i} \tag{3}$$

where S_i is the number of sub-trees, and is the number of routing switches per sub-tree at level i.

2.2   Constrains

The cost equation 1 is minimised subject to several constraints that ensure a high performance for the topology. This is achieved by setting the number of routing switches and ports per switch and per level to an adequate number that guarantees an overall minimum latency. The total number of processors is set as a constraint among the sub-trees which is defined in the following equation.

$$P = \prod_{i=1}^{n} S_i \rightarrow P = S_i \times S_{i+1} \times \cdots \times S_n \tag{4}$$

The connectivity constrains can be illustrated in the two following equations. The ratio between routing switches of different levels has to be a positive integer as it defines the number of ports per routing switch. The number of the routing switches per sub-tree has to increases from leafs to the root to satisfy the connectivity requirements of a fat tree.

$$\frac{R_{i+1}}{R_i} \in Z^+ \tag{5}$$

$$R_i \leq R_{i+1} \leq R_{i+2} \leq \ldots \leq R_n \tag{6}$$

The number of sub-trees per level is the most important constrain. The sub-trees and the routing switches are related to make sure that the numbers of ports per level are adequate enough to fully connect the number of sub-trees per level and hence minimise the delay in the network. This equation is based on queuing theory and can be simplifies into:

$$S_i \leq \frac{R_{i+1}}{\prod_{j=1}^{n} S_j} \tag{7}$$

One can also include another constraint to relate to the current technology which requires the maximum number of ports supported by a given switch at any level.

3   Performance Analysis

For the purpose of this research two simulators were implemented. The first one in C++ called m: Z-node that can simulate multiple fat tree topologies with multiple groups of levels and sub-trees along with the option of adjusting the properties of the channel links, routing, and applications patterns, and the second one called SimOpt in VB and Excel that takes a set of constraints and produces an optimal topology based on Eq. 1. Two configurations of XGFT [4] with different processors and number of levels were compared to their optimal versions obtained from our optimisation simulator. Figure 2(a) illustrates a two-level configuration of XGFT, which consists of 36 processing elements, divided into groups of 6, with each group connecting to an ancestor switch. The total number of ancestor switches for the first and second levels are 6 and 4 respectively. Figure 2(b) shows a three level configuration with 60 processors.

According to the constraints and the objective function discussed above, both configurations do not satisfy the requirements for high performance based on the given number of processors. Their optimised versions for the same number of processors and levels are shown in Fig. 5. However, for 60 processors the optimum configuration pays a smaller price for power consumption as illustrated in Fig. 6 compared to the non-optimised shown in Fig. 3, at the achievement of better performance, as we will demonstrate later.

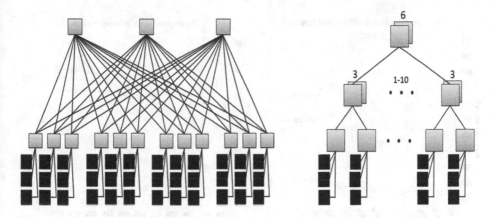

Fig. 5. a Optimal configuration for 36 processors. b Optimal configuration for 60 processors.

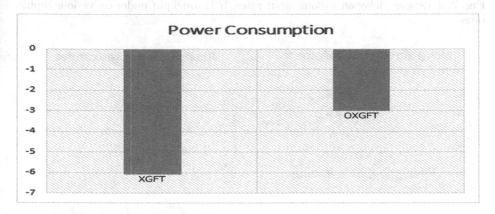

Fig. 6. Power reduction with 60 processors in XGFT and OXGFT

3.1   Discussion of the Optimal Configuration of XGFT

The optimal configuration for 36 processors with two levels, according to our assumptions, consists of a total of 12 switches for the first level, with each switch connected to 3 processors, and 3 switches for the second level (Fig. 5(a)).

XGFT configuration and optimal configuration on 36 processors were tested under various offered traffic load. Both the configurations performed similarly on a load of 5–50 %, with the optimum configuration achieving slightly lower message delay of −1.00 to −2.00 ns compared to XGFT (Fig. 7(a)). When the load increases to 60 % the difference in throughput between the two configurations becomes noticeable (Fig. 7(b)), and the message delay becomes significantly higher in XGFT. This is due to the lack of ports interconnections to service all the backed traffic at each level.

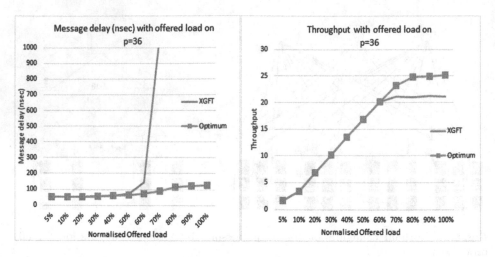

Fig. 7. a Message delay on various input rates. b Throughput under on various input rates.

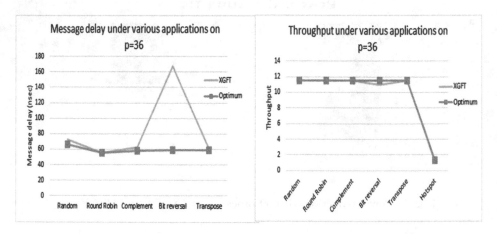

Fig. 8. a Message delay on 36 processors under different applications. b Throughput on 60 processors under various applications

Figure 8 indicates that the message delay is lower under all traffic patterns in the optimum configuration with the exception of Round Robin where both configurations have equal values. The throughput is equal on both cases except on bit reversal traffic where in the OXGFT configuration is slightly higher.

The XGFT configuration (Fig. 2(b) against optimum XGFT configuration on 60 processors was also tested, the configuration of the optimum XGFT consists of 20, 30 and 6 switches for levels 1, 2 and 3 respectively (Fig. 5(b)).While the XGFT configuration consists of the switching elements illustrated in Fig. 2(b).

Under various normalised loads (Fig. 9) , it is identified that even on higher number of processors the optimum configuration still overcomes the XGFT

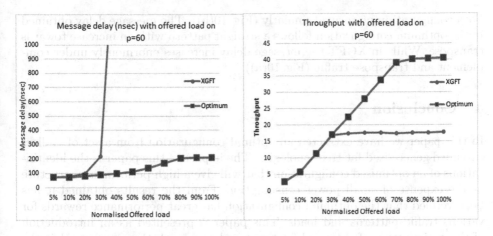

Fig. 9. a Message delay under various normalised input rates on 60 processors. b Throughput under various normalised input rates on 60 processors

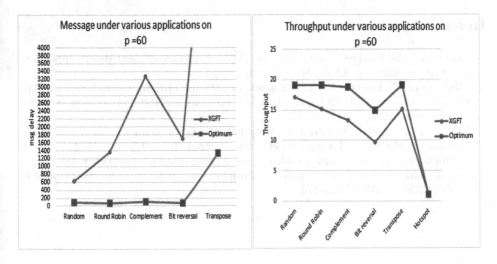

Fig. 10. a Message delay on 60 processors under various applications. b Throughput on 60 processor configuration under various applications.

configuration in all the different loads. The difference in the message delay between the two configurations is even higher. The message delay obtained in XGFT shows a significant increase in Complement traffic while the OXGFT (optimum) follows amore constant pattern against all traffics with an increase towards hotspot.

Based on the simulation results illustrated in Fig. 10, it is identified that the optimum configuration performs significantly better on 60 processors than the non-optimised XGFT. The throughput is higher in the optimum configuration under most of the traffic patterns, with the exception of hotspot traffic in which both configuration performed similarly (Fig. 10(b)). The message delay obtained in the optimum configuration follows a straight pattern with an increase towards transpose. While in XGFT the message delay increases enormously under complement and transpose traffic (Fig. 10(a)).

4 Conclusion

In this paper we have extracted an optimal configuration from a set of endless extended generalised fat tree topologies. The impact of this paper is the identifications of an optimised configuration that will give a high performance structure at the expense of small power consumption. From the results obtained it has been verified that the optimum configuration has great performance rewards for various traffic patterns and loads. This paper is presented as an introduction of the optimisation of extended fat tree topologies, with the aim to exploit the optimisation of all fat tree class topologies in future work.

References

1. Leiserson, C.E.: Fat-trees: universal networks for hardware efficient supercomputing. IEEE Trans. Comput. **C−34**(10), 892–901 (1985)
2. Minkenberg C., Luijten, R.P., Rodríguez, G.: On the optimum switch radix in fat tree networks. In: IEEE 12th International Conference on High Performance Switching and Routing, pp. 44–51 (2011)
3. Kariniemi, H., Nurmi, J.: Performance evaluation and implementation of two adaptive routing algorithms for XGFT networks. Comput. Inf. **23**(5–6), 415–435 (2004)
4. Kariniemi H.: On-line reconfigurable extended generalised fat tree xgft network-on-chip for multiprocessor system-on-chip. Tampere University of Technology Publication 614.2006. ISBN 952-15-1746-8

The Problems of Digital TV and Radio Broadcasting Systems Implementation

Viktor P. Dvorkovich[1] and Alexander V. Dvorkovich[2]([✉])

[1] Bauman University, Moscow, Russian Federation
[2] Moscow Power Engineering Institute, Moscow, Russian Federation
dvp@niircom.ru,
dvork.alex@gmail.com

Abstract. Implementation of digital broadcasting systems requires the solution of a set of problems, including national and international standardization, advances in video and channel coding, measurement methods and equipment for broadcasting systems. Different systems could be effective for highly populated and sparsely inhabited regions. New digital terrestrial multimedia broadcasting system may solve such problems. Some key characteristics of the RAVIS system are presented and the prospects of the system are discussed.

Keywords: Digital broadcasting · Multiplex · Video coding · Metrology · FEC coding · QEF

One of the most important tasks of modern digital broadcasting systems' implementation is to create favorable conditions for the advanced (breakthrough) development of world-class systems in leading areas of technology. The telecommunications is such direction, in which, inter alia, a significant increase in efficiency of use of limited natural resource – radio frequency spectrum – is realized on the basis of modern digital multimedia processing algorithms.

The solution of these problems should be systematic, pushing the development of highly effective methods for digital processing and transmission of video and audio information in real time and the implementation of appropriate transceiver systems.

Attention should be paid for solution of a set of problems, without which the effective development of digital technologies is impossible. The two major ones are the elaboration of appropriate standards and the creation of measurement systems to ensure the effectiveness of development, production and implementation of innovations.

Development of TV and radio broadcasting and implementation of digital broadcasting is a very complex problem in many countries. Digital broadcasting technology is advancing rapidly enough. Development of high-performance content processing systems and second generation broadcasting systems leads to the need to revise the plans for digital broadcasting employment, for example, in the part of the design of multiplexes.

V. Vishnevsky et al. (Eds.): DCCN 2013, CCIS 279, pp. 91–100, 2014.
DOI: 10.1007/978-3-319-05209-0_8, © Springer International Publishing Switzerland 2014

However, these statements are only a part of the existing challenges and practically do not affect the technical and political aspects of introduction of modern broadcasting systems.

For example, in the Russian Federation special attention was given to upgrading of the broadcasting network, including the transition to digital broadcasting, expansion of broadcasting in the country and abroad, better targeting and greater diversity of information services, development, implementation and deployment of new information products and technologies in the field of mass communications.

Elaboration of national specifications for TV broadcasting based on international standards can solve a lot of problems for developers, manufacturers and operating personnel. It is actual for modern multivariate broadcasting systems, in particular second generation DVB family — high-bitrate system for digital satellite broadcasting DVB-S2 [1], an improved system for digital terrestrial broadcasting DVB-T2 [2] and high-bitrate digital cable broadcasting system DVB-C2 [3].

For instance, an analysis of DVB-T2 materials can explain the need for the national specifications. DVB-T2 standard involves the use of various radio channel bandwidths 1.7, 5, 6, 7, 8, and 10 MHz. In accordance with this a different number of OFDM carriers, reference signals and so on are used. National specification may involve the use of only certain values of TV channels bandwidths, for example, 8 MHz. Thus, a set of parameters indicated by DVB-T2 standard, will never be used.

It should be mentioned that the development of China's national TV broadcasting standard DTMB have been done, taking into account the advantages and disadvantages of adopted TV broadcasting standards ATSC, DVB and ISDB.

Currently, discussions take place in the Russian Federation about the parameters of three multiplexes for digital television broadcasting — 8–9 programs in each. Indeed one such multiplex could be implemented with DVB-T broadcasting standard and video encoding according to MPEG-2 standard. Second generation broadcasting standard DVB-T2 and H.264/AVC video coding standard [4] already can provide multiplex with 24 standard TV programs or 8 high-definition television programs (Fig. 1).

Implementation of new video codec H.265/HEVC [5] will increase the number of high-definition programs in one channel to at least 12. Undoubtedly, this new encoder and the corresponding receiving devices will be widely used within 2–4 years.

Another major problem is the lack of metrological methods and equipment of broadcasting. The availability of metrological base is a pledge of creating of high-quality equipment and its effective usage. At the present stage of engineering development the evolution of metrology is associated with the creation of virtual measurement systems based on the use of computer programs that provide analysis and organization of systems for forming and processing of measurement information. There are serious successes of the Russian Federation

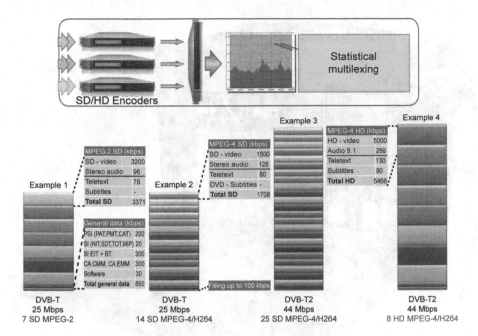

Fig. 1. Multiplexing of TV programs in DVB-T/DVB-T2 systems

scientific school in this direction [6,7]. The school was established by eminent Russian scientist Professor Mark I. Krivosheev.

Figure 2 shows the systems for TV measurements KI-TVM, KI-TVM- E (reference) and KI-TVC. They provide reference measurement signals generation and measurement of parameters of video and radio signals for analogue and digital television systems.

It should be noted particularly the need for the implementation of digital cable television broadcasting, that uses the frequency resource on a secondary basis.

There is one more important problem of TV broadcasting digitalization. It is associated with very irregular population of the Earth, except Europe and North America. Night shot of the Earth from space (Fig. 3) shows that, for example, only European part of the Russian Federation is lightened. It is impossible to cover such a large country with digital television broadcasting in UHF frequency band and broadband access using LTE, for example.

Moreover, modern television systems, providing transmission of a large number of programs in a single standard radio channel, couldn't be implemented in relatively small localities, for TV broadcasting in universities, for mobile reception in public and individual transport. In these cases, video data can be delivered in a narrow radio channel.

For the realization of these opportunities a real-time audiovisual information system RAVIS have been developed, providing very efficient use of VHF

Fig. 2. Systems for TV measurements KI-TVM, KI-TVM- E (reference) and KI-TVC

spectrum band and introduction of new multimedia services, including video broadcasting and improved sound quality.

The RAVIS system provides:

- essential growth (tenfold and more) of VHF spectrum bands utilization efficiency;
- broadcasting of video programs for mobile users, an opportunity of television broadcasting in sparsely populated localities with drastic cut in the broadcasting system cost;
- deployment of single-frequency networks (SFN) for mobile reception along highways and railways;
- implementation of the local radio public warning systems for civilians and organizations in emergency situations;
- essential (tenfold) decrease in power consumption of broadcasting facilities.

The main competitive advantages of the system for all categories of consumers are the following.

Fig. 3. Night shot of the earth from space (http://apod.nasa.gov/apod/ap001127.html)

For listeners:

- high quality sound broadcasting (stereo and multichannel);
- new multimedia services (video, text, supplement data, EPG, etc.);
- simple tuning using the station name, genre, etc.

For manufacturers:

- mass replacement of old analogue receivers;
- modernization of transmission facilities with preserving of infrastructure;
- overall growth of the market potential for transmitting and receiving equipment.

For broadcasters:

- decrease of power consumption per one program;
- increase of coverage;
- increase of the quality and the quantity of programs, the variety of services for customers;
- possibility of adaptation to the requirements of both large and small broadcasters.

For regulation authorities:

- growth of spectrum utilization efficiency;
- possibility of preserving of frequency allocations, simple licensing procedures;
- coordination in the framework of the international-recognized system.

Two VHF frequency bands are used for FM sound broadcasting in former USSR countries (Fig. 4a, b):

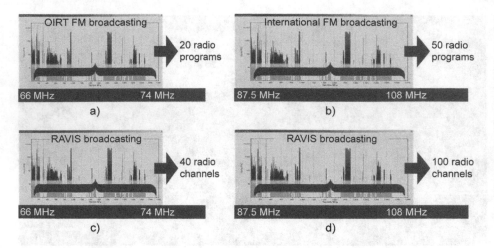

Fig. 4. The effectiveness of VHF frequency bands utilization using the RAVIS system

- 66–74 MHz (OIRT FM band) and
- 87.5–108 MHz (international FM band).

Implementation of the RAVIS system allows the usage of up to 140 radio channels with 200 kHz bandwidth in these frequency bands (Fig. 4c, d). It is possible to transmit up to 20 stereo sound programs with CD quality or up

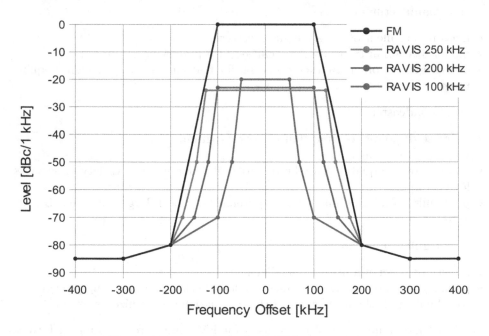

Fig. 5. Spectrum masks for RAVIS transmission

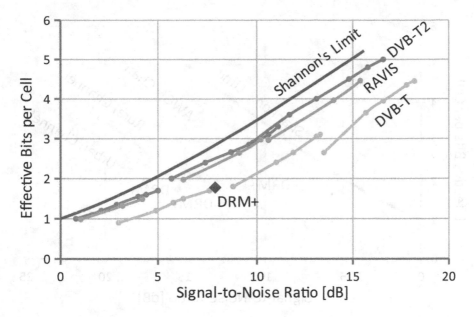

Fig. 6. The comparison of error-rate performance of the systems RAVIS, DVB-T, DVB-T2 and DRM+ (BER = 10^{-4})

to 8 multichannel (5.1) sound programs, or video program accompanied with multilingual sound in each radio channel. In addition, the necessary broadcasting coverage requires the use of significantly lower transmission power compared to analogue FM broadcasting.

Figure 5 shows spectrum masks for RAVIS transmission and spectrum mask for analogue FM transmission according to ETSI ES 302 018-1 [8]. It illustrates that the RAVIS system could be used without changing of frequency allocations.

Russian Federation Patent [9] lays in the base of the RAVIS system. Russian Federation national standard on the system [10] has been adopted. Four additional nation standards are under adoption procedure. These standards concerns:

- digital modulator,
- test receiver,
- content builder and
- norms and methods of metrology supplement.

The RAVIS system is recognized at international level. A set of international reports includes the description and parameters of the system [11–15]. Implemented in FEC coding at bitrates up to 900 kbps in the presence of Gaussian noise error-rate performance close to the Shannon bound, as it is for television standard DVB-T2 (Fig. 6).

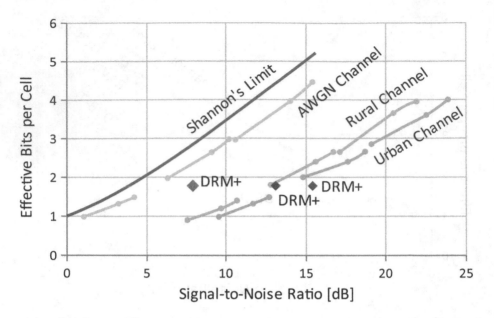

Fig. 7. RAVIS and DRM+ performance for Gauss, urban and suburban channels (BER $= 10^{-4}$)

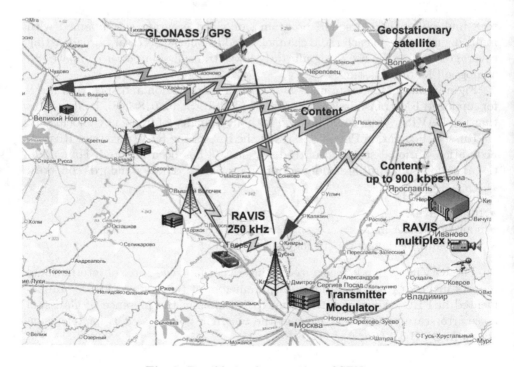

Fig. 8. Possible implementation of SFN

However, in contrast to television and broadband systems, used FEC coding methods, narrow bandwidth of the system and utilization of VHF frequency bands provides quasi-error-free information reception under multipath environment for urban and suburban mobile reception (receiver speed — 60 and 150 km/h, Fig. 7).

RAVIS receivers provide automatic emergency notifications. Along with the main service channel (MSC) the system has two additional logical channels with increased error-rate performance: 12 kbps low bitrate channel (LBC, for example, for sound notifications) and 5 kbps reliable data channel (RDC, for example, for digital notifications and telemetry).

The RAVIS system practically has no analogs in its capabilities. It must be emphasized that one of the most important advantages of the system is the possibility of SFN implementation along highways and railways with transmitters' synchronization using GLONASS/GPS signals and content transmission via fiber optic cables or geostationary satellites (Fig. 8).

This year the development of prototypes of equipment for RAVIS system have been finalized — video encoders, multi-program sound encoders, multiplexers, modulators, power amplifiers and receivers. The next planned works are equipment production and implementation of several fragments of RAVIS broadcasting network — in large and small cities, SFNs along highways and railways.

References

1. ETSI EN 302 307 V1.3.1: Digital video broadcasting (DVB); Second generation framing structure, channel coding and modulation systems for broadcasting, interactive services, news gathering and other broadband satellite applications (DVB-S2), March 2013
2. ETSI EN 302 755 V1.3.1: Digital video broadcasting (DVB); Frame structure channel coding and modulation for a second generation digital terrestrial television broad-casting system (DVB-T2), April 2012
3. ETSI EN 302 769 V1.2.1: Digital video broadcasting (DVB); Frame structure channel coding and modulation for a second generation digital transmission system for cable systems (DVB-C2), April 2011
4. ITU-T Recommendation H.264: Advanced video coding for generic audiovisual services (ISO/IEC 14496–10:2012), April 2013
5. ITU-T Recommendation H.265: High efficiency video coding (ISO/IEC 23008–2:2013), April 2013
6. Dvorkovich, V.P., Dvorkovich, A.V.: Digital Video Information Systems (Theory and Practice), p. 1008. Technosphere, Moscow (2012). ISBN 978-5-94836-336-3
7. Basiy, V.T., Dvorkovich, V.P., Makarov, D.G., Latrygin, N.M.: Monitoring of quality characteristics of television channels. Trudy NIIR **3**, 26–33 (2007)
8. ETSI EN 302 018–1 V. 1.2.1: Electromagnetic compatibility and radio spectrum matters (ERM); Transmitting equipment for the frequency modulated (FM) sound broadcasting service; Part 1: Technical characteristics and test methods, March 2006
9. Russian Federation Patent 2441321: The method for mobile narrowband digital multimedia radio broadcasting. Priority 26/07/2010 (2010)

10. Russian Federation National Standard GOST R 54309-2011: Realtime audiovisual information system (RAVIS). Framing structure, channel coding and modulation for digital terrestrial narrowband broadcasting system for VHF band. Technical specification (2011)
11. Report ITU-R BT.2049-5: Broadcasting of multimedia and data applications for mobile reception, May 2011
12. Report ITU-R BS.2214: Planning parameters for terrestrial digital sound broadcasting systems in VHF bands, May 2011
13. ECC Report 117: Managing the transition to digital sound broadcasting in the frequency bands below 80 MHz, September 2010
14. ECC Report 141: Future possibilities for the digitalisation of Band II (87.5–108 MHz), May 2010. Technical supplement to ECC Report 141, April 2012
15. ECC Report 177: Possibilities for future terrestrial delivery of audio broadcasting services. April 2012

Cross-Layer Adaptation-Based Video Downlink Transmission over LTE: Survey

Tatiana Efimushkina$^{(\boxtimes)}$ and Moncef Gabbouj

Department of Signal Processing, Tampere University of Technology,
Korkeakoulunkatu 10, 33720 Tampere, Finland
{tatiana.efimushkina,moncef.gabbouj}@tut.fi
http://www.tut.fi

Abstract. The recent technological growth in the field of video communications has caused a rise of video applications along with an emergence of new large screen mobile devices. Note that unlike the second and the third generation wireless technologies, the Long-Term Evolution (LTE) is able to support high data rate video applications based on its increased capacity. A Cross-Layer Adaptation (CLA) approach has become very popular due to its ability to achieve an acceptable video quality and satisfy delay requirements over the error-prone LTE networks. The CLA design implies the interaction between the network layers that results in adaptation of the source compression and channel coding toward an optimized video delivery. This paper presents a survey on the recent CLA advances which claim to provide considerable improvements for video communication over LTE. State-of-the-art CLA techniques are studied based on the type of adaptation, distortion estimation and optimization problem formulation and solution.

Keywords: H.264 · Cross-layer · Video communications · End-to-end distortion estimation · LTE · Optimization problem

1 Introduction

The Long Term Evolution (LTE) is a wireless technology devised by the 3rd Generation Partnership Program (3GPP) to meet the requirements of the emerging video applications such as video-on-demand, video conferencing and distant learning. Note that the second and third-generation (2G and 3G) wireless technologies, such as Global System for Mobile Communications (GSM), Universal Mobile Telecommunication System (UMTS), High Speed Packet Access (HSPA), etc., have not been able to support high data rate video applications given their capacity limitations. LTE achieves a greatly enhanced user experience by providing high peak throughputs and low latencies ubiquitously along with an interoperability and service continuity with the existing UMTS networks [1].

LTE adopted an efficient multiple access technique called the Orthogonal Frequency Division Multiple Access (OFDMA) technology for downlink transmission (from evolved Node B (eNB) to User Equipment (UE)) that improves

V. Vishnevsky et al. (Eds.): DCCN 2013, CCIS 279, pp. 101–113, 2014.
DOI: 10.1007/978-3-319-05209-0_9, © Springer International Publishing Switzerland 2014

the spectrum efficiency and facilitates flexible user resource allocation [2]. Note that the video compression research direction does not stand still. The latest High Efficiency Video Coding (HEVC) standard claims to double data compression ratio compared to its predecessor H.264/AVC at the same video quality level, but it is still partly under development [3]. The H.264/AVC is the most commonly used video compression standard nowadays that has achieved significant improvements in rate-distortion efficiency compared to the existing ones [4]. Moreover, Scalable Video Coding (SVC) represents an efficient extension of the H.264/AVC standard that easily adapts to different preferences of the users, including network conditions and devices capabilities [5].

The given drastic technological progress has driven the video communications over LTE to be a popular research topic nowadays [6–13]. Indeed, limited radio resources, dynamics in network conditions, presence of wireless channel errors and high user demands along with the strict requirements of video traffic to variable bit rates and low delay imposes challenges on video transmission over wireless networks. In the given scenario of video delivery over LTE the layer-separated design can no longer be optimal, whereas a Cross-Layer Adaptation (CLA) principle is getting more attention. The first application of the cross-layer design refers to Adaptive Modulation and Coding (AMC) technology that has gained a huge increase in throughput under varying network conditions [14]. The successful breakthrough of AMC has resulted in application of a cross-layer principle into various areas, including joint source-channel coding optimization, queuing and scheduling, etc.

Our primary focus in this paper is on recent advances of CLA approach in downlink video transmission over LTE [6–12], which are studied in detail, and we would like to refer the reader to [13] to get more information on the uplink CLA-based video transmission. In the next Section the generic cross-layer video communication system is discussed. Section 3 provides a thorough comparison between the most commonly used distortion estimation algorithms, whereas the network channel estimation is declared in Sect. 4. In Sect. 5 optimization problems and their solutions are covered. Lastly, conclusion and future research goals are presented.

2   CLA Video Transmission System

Cross-layer adaptation approach aims optimized video delivery by jointly using information from various network protocol layers. According to the recent research on cross-layer design of LTE networks there are three types of adaptation principles:

1. The first one adapts MAC/PHY parameters based on the knowledge of the source data, e.g. resource management is performed based on the predefined rate-distortion characteristics of video towards optimizing user Quality of Experience (QoE) [8,11,12].

2. The second one is an application layer bit rate adaptation that takes into account the mechanisms provided by the lower levels (MAC/PHY), such as resource allocation or error control, to allow the streaming service adapt to varying network conditions [15].

3. The third principle deals with joint adaptation of both application layer and MAC/PHY parameters [6,7,9,10].

Obviously, joint adjusting of source and transmission parameters comes along with the computational complexity increase; however, according to experimental results declared in [16] the joint CLA approach shows the best gains in terms of decoded video quality compared to the first two techniques especially when the channel conditions deteriorate. For the details on simulation address [16].

Figure 1 demonstrates the high-level overview of the generic cross-layer based real-time video transmission system with joint parameters adaptation. Let us go through the main conceptual components of the system. In a cross-layer framework there are two case scenarios of operating on input video data: either it can be encoded-on-the-fly, and, accordingly, the video coding parameters are easily controllable (applied in real-time video transmission, e.g. in [6–10]), or it can be pre-encoded and stored at a media server, so the basic characteristics of video packets are known in advance for the whole video sequence (applied for video-on-demand services, in [11,12]).

The cross-layer schemes mentioned above give their preferences to block-based motion compensated video compression standards either MPEG-4, or MPEG-4 Part 10 aka H.264/AVC. Scalable video coding is not considered due to the wider acceptance of a single layer H-264/AVC video encoding and less overhead, however, an easily adaptable SVC technology is getting popularity in a cross-layer design of video communications over LTE [17].

The video encoder supplies the controller with the rate and distortion information of the video encoded sequence. Both the bit rate and quality of the video stream can be directly affected by the selection of the **source coding**

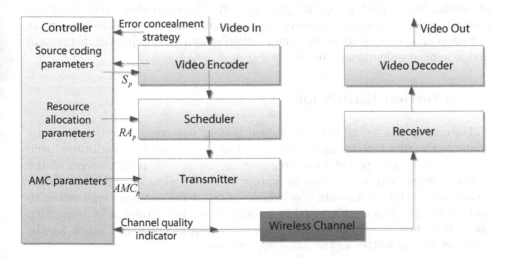

Fig. 1. The generic cross-layer real-time video transmission system.

parameters S_p, such as the coding mode (intra, inter or skip) and quantization step-size. The detailed explanation of the video compression routine is presented in [4].

After getting the selected source coding parameters from the controller, video is encoded and packetized. Scheduler performs allocation of the Resource Blocks (RBs) to the user based on the **resource allocation parameters** RA_p specified by the controller. Transmitter codes the video packets according to the chosen **AMC parameters** AMC_p, such as modulation (QPSK, 16-QAM and 64-QAM) and a coding rate (1/2, 3/4, 9/16), received from the controller that allows enhancing the throughput of the system. Indeed, in a good quality channel larger constellation sizes and higher channel coding rate are selected to gain in the transmission rate. Otherwise, in poor network conditions to decrease the losses of the packets transmission rate is reduced.

Note that in the LTE technology after receiving a downlink signal the UE obtains the Channel Quality Indicator (CQI) evaluated from its Signal to Noise Ratio (SNR) measurements and then reports it to eNB via upload control channel. More information on sub-band, wideband CQI reports, including aperiodic and periodic modes can be found in [18]. At the receiver, the demodulated bit stream is processed by the channel decoder, which performs error detection and correction.

The video decoder is responsible for reconstructing the video sequence for playback. In case some of the encoded information is lost, the video decoder conceals any lost information based on one of the selected error concealment techniques, the broad overview of which is presented in [19].

In the cross-layer system, the controller is the most important component, which is responsible for adapting the parameters of different layers using the knowledge of the error concealment strategy, the source content and CQI. The adaptation is performed by solving the multi-dimensional optimization problem of maximizing or minimizing the given objective function that varies from one research paper to another. However, the estimated end-to-end video quality at the encoder is integrated as the main criterion into the optimization problem and is applied ubiquitously [6–13].

3 Distortion Estimation

Video quality can be affected by source coding parameters that control the amount of quality loss during the encoding process, the network variations, such as delay, jitter and packet loss rate, that are caused by packetization of the video sequence and an error-prone nature of the wireless channel, and error-concealment strategy exploited by decoder. One of the common approaches to address the problem of video quality estimation at the encoder is by measuring the expected end-to-end distortion, the general representation of which for the k packet can be written in the following way [16]:

$$E[D_k] = (1 - p_k)E[D_{r,k}] + p_k E[D_{l,k}], \tag{1}$$

where p_k is the kth packet loss probability, $E[D_{r,k}]$ is the expected distortion if the packet is received correctly, which includes source coding distortion as well as error propagation, and $E[D_{l,k}]$ is the expected distortion if the packet is lost, which accounts for the distortion due to error concealment strategy. Even the simple temporal error-concealment scheme that replaces the pixel values of a block contained on lost packet with pixel values from the same location in previous frame along with the predictive coding create complex dependencies between packets. Moreover, the varying packet loss probability in time varying channels causes the distortion process to be a non-stationary random process. All of these issues jointly make an expected distortion computation a challenging task. Although, there are a number of algorithms for expected video distortion estimation in the literature (address the bibliography of [16]), we would like mostly to specify the ones that are currently used by the state-of-the-art cross-layer schemes [6–13] when addressing downlink video communications over LTE.

The cross-layer techniques [6,7] employ Recursive Optimal per-Pixel Estimate (ROPE) algorithm at the controller that can trace the error propagation from previous frames and enables accurate estimation of video distortion [20]. ROPE uses Mean Squared Error (MSE) as a metric to calculate the first and second moments of the expected pixel's distortion given a probability of packet loss. However, nonlinear clipping that contributes to the transmission distortion is neglected. Moreover, ROPE is said to be very sensitive to the approximation errors caused by subpixel motion compensation and other pixel averaging operations.

The video estimation scheme that can overcome this problem is proposed in [21], and is exploited by the cross-layer scheme [9]. The proposed concise model categorizes the distortion into source, error-propagated and error-concealment distortion items, which are further separated into several small items that can be found either directly or recursively. Besides, the model can be extended to block-level implementation, which results in reduction of computational complexity and memory cost.

Table 1 shows the comparison of three end-to-end distortion estimation methods, including ROPE, method in [21] and Error-Resilient Rate Distortion Optimization (ER-RDO) technique, that are used for evaluation of the H.264 error-resilient video coding performance. The details of simulation are declared in [21]. The researchers in [22,23] proposed an ER-RDO distortion estimation algorithm by simulating K independent decoders (here $K = 30,500$) at the encoder during the encoding process and then averaging the given simulated distortion. The algorithm is based on the Law of Large Numbers (LLN), and the result asymptotically approaches the expected end-to-end distortion when K goes to infinity. ER-RDO has been included in H.264 test model for mode decision purpose. Among the main disadvantages of this method is first high computational complexity and large memory requirements, and second high variation of the estimated distortion result at various encoders given identical input parameters, such as packet loss, video sequence, due to the random nature of the produced error events.

Table 1. Comparison of the main end-to-end distortion estimation methods applied to error-resilient video coding [21].

Sequence	Method	PSNR at different loss rate [dB]			Running time [s]
		3 %	10 %	20 %	
Foreman, 64 kbps	ROPE	30.35	27.51	25.54	34.32
	[21]	30.31	27.60	25.58	22.44
	ER-RDO (30)	30.04	27.28	25.35	88.70
	ER-RDO (500)	30.21	27.46	25.50	1317.38
Paris, 144 kbps	ROPE	27.47	25.88	24.79	582.63
	[21]	27.51	25.93	24.84	387.88
	ER-RDO (30)	27.01	25.65	24.35	968.75
	ER-RDO (500)	26.71	25.59	24.54	7243.56

As demonstrated in Table 1, the method in [21] outperforms ER-RDO for various video sequences at different loss rates. ROPE and method in [21] show quite similar behavior in terms of Peak Signal to Noise Ratio (PSNR), but the method in [21] benefits in terms of computational complexity that is the smallest among all the studied methods.

Although the algorithms declared in [24,25] have not been adopted in LTE CLA-based downlink video communication yet, they comprise the state-of-the-art expected end-to-end distortion estimation approaches. The idea in [24] is to quantify the effects of the following individual terms on transmission distortion: residual concealment error, Motion Vector (MV) concealment error, propagation error and clipping noise, and correlations between any two of them (RMPC). The given pixel-level algorithm is based on the integer pixel MV assumption, and therefore, does not show accurate results in state-of-the-art video codecs, including H.264 and HEVC, that use fractional pixel motion compensation with interpolation filtering. Note that the distortion estimation for pixels under filtering operations requires the computation of the second moment of a weighted sum of random variables. The given problem is solved in [25], where the extension of RMPC algorithm known as ERMPC to subpixel-level distortion estimation without significantly increasing complexity is introduced.

In order to visualize the advantages of the latter approach address Table 2, which demonstrates the average PSNR gain (in dB) of ERMPC over other distortion estimation algorithms, including RMPC, ERRDO (30) and ROPE for different video sequences. The above mentioned distortion estimation algorithms are used for mode decision in H.264 environment (refer to [25] for the details of simulation). It can be seen from Table 2 that ERMPC outperforms RMPC in terms of average PSNR gain by 0.25 dB for Mobile video sequence (packet loss = 2 %), it achieves an average PSNR gain of 1.34 dB over ERRDO for Foreman video sequence (packet loss = 1 %), and shows 3.18 dB increase in average PSNR gain over ROPE for Mobile sequence (packet loss = 0.5 %).

The simplified way of finding the estimate of video distortion D is used in [11] that implies a complex function of both bit rate R of a resource block and

Table 2. Average PSNR gain of ERMPC over different video distortion estimation algorithms in error-resilient video coding [25].

Sequence	Foreman				Mobile			
Packet loss	5%	2%	1%	0.5%	5%	2%	1%	0.5%
ERMPC vs RMPC, (dB)	0.08	0.13	0.21	0.17	0.20	**0.25**	0.21	0.21
ERMPC vs ERRDO(30), (dB)	0.64	1.07	**1.34**	1.24	0.50	0.82	0.56	0.54
ERMPC vs ROPE, (dB)	1.59	1.37	1.41	1.42	1.11	1.89	2.79	**3.18**

Bit Error Rate (BER) P_{BER}:

$$D = 2^{-2R}(1 - P_{BER})^R + (1 - (1 - P_{BER})^R), \qquad (2)$$

where P_{BER} is approximated based on the received SNR parameter and AMC parameters. Although the given approximation may benefit in simplicity, its accuracy is under the question due to the fact that the error-concealment and error-propagation are not taken into account.

The cross-layer schemes [8,10] use Mean Opinion Score (MOS) metric for user-perceived quality measurements. In [8] the video quality is firstly measured by the Video Structural SIMilarity (VSSIM) index due to the fact that human eyes are highly adapted to the structural distortion rather than the pixel-based distortion, and then the linear mapping between VSSIM and MOS is done. In its turn in [10] a non-linear mapping model based on a hyperbolic tangent function between PSNR and MOS is proposed. First of all, the distortion can be found in terms of MSE as the sum of two expected distortions

$$D = D_s + D_l. \qquad (3)$$

Here, D_s is the distortion due to compression of video sequence, which can be approximated as the following exponential function

$$D_s = a/(e^{R/b} - 1), \qquad (4)$$

where a, b are model parameters and R is a source bit rate. D_l is generated by the time-varying and error-prone characters of wireless channels $D_l = cP_{BER}$, where c is independent of the source bit rate. Therefore, the PSNR of the estimated video distortion can be expressed:

$$D_{PSNR} = 10 \log_{10} \frac{255^2}{\frac{a}{e^{R/b}-1} + cP_{BER}}, \qquad (5)$$

after which the non-linear mapping is performed based on a hyperbolic tangent function [10], whereas the traditional relation is shown in Table 3.

To summarize it up, techniques that use simple approximation of the end-to-end video distortion [8,10,11], e.g. based on coding rate and packet loss are not accurate, and therefore, lose in solving the task of joint adaptation of various parameters to the time-varying channel conditions. The more advanced recursive

Table 3. Traditional relation between PSNR and MOS [10].

PSNR (dB)	<19.9	20–24.9	25–30.9	31–36.9	>37
MOS	1	2	3	4	5
Quality	Bad	Poor	Fair	Good	Excellent
Impairment	Very annoying	Annoying	Slightly annoying	Perceptible but not annoying	Imperceptible

techniques exploited in [6,7,9] that account for error concealment and error propagation are computationally more expensive, but are the candidates for the accurate and precise video quality estimation. Note that research on end-to-end video distortion estimation at the encoder is ongoing, and among the recent advances the contributions in [24,25] have demonstrated accurate results and significant improvements in terms of PSNR when applying for error-resilient video coding.

4   Network Channel Estimation

According to theoretical representation of the real-time video transmission system demonstrated in Fig. 1, every Time Transmission Interval (TTI) the controller is responsible for adapting the cross-layer parameters based on the CQI data. CQI reports are sent from the UE and imply quantized and scaled measures of the experienced by the receivers SNR [26]. Indeed, the wireless LTE channel is known to suffer from multi-path fading and shadowing effects that lead to random errors and decrease in SNR. Therefore, in order to optimize the system's performance it should be adapted to the varying network conditions. This adaptation is implemented at the controller by optimizing the estimated distortion parameter, which in its turn depends on the channel packet loss.

Most of the schemes [7,8,10–12] approximate the channel packet loss P_{BER} due to wireless channel errors based on the AMC scheme with a and b parameters and channel SNR γ received via CQI reports as follows

$$P_{BER} = \frac{a}{e^{\gamma * b}}. \tag{6}$$

However, in [6] the packet loss which is assumed to appear due to random wireless channel errors and network congestion is evaluated based on queuing analysis. The queue is modeled as M/G/1, where service time is formulated as Geometric distribution. Packet loss rate is then obtained based on the drop probability due to exceeding the specified period for playback that is calculated by the tail distribution of the waiting time. The paper claims that the downlink network conditions can be derived by the queuing analysis at the sender without the need of feedback in terms of CQI reports from the user side, which can overcome the problem of delay in reaction that exists in CLA methods.

The authors in [9] that proposed a state-of-the-art CLA-based real-time video transmission system consider a general form of wireless channels using an outage probability. In this system that utilize capacity achieving codes, the outage probability is equal to the packet error probability [27]. The channel mutual information, which can be described by its Probability Density Function (PDF) $\rho(I)$, is calculated

$$I = \log_2 (1 + \gamma), \tag{7}$$

where γ is the channel SNR. After receiving from the UE the parameters that describe the PDF of the channel mutual information, the outage probability, when utilizing r bits per channel use, is found

$$P_0(r) = P(I < r) = \int_0^r \rho(I)dI. \tag{8}$$

5   Optimization Problem

Based on the estimated video distortion, the cross-layer optimization problem is formulated, the general objective of which is to minimize the expected video distortion D under the given packet delay constraint [7]:

$$\min_{\{S_p, AMC_p\}} D(S_p, AMC_p), \text{s.t.} : C(S_p, AMC_p) \leq C_0, T(S_p, AMC_p) \leq T_0, \tag{9}$$

where C, C_0 corresponds to the total and maximum allowable resource consumption, and T, T_0 are the total delay and end-to-end delay constraint, respectively.

The optimization problem in [7] is solved as a problem of finding the shortest path in a weighted directed acyclic graph (DAG) [28]. In cross-layer technique [6] the best video coding parameters S_p and AMC_p for each video slice are chosen so as to maximize the total number of supportable users, given the video quality requirement, the video transmission delay constraint, and a radio resource budget. The detailed algorithm of the proposed optimization solution is presented in [6], and allows finding optimal parameters for each slice over all possible combinations with the considerations of distortion constraint and the minimum number of RBs RA_p for transmission. Due to high computational costs of the optimization solution algorithm the optimization is performed one frame at a time. In [11] the resource allocation of resource blocks for multiple users is formulated as a combinatorial optimization problem that target optimized video quality within delay bounds. The optimization problem, involving multiple constraints and an exponential search space, is solved by exploiting the genetic algorithm (GA) based heuristic approach [29]. The goal of the cross-layer scheme in [8] is to allow more users to join the system while keeping the target mean MOS for all users in a cell. The optimization problem is formed so as to maximize the average utility for all users, but with target mean MOS, and is solved by the greedy search algorithm [8]. In [10] firstly the scheme determines the AMC_p based on mapping the received SNR to the transmission mode. Then, the optimization problem is formulated that aims at finding the set of S_p and RA_p so that

Table 4. A short summary of the survey on downlink video transmission over LTE networks.

Paper, Year	Type of CLA scheme	Network channel	Distortion estimation	Optimization problem	Weak point
[6], 2012	(3) RA_p, AMC_p, S_p are jointly adapted	Packet loss is derived by means of queuing analysis	ROPE	System capacity is max. given video quality requirem., **Solution:** full search algorithm	RB allocation is channel unaware
[7], 2010	(3) RA_p, AMC_p, S_p, are jointly adapted	Packet loss is obtained based on CQI	ROPE	D is minimized under the given packet delay, **Solution:** DAG	A lack of experiment. analysis
[8], 2011	(1) RA_p are adapted based on utility func. of video	Average data rate is estimated based on CQI	MOS	The average utility is maximized with target MOS, **Solution:** algorithm is present	Error concealment, delay bounds are not considered
[9], 2012	(3) AMC_p, S_p are jointly adapted	Outage prob. is found based on channel pdf	[21]	The estimated distortion is minimized, **Solution:** is present	A single user scenario is assumed, delay is neglected
[10], 2013	(3) S_p, RA_p are jointly adapted	packet loss is approx. based on CQI	MOS	Max. video quality under the given delay, **Solution:** PSO	Inaccurate D, which is an aver. estimate
[11], 2012	(1) RA_p are adapted based on video R-D	Packet loss is approx. based on CQI	D is approx. based on bit rate and packet loss	Max. video quality under the given delay, **Solution:** GA	Computationally complex scheme
[12], 2011	(1) RA_p are adapted	Packet loss is approx. based on CQI	No D, transmission rate of user is estimated	Max. transm. rate given the target threshold, **Solution:** is present	Delay is neglected

to maximize the user-perceived video quality as well as ensure fairness among users. The optimization problem is solved by the Particle Swarm Optimization (PSO) method developed in [30]. In [12] the optimization problem that aims at maximizing the transmission rate of the users given the power constraint, and the minimum required transmission rate is formulated. Its suboptimal solution is presented in the paper.

A short comparative summary of the survey that covers the main downlink video transmission advances over LTE [6–12] is shown in Table 4. Here the state-of-the-art CLA-based schemes are comparatively analyzed by various characteristics, including the type of adaptation (Sect. 2), network channel estimation, expected end-to-end distortion estimation, optimization problem formulation and solution, and main weakness of the paper.

6   Conclusion

The cross-layer approach is an efficient way to address the complex problem of video communication over LTE networks. Indeed, the joint consideration of key parameters from different layers leads to a more precise end-to-end video quality estimation that after optimization results in an overall system performance improvements. In this paper, we have reviewed recent advances in existing cross-layer schemes that attempt to improve downlink video transmission over LTE, however there is a number of important research issues in this area that deserve more attention, one of which is the feedback mechanism. Note that the CQI feedback delay may cause a wrong selection of the parameters that may not match current channel conditions. A higher feedback frequency leads to a design trade-off between the performance gain and the signaling cost in cross-layer design.

It is worth mentioning that all the above mentioned CLA-based approaches adopt the theoretical representation of the video transmission system, the major and most important part of which is a controller. However, in real life LTE network deployment there are different modules, including the media server, eNB and UE, but no such processing entity, as controller. Another shortcoming of the existing CLA approaches is an inability to analyze the downlink video transmission simultaneously at the eNB and the UE. Therefore, our future research will comprise development of the two-phase analytical model in discrete time that presents an efficient mathematical tool to evaluate the behavior of the overall system (MS, eNB, UE with CQI) and estimate the main performance measures of the video transmission over LTE networks.

References

1. 3GPP TR 36.913 v8.0.0: Requirements for Further Advancements for E-UTRA (LTE-Advanced). Release 8 (2008)
2. 3GPP TS 36.201 v1.0.0: LTE Physical Layer General Description (2012)
3. Sullivan, G., Ohm, J., Han, W., Wiegand, T.: Overview of the High Efficiency Video Coding (HEVC) standard. J. IEEE Trans. Circ. Syst. Video Technol. 22(12), 1649–1667 (2012). doi:10.1109/TCSVT.2012.2221191
4. Wiegand, T., Sullivan, G., Bjontegaard, G., Luthra, A.: Overview of the H.264/AVC video coding standard. J. IEEE Trans. Circ. Syst. Video Technol. 13(7), 560–576 (2003). doi:10.1109/TCSVT.2003.815165
5. Schwarz, H., Marpe, D., Wiegand, T.: Overview of the scalable video coding extension of the H.264/AVC standard. J. IEEE Trans. Circ. Syst. Video Technol. 17(9), 1103–1120 (2007). doi:10.1109/TCSVT.2007.905532
6. Wu, D., Ci, S., Zhang, W., Zhang J.: Cross-layer rate adaptation for video communications over LTE networks. In: IEEE Global Communications Conference, Anaheim, CA, pp. 5056–5061 (2012). doi:10.1109/GLOCOM.2012.6503884
7. Luo, H., Ci, S., Wu, D., Wu, J., Tang, H.: Quality-driven cross-layer optimized video delivery over LTE. J. IEEE Commun. Mag. 48(2), 102–109 (2010). doi:10.1109/MCOM.2010.5402671
8. Shehada, M., Thakolsri, S., Despotovic, Z., Kellerer, W.: QoE-based cross-layer optimization for cideo delivery in long term evolution mobile networks. In: 14th Wireless Personal Multimedia Communications, Brest, France, pp. 1–5 (2011)
9. Pejoski, S., Kafedziski, V.: Cross-layer framework for real time H.264/AVC video transmission over wireless channels using outage probability. In: IX International Symposium on Telecommunications BIHTEL, Sarajevo, Bosnia and Herzegovina, pp. 1–6 (2012)

10. Ju, Y., Lu, Z., Ling, D., Wen, X., Zheng, W., Ma W.: QoE-based cross-layer design for video applications over LTE. J. Multimedia Tools Appl. doi:10.1007/s11042-013-1413-0 (2013). (Springer, New York)

11. Cheng, X., Mohapatra, P.: Quality-optimized downlink scheduling for video streaming applications in LTE networks. In: Communications Software, Services and Multimedia Symposium, Globecom, Anaheim, CA, pp. 1914–1919 (2012)

12. Karachontzitis, S., Dagiuklas, T., Dounis, L., et al.: Novel cross-layer scheme for video transmission over LTE-based wireless systems. In: IEEE International Conference on Multimedia Expo, Barcelona, pp. 1–6 (2011). doi:10.1109/ICME.2011.6012174

13. Essaili, A., Zhou, L., Shroeder, D., Steinbach, E., Kellerer, W.: QoE-driven live and on-demand LTE uplink video transmission. In: 13th Workshop on Multimedia Signal Processing, Hangzhou, pp. 1–6 (2007). doi:10.1109/MMSP.2011.6093821

14. Goldsmith, A., Chua, S.: Adaptive coded modulation for fading channels. J. IEEE Trans. Commun. **46**(5), 595–602 (1998)

15. Oyman, O., Foerster, J., Tcha, Y., Lee, S.: Toward enhanced mobile video services over WiMAX and LTE. J. IEEE Commun. Mag. **48**(8), 68–76 (2010). doi:10.1109/MCOM.2010.5534589

16. Pahalawatta, P., Katsaggelos, A.: Review of content-aware resource allocation schemes for video streaming over wireless networks. J. Wirel. Commun. Mob. Comput. **7**, 131–142 (2007). doi:10.1002/wcm469

17. Radhakrishnan, R., Tirouvengadam, B., Nayak, A.: Cross layer design for efficient video streaming over LTE using scalable video coding. In: IEEE International Conference on Communications, Ottawa, ON, pp. 6509–6513 (2012). doi:10.1109/ICC.2012.6364725

18. Khan, F.: LTE for 4G Mobile Broadband. Cambridge University Press, Cambridge (2009). doi:10.1017/CBO9780511810336

19. Wang, Y., Zhu, Q.: Error control and concealment for video communication: a review. J. Proc. IEEE **86**(5), 974–997 (1998). doi:10.1109/5.664283

20. Zhang, R., Regunathan, L., Rose, K.: Video coding with optimal inter/intra-mode switching for packet loss resilience. IEEE J. Sel. Areas Commun. **18**(6), 966–976 (2000). doi:10.1109/49.848250

21. Zhang, Y., Gao, W., Lu, Y., Huang, Q., Zhao, D.: Joint source-channel rate-distortion optimization for H.264 video coding over error-prone networks. J. IEEE Trans. Multimedia **9**(3), 445–454 (2007). doi:10.1109/TMM.2006.887989

22. Stockhammer, T., Wiegand, T., Wenger, S.: Optimized transmission of H.261/JVT coded video over packet-lossy networks. In: IEEE ICIP, pp. 173–176 (2002)

23. Stockhammer, T., Hannuksela, M., Wiegand, T.: H.264/AVC in wireless environments. J. IEEE Trans. Circ. Syst. Video Technol. **13**(7), 657–673 (2003)

24. Chen, Z., Wu, D.: Prediction of transmission distortion for wireless video communication, part I: analysis. J. IEEE Trans. Image Process. **21**(3), 1123–1137 (2011). doi:10.1109/TIP.2011.2168411

25. Chen, Z., Pahalawatta, P.V., Tourapis, A.M., Wu, D.: Improved estimation of transmission distortion for error-resilient video coding. J. IEEE Trans. Circ. Syst. Video Technol. **22**(4), 636–647 (2012). doi:10.1109/TCSVT.2011.2171262

26. Kolehmainen, N., Puttonen J., Kela, P., Ristaniemi, T., Henttonen, T., Moisio, M.: Channel quality indication reporting schemes for UTRAN long term evolution downlink. In: IEEE Vehicular Technology Conference, Singapore, pp. 2522–2526 (2008). doi:10.1109/VETECS.2008.555

27. Maani, E., Pahalawatta, P., Berry, R., Pappas, T., Katsaggelos, A.: Resource allocation for downlink multiuser video transmission over wireless lossy networks. J. IEEE Trans. Image Process. **17**(9), 1663–1671 (2008). doi:10.1109/TIP.2008. 2001402

28. Schuster, G., Katsaggelos, A.: Rate-Distortion Based Video Compression: Optimal Video Frame Compression and Object Boundary Encoding. Springer, New York (1997). doi:10.1007/978-1-4757-2566-7

29. Blum, C., Roli, A.: Metaheuristics in combinatorial optimization: overview and conceptual comparison. J. ACM Comput. Surv. **35**(3), 268–308 (2003). doi:10. 1145/937503.937505

30. Kennedy, J., Eberhart, R.: Particle swarm optimization. In: IEEE International Conference on Neural Networks IV, Perth, WA, pp. 1942–1948 (1995). doi:10. 1109/ICNN.1955.488968

A Discrete Waiting Time Model
for Optical Signals

Laszlo Lakatos[1](✉) and Dmitry Efrosinin[2]

[1] Eötvös Loránd University, Pázmány P. s. 1/C, Budapest 1117, Hungary
[2] Johannes Kepler University Linz, Altenbergerstrasse 69, 4040 Linz, Austria
lakatos@inf.elte.hu, dmitry.efrosinin@jku.at
http://www.elte.hu, http://www.jku.at

Abstract. We consider a discrete time queueing system where the service of a customer may start at the moment of arrival or at moments differing from it by the multiples of a given cycle time. One finds the distribution of waiting time and its mean value. These results give possibility for the numerical optimization of cycle length. The original model was raised in connection with the landing process of airplanes, but it appears to be an exact model to describe the functioning of a node at the transmission of optical signals.

1 Introduction

We propose to consider a single-server queueing system, where an entering customer may be accepted for service either at the moment of arrival or at moments differing from it by the multiples of a given so-called cycle time. In order to illustrate the problem we give a practical example.

Optical signals enter a node and they should be transmitted according to the FCFS rule. This information cannot be stored, if it cannot be serviced at once is sent to a delay line and returns to the node after having passed it. Clearly, the signal can be transmitted from the node at the moment of its arrival or at a moment that differs from it by a multiple of time necessary to pass the delay line.

The original problem was raised in connection with the landing process of airplanes [4], later it appeared to be an exact model for the transmission of optical signals where because of lack of optical RAM the fiber delay lines are used. At the beginning the system was studied characterizing it by the number of present customers [4,6], later Koba [1–3] found sufficient condition for the stability of GI/G/1 system and gave the system of equations determining the waiting time's ergodic distribution. Rogiest et al. [7,8] describes the application of model for the transmission of optical signals.

The author 'L. Lakatos' was supported by the TAMOP-4.2.2.C-11/1/KONV-2012-0001 project. The project has been supported by the European Union, co-financed by the European Social Fund.

V. Vishnevsky et al. (Eds.): DCCN 2013, CCIS 279, pp. 114–123, 2014.
DOI: 10.1007/978-3-319-05209-0_10, © Springer International Publishing Switzerland 2014

First this system was considered from the viewpoint of the number of present customers [4]. By using Koba's results [1] in [5] we investigated the distribution and characteristics of waiting time for the continuous model. In the present paper we do the same for the discrete time case.

2   The Description of System: Number of Customers

We describe the queueing system and give some results without details concerning the number of customers in it.

We are going to consider the discrete time version of the above described queueing system. Let us divide the cycle time T into n equal parts and assume that for a time slice T/n a new customer arrives with probability r (so there is no entry with probability $1 - r$), and the service of actual customer (if for this time slice it takes place) is continued with probability q and terminated with probability $1 - q$. From these assumptions follows both the interarrival and service times have geometrical distributions. The main result is given in the following:

Theorem 1. *Let us consider a discrete queueing system in which both the inter-arrival and service time distributions are geometrical, the service of a customer may be started upon arrival or (in case of busy server or waiting queue) at moments differing from it by the multiples of a cycle time T equal to n time units. Let us define an embedded Markov chain whose states correspond to the number of customers in the system at moments $t_k - 0$, where t_k is the moment of beginning of service of the k-th one. The matrix of transition probabilities has the form*

$$\begin{bmatrix} a_0 & a_1 & a_2 & a_3 & \ldots \\ a_0 & a_1 & a_2 & a_3 & \ldots \\ 0 & b_0 & b_1 & b_2 & \ldots \\ 0 & 0 & b_0 & b_1 & \ldots \\ \vdots & \vdots & \vdots & \vdots & \ddots \end{bmatrix}$$

its elements are determined by the generating functions

$$A(z) = \sum_{i=0}^{\infty} a_i z^i = \tag{1}$$
$$= \frac{(1-r)(1-q)}{1-q(1-r)} + z\frac{r(1-q)}{1-q(1-r)} + z\frac{rq(1-r+rz)^n(1-q^n)}{[1-q(1-r)][1-q^n(1-r+rz)^n]},$$

$$B(z) = \sum_{k=1}^{\infty} b_i z^i = \frac{1-(1-r)^n(1-r+rz)^n}{1-(1-r)(1-r+rz)}\frac{r(1-r+rz)}{1-(1-r)^n} + \tag{2}$$
$$+\frac{1-q^n(1-r)^n(1-r+rz)^n}{1-q(1-r)(1-r+rz)}\frac{rq(1-r+rz)[(1-r+rz)^n-1]}{[1-(1-r)^n][1-q^n(1-r+rz)^n]}.$$

The generating function of ergodic distribution $P(z) = \sum\limits_{i=0}^{\infty} p_i z^i$ has the form

$$P(z) = p_0 \frac{zA(z) - B(z) + \frac{rz}{(1-r)(1-q)}[A(z) - B(z)]}{z - B(z)},$$

where

$$p_0 = \frac{1 - B'(1)}{1 - B'(1) + A'(1) + \frac{r}{(1-r)(1-q)}[A'(1) - B'(1)]}. \tag{3}$$

The condition of existence of ergodic distribution is the fulfilment of inequality

$$\frac{rq}{1 - q^n} \frac{1 - q^n(1 - r)^n}{1 - q(1 - r)} < (1 - r)^n.$$

Proof. The transition probabilities for the embedded chain are determined by (1) and (2). (1) gives the generating function of transition probabilities if at the beginning of service there is no further customer in the system, (2) determines the generating function if at this moment there are at least two customers.

Let us denote the ergodic distribution by p_i $(i = 0, 1, 2, \ldots)$ and introduce the generating function $P(z) = \sum\limits_{i=0}^{\infty} p_i z^i$. For p_i we have the system of equations

$$p_0 = p_0 a_0 + p_1 a_0,$$

$$p_j = p_0 a_j + p_1 a_j + \sum_{i=2}^{j+1} p_i b_{j-i+1},$$

from which

$$P(z) = \sum_{j=0}^{\infty} p_j z^j = p_0 A(z) + p_1 A(z) + \sum_{j=0}^{\infty} \sum_{i=2}^{j+1} p_i b_{j-i+1} z^j$$

or

$$P(z) = \frac{p_0[zA(z) - B(z)] + p_1 z[A(z) - B(z)]}{z - B(z)}.$$

By using the first equation p_1 can be expressed via p_0

$$p_1 = \frac{1 - a_0}{a_0} p_0 = \frac{r}{(1 - r)(1 - q)} p_0.$$

We can find p_0 from the condition $P(1) = 1$

$$p_0 = \frac{1 - B'(1)}{1 - B'(1) + A'(1) + \frac{r}{(1-r)(1-q)}[A'(1) - B'(1)]}.$$

The chain is irreducible, so $p_0 > 0$. By using (1) and (2) we have

$$A'(1) = \frac{r}{1 - q(1 - r)} + \frac{nr^2 q}{[1 - q(1 - r)](1 - q^n)},$$

$$B'(1) = 1 - \frac{nr(1-r)^n}{1-(1-r)^n} + \frac{nr^2q[1-q^n(1-r)^n]}{(1-q^n)[1-(1-r)^n][1-q(1-r)]},$$

and we obtain

$$\left(1 + \frac{r}{(1-r)(1-q)}\right)A'(1) - \frac{r}{(1-r)(1-q)}B'(1) =$$

$$= \frac{nr^2q}{(1-q^n)[1-q(1-r)]} + \frac{nr^2(1-r)^n}{(1-r)[1-(1-r)^n][1-q(1-r)]} > 0.$$

Since in the denominator of (3) the value of expression apart from $1 - B'(1)$ is positive, the condition $1 - B'(1) > 0$ must be fulfilled. This leads to the expression

$$\frac{nr(1-r)^n}{1-(1-r)^n} - \frac{nr^2q[1-q^n(1-r)^n]}{(1-q^n)[1-(1-r)^n][1-q(1-r)]} > 0.$$

From it we obtain the stability condition

$$\frac{rq}{1-q^n}\frac{1-q^n(1-r)^n}{1-q(1-r)} < (1-r)^n.$$

3   Waiting Time

We consider the above described queueing system and we will use Koba's results [1] to find the waiting time distribution. We shortly repeat these results.

Let t_n denote the time of arrival of the n-th customer; its service will begin at the moment $t_n + T \cdot X_n$, where X_n is a nonnegative integer. Let $Z_n = t_{n+1} - t_n$, and S_n be the service time of n-th customer. Furthermore, let $X_n = i$, if

$$(k-1)T < iT + S_n - Z_n \le kT \qquad (k \ge 1),$$

then $X_{n+1} = k$, and if $iT + S_n - Z_n \le 0$, then $X_{n+1} = 0$. Hence, X_n is a homogeneous Markov chain with transition probabilities p_{ik}, where

$$p_{ik} = P\{(k-i-1)T < S_n - Z_n \le (k-i)T\}$$

if $k \ge 1$, and

$$p_{i0} = P\{S_n - Z_n \le -iT\}.$$

Introduce the notations

$$f_j = P\{(j-1)T < S_n - Z_n \le jT\}, \tag{4}$$

$$p_{ik} = f_{k-i} \quad \text{if} \quad k \ge 1, \quad p_{i0} = \sum_{j=-\infty}^{-i} f_j = \hat{f}_i. \tag{5}$$

The ergodic distribution of this chain satisfies the system of equations

$$p_j = \sum_{i=0}^{\infty} p_i p_{ij} \quad (j \ge 0), \quad \sum_{j=0}^{\infty} p_j = 1.$$

Theorem 2. *Let us consider the above described system and introduce a Markov chain whose states correspond to the waiting time (in the sense that the waiting time is the number of actual state multiplied by T) at the arrival time of customers. The matrix of transition probabilities for this chain is*

$$
\begin{bmatrix}
\sum\limits_{j=-\infty}^{0} f_j & f_1 & f_2 & f_3 & f_4 & \cdots \\
\sum\limits_{j=-\infty}^{-1} f_j & f_0 & f_1 & f_2 & f_3 & \cdots \\
\sum\limits_{j=-\infty}^{-2} f_j & f_{-1} & f_0 & f_1 & f_2 & \cdots \\
\vdots & \vdots & \vdots & \vdots & \vdots & \ddots
\end{bmatrix}
$$

its elements are defined by (4) and (5). The generating function of the ergodic distribution is

$$
P(z) = \left[1 - \frac{rq[1-(1-r)^n]}{(1-q)(1-q^n)(1-r)^n} \right] \times \tag{6}
$$

$$
\times \frac{\dfrac{1-q}{1-q(1-r)} - \dfrac{(1-q)[1-(1-r)^n]}{1-q(1-r)}\dfrac{z}{z-(1-r)^n}}{1 - \dfrac{rq(1-q^n)}{1-q(1-r)}\dfrac{z}{1-q^n z} - \dfrac{(1-q)[1-(1-r)^n]}{1-q(1-r)}\dfrac{z}{z-(1-r)^n}},
$$

the condition of existence of ergodic distribution is

$$
\frac{rq[1-(1-r)^n]}{(1-q)(1-q^n)(1-r)^n} < 1. \tag{7}
$$

Proof. For the system we have

$$
P\{Z = j\} = (1-r)^{j-1}r \qquad \text{and} \qquad P\{S = j\} = q^{j-1}(1-q).
$$

$S - Z$ has the distribution

$$
\sum_{i=1}^{\infty} (1-r)^{i-1} r q^{i-1+j}(1-q) = \frac{r(1-q)q^j}{1-q(1-r)} \qquad (j = 1, 2, \ldots)
$$

if $S - Z > 0$ and

$$
\sum_{i=1}^{\infty} q^{i-1}(1-q)(1-r)^{i-1+j} r = \frac{r(1-q)(1-r)^j}{1-q(1-r)} \qquad (j = 0, 1, 2, \ldots)
$$

if $S - Z \leq 0$. By using these values, we obtain

$$
f_j = \sum_{k=(j-1)n+1}^{jn} \frac{r(1-q)q^k}{1-q(1-r)} = \frac{rq(1-q^n)}{1-q(1-r)} q^{(j-1)n}
$$

for the positive jumps and

$$f_{-j} = \sum_{k=jn}^{(j+1)n-1} \frac{r(1-q)(1-r)^k}{1-q(1-r)} = \frac{(1-q)[1-(1-r)^n]}{1-q(1-r)}(1-r)^{jn}$$

for the negative jumps. Furthermore, we have

$$p_{j0} = \sum_{k=-\infty}^{-j} f_k = \sum_{k=j}^{\infty} \frac{(1-q)[1-(1-r)^n]}{1-q(1-r)}(1-r)^{kn} = \frac{(1-q)(1-r)^{jn}}{1-q(1-r)} = \hat{f}_j.$$

By using these transition probabilities for the equilibrium distribution we have the system of linear equations

$$p_0 = p_0\hat{f}_0 + p_1\hat{f}_1 + p_2\hat{f}_2 + p_3\hat{f}_3 + \cdots$$
$$p_1 = p_0 f_1 + p_1 f_0 + p_2 f_{-1} + p_3 f_{-2} + \cdots$$
$$p_2 = p_0 f_2 + p_1 f_1 + p_2 f_0 + p_3 f_{-1} + \cdots$$
$$\vdots$$

Multiplying the j-th equation by z^j, summing up from zero to infinity, for the generating function $P(z) = \sum_{j=0}^{\infty} p_j z^j$ we have

$$P(z) = P(z)F_+(z) + \sum_{j=1}^{\infty} p_j z^j \sum_{i=0}^{j-1} f_{-i} z^{-i} + \sum_{j=0}^{\infty} p_j \hat{f}_j,$$

where

$$F_+(z) = \sum_{i=1}^{\infty} f_i z^i.$$

By using the transition probabilities f_j we have

$$F_+(z) = \sum_{j=1}^{\infty} f_j z^j = \sum_{j=1}^{\infty} \frac{rq(1-q^n)}{1-q(1-r)} q^{(j-1)n} z^j = \frac{rq(1-q^n)}{1-q(1-r)} \frac{z}{1-q^n z},$$

$$\sum_{i=0}^{j-1} f_{-i} z^{-i} = \sum_{i=0}^{j-1} \frac{(1-q)[1-(1-r)^n]}{1-q(1-r)} \left(\frac{(1-r)^n}{z}\right)^i =$$
$$= \frac{(1-q)[1-(1-r)^n]}{1-q(1-r)} \frac{z}{z-(1-r)^n} \left[1 - \left(\frac{(1-r)^n}{z}\right)^j\right],$$

$$\sum_{j=1}^{\infty} p_j z^j \sum_{i=0}^{j-1} f_{-i} z^{-i} =$$

$$= \frac{(1-q)[1-(1-r)^n]}{1-q(1-r)} \frac{z}{z-(1-r)^n} \sum_{j=1}^{\infty} p_j z^j \left[1 - \left(\frac{(1-r)^n}{z}\right)^j\right] =$$

$$= \frac{(1-q)[1-(1-r)^n]}{1-q(1-r)} \frac{z}{z-(1-r)^n} [P(z) - P((1-r)^n)],$$

$$\sum_{i=0}^{\infty} p_i \hat{f}_i = \frac{1-q}{1-q(1-r)} \sum_{i=0}^{\infty} p_i (1-r)^{in} = \frac{1-q}{1-q(1-r)} P((1-r)^n).$$

So, the expression for the generating function may be written in the form

$$P(z)\left[1 - \frac{rq(1-q^n)}{1-q(1-r)} \frac{z}{1-q^n z} - \frac{(1-q)[1-(1-r)^n]}{1-q(1-r)} \frac{z}{z-(1-r)^n}\right] =$$

$$= P((1-r)^n)\left[\frac{1-q}{1-q(1-r)} - \frac{(1-q)[1-(1-r)^n]}{1-q(1-r)} \frac{z}{z-(1-r)^n}\right].$$

This expression contains the unknown value $P((1-r)^n)$, it can be found from the condition $P(1) = 1$. It is equal to

$$P((1-r)^n) = 1 - \frac{rq[1-(1-r)^n]}{(1-q)(1-q^n)(1-r)^n},$$

so, finally, the generating function takes on the form (6).

The chain is irreducible and aperiodic, in order to get the stability condition we find p_0 from the generating function

$$p_0 = \left[1 - \frac{rq[1-(1-r)^n]}{(1-q)(1-q^n)(1-r)^n}\right] \frac{1-q}{1-q(1-r)}.$$

It is positive if

$$\frac{rq[1-(1-r)^n]}{(1-q)(1-q^n)(1-r)^n} < 1, \tag{8}$$

i.e. it is equivalent to the stability condition in the case of the number of customers.

We obtained the distribution of waiting time measured in cycles. From the generating function we can get its mean value, it is equal to

$$\bar{C} = P'(1) = \frac{rq[1-q^n(1-r)^n]}{(1-q^n)\{(1-q)(1-q^n)(1-r)^n - rq[1-(1-r)^n]\}}. \tag{9}$$

This expression gives the mean value in cycle numbers, in order to get in time it must be multiplied by the length of a cycle n.

4   Optimization of the Retrial Cycle

We consider the problem of minimization of waiting time. The mean number of cycles is determined by (9) and we look for

$$\bar{C} = \bar{C}(n) \Rightarrow \min_{n \in \mathbf{N}}. \tag{10}$$

Figure 1(a, b) shows the behaviour of $\bar{C}$ depending on the length of cycle n for different values of probabilities r and q. Of course, the values of r and q are chosen to satisfy the stability condition (7). From the figures it is clear the concave structure of function $\bar{C}$ for the different values of r and q.

On Fig. 1(a) the probability to continue the service is fixed $q = 0.66$, the parameter to enter a customer for a time interval changes

$$r = 0.05;\ 0.06;\ 0.07;\ 0.09;\ 0.15.$$

The minimum mean number of cycles is reached for $n^* = 5,\ 5,\ 5,\ 4,\ 3$, respectively, i.e. decreases if the working load or the probability of arrivals increases, or the left side of stability condition is greater. On Fig. 1(b) the parameter $r = 0.07$ is fixed, and the probability to continue the service takes on the values

$$q = 0.50;\ 0.60;\ 0.66;\ 0.75;\ 0.85.$$

In these cases the optimal values are $n^* = 4,\ 4,\ 5,\ 5,\ 6$, respectively.

For fixed values of r and q the mean number of cycles first decreases and achieving some optimal value it increases. In case of small n till the beginning of service of the next customer a large number of cycles is required (the service time is significantly greater than the length of cycle). Approaching the optimal value n^* the number of cycles decreases, leaving it the length of cycle becomes greater and greater, and the waiting time will mainly be determined by the length of cycle, not by the service time. It makes the waiting time large, consequently the number of required cycles will grow.

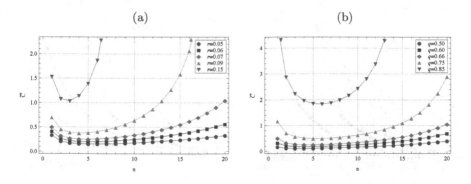

Fig. 1. Optimization of $\bar{C}$

Besides the mean number of cycles $\bar{C}$ different alternative performance characteristics of the system can be evaluated, e.g.:

Utilization of the system

$$\bar{U} = 1 - p_0;$$

Mean waiting and sojourn time

$$\bar{W} = n\bar{C}, \ \bar{S} = \bar{W} + \frac{1}{1-q};$$

Mean number of customers in orbit and system

$$\bar{Q} = r\bar{W}, \ \bar{N} = r\bar{S}.$$

Figure 2(a–d) illustrates respectively the functions $\bar{U}, \bar{C}, \bar{W}$ and $\bar{N}$ for varying r and q for optimal value n^* which minimizes expression (10). Note that the value of system parameters are chosen to suite the stability condition (8). The curves in pictures exhibit indented structure as r increases. It could be explained by decreasing of optimal cycle length n^* as r increases. In presented examples the increasing of r above some threshold level lead to the sufficient diminishing of the system utilization, waiting and sojourn times. Therefore it is possible to formulate and solve some local optimization problems for the mean performance measures with respect to the system parameters.

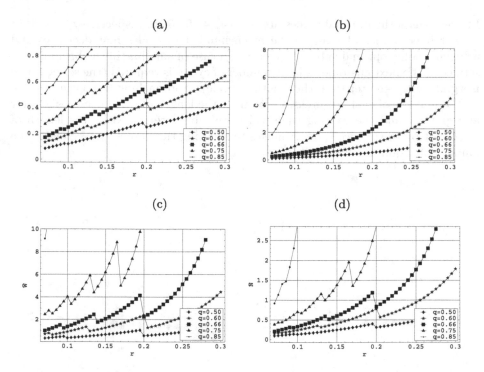

Fig. 2. $\bar{U}$ (a), $\bar{C}$ (b), $\bar{W}$ (c) and $\bar{N}$ (d) versus r and q

References

1. Koba, E.V.: On a GI/G/1 queueing system with repetition of requests for service and FCFS service discipline. Dopovidi NAN Ukrainy **6**, 101–103 (2000) (in Russian)
2. Koba, E.V.: On a GI/G/1 retrial queueing system with a FIFO queueing discipline. Theory Stoch. Proc. **8**(24), 201–207 (2002)
3. Koba, E.V., Pustovaya, S.V.: Lakatos-type queueing systems, their generalization and application. Kibernetika i Sistemnyi Analiz **3**, 78–90 (2012) (in Russian)
4. Lakatos, L.: On a simple continuous cyclic-waiting problem. Annales Univ. Sci. Budapest. Sect. Comp. **14**, 105–113 (1994)
5. Lakatos, L., Efroshinin, D.: Some aspects of waiting time in cyclic-waiting systems. In: Dudin, A., Klimenok, V., Tsarenkov, G., Dudin, S. (eds.) BWWQT 2013. CCIS, vol. 356, pp. 115–121. Springer, Heidelberg (2013)
6. Lakatos, L., Szeidl, L., Telek, M.: Introduction to Queueing Systems with Telecommunication Applications. Springer, Heidelberg (2013)
7. Rogiest, W., Laevens, K., Fiems, D., Bruneel, H.: Analysis of a Lakatos-type queueing system with general service times. In: Proceedings of ORBEL 20. Quantitative Methods in Decision Making, pp. 95–97. Ghent University, Ghent (2006)
8. Rogiest, W., Laevens, K., Walraevens, J., Bruneel, H.: Analyzing a degenerate buffer with general interarrival and service times in discrete time. Queueing Syst. **56**, 203–212 (2007)

Analytical Model of the TCP Reno Congestion Control Procedure through a Discrete-Time Markov Chain

Vladimir Kokshenev$^{(\boxtimes)}$ and Sergey Suschenko

Tomsk State University, Tomsk, Russia
vladimir_finf@mail.ru, ssp@inf.tsu.ru

Abstract. An analytical model of the Reno congestion control procedure for Transmission Control Protocol is presented, and its theoretical predictions are compared with real TCP traces and some well-known results [1]. The model is based on Discrete-Time Markov Chain, and it covers slow start, congestion avoidance, fast recovery, fast retransmit, cumulative and selective acknowledgments, timeouts with exponential back-off and appropriate byte counting features of TCP. The model provides a way to estimate Reno TCP performance as a function of round trip time and loss rate for bulk transfer TCP flow. The model allows to compare SACK performance with traditional cumulative acknowledgments. The model can be used to find conditions beneficial for advertised window adjustment to improve performance.

Keywords: TCP · Reno · Throughput · Analytical model · Discrete-time Markov chain

1 Introduction

TCP is the most widely used transport layer protocol in the Internet [23]. The performance of many network services and applications are influenced by TCP performance. The protocol was introduced in 1970s, and since then a lot of new standards, proposals, modifications and analytical models were developed and discussed.

Congestion control (CC) is a feature that each TCP implements in attempt to avoid network overloading by excessive traffic. As specified in [2] TCP uses windowing mechanism to implement it. The main idea of congestion control is to reduce or increase the offered sending rate depending on network conditions. Congestion control procedures base their decisions on different types of feedback received from a network: loss rate, propagation delay, explicit congestion signaling, etc. A lot of congestion control algorithms were developed to address different network types and conditions: Tahoe, Reno, NewReno, Vegas, Hybla, CTCP, Illinois, BIC, CUBIC, Westwood+, H-TCP, High Speed TCP, Scalable TCP, Veno, YeAH, FAST, etc.

V. Vishnevsky et al. (Eds.): DCCN 2013, CCIS 279, pp. 124–135, 2014.
DOI: 10.1007/978-3-319-05209-0_11, © Springer International Publishing Switzerland 2014

Based on feedback type used in congestion control decisions all TCP implementations can be divided into three main categories: loss-based, delay-based and hybrid algorithms. Loss-based algorithms are the most widely used. Tahoe, Reno, NewReno, BIC, CUBIC, HSTCP are members of loss-based congestion control class. Vegas and FAST TCP are examples of delay-based congestion control approach. CTCP, Illinois and Westwood+ can be counted as hybrid algorithms. They base their decisions on both loss rate and delay of the network path.

Being an important networking area, TCP congestion control is an integral part of any operating system. Linux OS supports as many as thirteen congestion control algorithms [17]: Reno/NewReno, Vegas, Veno, Westwood+, BIC, CUBIC, HSTCP, Hybla, Scalable, Illinois, YeAH, HTCP, LP, with CUBIC as a default method since kernel version 2.6.18. At the same time FreeBSD implements six congestion control methods: Reno/NewReno (default in FreeBSD 9.0-release), CUBIC, HTCP, Vegas, CHD, HD. And Microsoft operating systems starting from Vista and Windows Server 2008 use just two congestion control variants: Reno/NewReno and CTCP.

TCP Reno and NewReno are only congestion control algorithms that reached Standard Track category in RFC series [5,18]. And they are the most widely implemented in modern operating systems and network equipment.

2   The Reno Protocol

Reno protocol is an enhanced version of TCP Tahoe implementation. It uses three congestion window (CWND) management algorithms [3,5]: Slow Start (SS), Congestion Avoidance (CA) and Fast Recovery (FR). For sake of simplicity we measure window size in segments.

Being in slow start mode, TCP increases congestion window by one for each positive acknowledgment received. Therefore CWND doubles each round trip time (RTT) cycle. Slow start threshold variable (SSTHRESH) defines when to switch from slow start to congestion avoidance mode. Protocol acts in slow start mode if CWND is less than SSTHRESH.

Protocol switches into congestion avoidance when CWND reaches SSTHRESH value. Being in congestion avoidance state, TCP increments congestion window by 1/CWND for each positive acknowledgment received. A window of size CWND segments will generate at most CWND acknowledgments in one RTT, so therefore an increment of 1/CWND per acknowledgment will increase CWND by at most one segment in one RTT [4].

TCP Reno supports two loss detection methods: timeout (TO) and duplicate acknowledgment (DupACK) based analysis. The sender resets the already running retransmission timer and starts a new one each time a new segment is transmitted. The timer is set for the retransmission timeout (RTO) value supplied by the RTT estimation procedure. All unacknowledged segments are counted as lost if this timer expires. In addition TCP performs several actions when RTO timer expiration happens: SSTHRESH parameter is set to CWND/2,

CWND variable is set to 1, protocol performs retransmission, and slow start is re-initiated.

The other option to detect losses is to analyze incoming acknowledgments. When the number of consecutive duplicate acknowledgments reaches threshold value the closest unacknowledged segment is counted as lost. By default the threshold value is equal to 3. Duplicate acknowledgements are segments with same values of Acknowledgment Number field in TCP header. When a loss is detected by DupACKs, Reno protocol performs immediate retransmission, which is also called Fast Retransmission. In addition it changes values of SSTHRESH and CWND variables to CWND/2 and CWND/2 + 3 accordingly, and switches into fast recovery mode.

Being in fast recovery mode, TCP temporary increases CWND by one segment for each duplicate acknowledgment received. And protocol continues transmission if general rules allow [3, 5]. When acknowledgment for the retransmitted segment is received, CWND is reset back to SSTHRESH, and protocol switches from fast recovery to congestion avoidance mode.

Modern TCP implementations in addition to modes and features specified above use Karn algorithm [14, 19], selective acknowledgments [12, 13, 20], appropriate byte counting [21] and other advanced techniques.

The idea of Karn algorithm is in RTO exponential back-off when retransmission timer expiration happens. When the protocol receives acknowledgment for retransmitted data the back-off is canceled. This measure improves TCP adaptation to sudden RTT changes but can affect performance and responsiveness in case of timeout expiration due to loss events.

Traditional TCP acknowledges the last continuous segment received successfully by cumulative acknowledgment. Therefore TCP may experience poor performance when multiple packets are lost from one window of data [12]. Selective acknowledgments (SACK) allow the receiver to acknowledge discontinuous blocks of segments that were received correctly. Thus SACK can help to overcome this limitation.

Appropriate Byte Counting (ABC) modifies the algorithm for increasing TCP's congestion window. Rather than increasing a TCP's congestion window based on the number of acknowledgments that arrive at the data sender, the congestion window is increased based on the number of bytes acknowledged by the arriving acknowledgments [21]. The algorithm mitigates the impact of delayed acknowledgments feature that TCP should implement according to [22].

3   Analytical Models of TCP Reno

Reno algorithm is frequently referenced as Standard TCP, and it is the most widely implemented congestion control method [3]. A lot of researches dedicated to Reno performance were done in past years. Some of the most noticeable TCP analytical models are shown in Table 1.

Though all presented researches offer Reno/NewReno analytical models, only [8, 10, 11] count all its phases. TCP SACK option was implemented only in [10].

Table 1. TCP analytical models.

Model	CC Algorithm	SS	CA	FR	TO	Karn	SACK	Fast Retr.
[1]	Reno	−	+	−	+	+	−	+
[6]	Tahoe	+	+	−	+	−	−	−
	Reno	+	+	−	−	−	−	+
[7]	OldTahoe	+	+	−	+	+	−	−
	Tahoe	+	+	−	+	+	−	+
	Reno	+	+	+	+	+	−	+
	NewReno	+	+	+	+	+	−	+
[8]	NewReno	+	+	+	+	−	−	+
[9]	Reno	−	+	−	+	+	−	+
[10]	Reno	+	+	+	+	+	+	+
	Vegas	+	+	+	+	+	+	+
[11]	Reno	+	+	+	+	−	−	+

Retransmission timeout was covered in all specified papers, but not all of them count exponential back-off behavior. The most complete model is presented in [10]. Our model covers all congestion control phases and includes Karn algorithm, SACK and fast retransmit mechanisms, and appropriate byte counting technique. It has several differences from [10]:

- We use separate Markov chain states for congestion avoidance phase for each value of SSTHRESH variable. From one side this gives an opportunity to get more statistics, but from the other side it makes the Markov chain bigger.
- We suggest bulk transfer TCP flow model and [10] provides model for bursty flows.
- We use Discrete-Time Markov Chain (DTMC) as opposed to Continuous-Time Markov Chain (CTMC) used in [10].

It is not a trivial task to compare our model with models proposed in [6–8, 10, 11], but [1] gives a simple formula, so therefore it can be counted as an easy option to compare with.

4   TCP Reno Analytical Model

We define the throughput (B) of TCP connection as:

$$B = \frac{E[A] - E[B]}{RTT} MSS, \qquad (1)$$

where A — is the number of segments sent during RTT cycle, B — is the number of retransmissions in RTT cycle, RTT — is the round trip time, MSS — is the maximum segment size.

Values of $E[A]$ and $E[B]$ are calculated based on sender window size (CWND) and logic behind slow start, congestion avoidance and fast recovery phases.

We use two-dimensional Discrete-Time Markov Chain to model TCP sender behavior. Statistical cycle duration is equal to round trip time. We suggest that RTT is constant and each segment has a size equal to MSS. It is also suggested that time required to send the whole receiver's advertised window (W) is less than RTT. The latter statement can be justified by results published in [15,16]. The first dimension of DTMC is for CWND variable, and the second one is for SSTHRESH.

The model doesn't count losses in reverse direction: it is assumed that TCP acknowledgments are never lost. We denote the probability to transfer a segment successfully from sender to receiver as F, so therefore $(1 - F)$ is the probability to lose a segment. The model suggests uncorrelated losses.

In order to model timeout phase we use S as a value of retransmission timeout expressed in cycles. It is based on RTT and RTO values expressed in seconds:

$$S = \left\lfloor \frac{RTO}{RTT} \right\rfloor, \tag{2}$$

We use additional parameter MK to limit timeout exponential back-off. The values of timeout can now be expressed as $2^k S$, $k = \overline{0, MK}$. Where k can be treated as a number of consecutive timeout expiration events without successful acknowledgments between them.

Additional parameter ML is used to model SACK option. It denotes the number of losses TCP is capable to detect being in slow start or congestion avoidance phases, and recover in fast recovery phase. To model cumulative acknowledgments ML value equal to 1 must be used.

Transition probabilities (π_{in}^{jm}) for the DTMC are listed below:

$$\pi_{in}^{jm} = \begin{cases} F^i, & i = 2^k, k = \overline{0, \lceil \log_2 n \rceil - 1}; \ n = \overline{2, \lfloor W/2 \rfloor}; \\ & j = \min(2i, n); \ m = n; \\ 1 - F^i, & i = \overline{1,3}; \ n = \overline{2,3}; \ j = i; \ m = \lfloor W/2 \rfloor + 1; \\ 1 - F^i, & i = \overline{1,2}; \ n = \overline{4, \lfloor W/2 \rfloor}; \ j = i; \ m = \lfloor W/2 \rfloor + 1; \\ \sum_{x=0}^{2} C_i^x F^x (1 - F)^{i-x}, & i = 2^k, k = \overline{2, \lceil \log_2 n \rceil - 1}; \ n = \overline{2, \lfloor W/2 \rfloor}; \ j = i; \\ & m = \lfloor W/2 \rfloor + 1; \\ 1 - \sum_{x=\max(3, i-ML)}^{i} C_i^x F^x (1 - F)^{i-x} - & i = 2^k, k = \overline{2, \lceil \log_2 n \rceil - 1}; \ n = \overline{2, \lfloor W/2 \rfloor}; \ j = 1; \\ - \sum_{x=0}^{2} C_i^x F^x (1 - F)^{i-x}, & m = \max(2, \lfloor i/2 \rfloor); \\ \sum_{x=\max(3, i-ML)}^{i-1} C_i^x F^x (1 - F)^{i-x}, & i = 2^k, k = \overline{2, \lceil \log_2 n \rceil - 1}; \ n = \overline{2, \lfloor W/2 \rfloor}; \ j = i; \\ & m = 1; \\ F^i, & i = \overline{n, W-1}; \ n = \overline{2, \lfloor W/2 \rfloor}; \ j = i+1; \ m = n; \\ F^W, & i = W; \ n = \overline{2, \lfloor W/2 \rfloor}; \ j = W; \ m = n; \\ \sum_{x=0}^{2} C_i^x F^x (1 - F)^{i-x}, & i = \overline{\max(4,n), W}; \ n = \overline{2, \lfloor W/2 \rfloor}; \ j = i; \\ & m = \lfloor W/2 \rfloor + 1; \\ 1 - \sum_{x=\max(3, i-ML)}^{i} C_i^x F^x (1 - F)^{i-x} - & i = \overline{\max(4,n), W}; \ n = \overline{2, \lfloor W/2 \rfloor}; \ j = 1; \\ - \sum_{x=0}^{2} C_i^x F^x (1 - F)^{i-x}, & m = \max(2, \lfloor i/2 \rfloor); \\ \sum_{x=\max(3, i-ML)}^{i-1} C_i^x F^x (1 - F)^{i-x}, & i = \overline{\max(4,n), W}; \ n = \overline{2, \lfloor W/2 \rfloor}; \ j = i; \ m = 1; \end{cases}$$

$$\pi_{in}^{jm} = \begin{cases} 1, & i = \overline{4, W}; \; n = 1; \; j = \lfloor i/2 \rfloor; \; m = \max(2, \lfloor i/2 \rfloor); \\ 1, & i = \overline{1, W}; \; n = \overline{\lfloor W/2 \rfloor + 1, \lfloor W/2 \rfloor + S - 1}; \; j = i; \; m = n + 1; \\ 1, & i = \overline{1, W}; \; n = \lfloor W/2 \rfloor + S; \; j = W + 1; \; m = \max(2, \lfloor i/2 \rfloor); \\ F, & i = W + 1; \; n = \overline{2, \lfloor W/2 \rfloor}; \; j = i + 1; \; m = n; \\ 1 - F, & i = W + 1; \; n = \overline{2, \lfloor W/2 \rfloor}; \; j = i; \; m = \lfloor W/2 \rfloor + 1; \\ F^2, & i = W + 2; \; n = \overline{2, 3}; \; j = 3; \; m = n; \\ F^2, & i = W + 2; \; n = \overline{4, \lfloor W/2 \rfloor}; \; j = 4; \; m = n; \\ 2(1 - F)F, & i = W + 2; \; n = \overline{2, \lfloor W/2 \rfloor}; \; j = 2; \; m = \lfloor W/2 \rfloor + 1; \\ (1 - F)^2, & i = W + 2; \; n = \overline{2, \lfloor W/2 \rfloor}; \; j = i; \; m = \lfloor W/2 \rfloor + 1; \\ F, & k = \overline{2, MK}; i = W + 2k - 1; \; n = 2; \; j = i + 1; \; m = n; \\ 1 - F, & k = \overline{2, MK}; i = W + 2k - 1; \; n = 2; \; j = i; \; m = \lfloor W/2 \rfloor + 1; \\ F^2, & k = \overline{2, MK}; i = W + 2k; \; n = 2; \; j = 3; \; m = n; \\ 2(1 - F)F, & k = \overline{2, MK}; i = W + 2k; \; n = 2; \; j = 2; \; m = \lfloor W/2 \rfloor + 1; \\ (1 - F)^2, & k = \overline{2, MK}; i = W + 2k; \; n = 2; \; j = i; \; m = \lfloor W/2 \rfloor + 1; \\ 1, & k = \overline{1, MK}; i = \overline{W + 2k - 1, W + 2k}; \\ & n = \overline{\lfloor W/2 \rfloor + 1, \lfloor W/2 \rfloor + 2^k S - 1}; \; j = i; \; m = n + 1; \\ 1, & k = \overline{1, MK - 1}; i = \overline{W + 2k - 1, W + 2k}; \; n = \lfloor W/2 \rfloor + 2^k S; \\ & j = W + 2k + 1; \; m = 2; \\ 1, & i = \overline{W + 2MK - 1, W + 2MK}; \; n = \lfloor W/2 \rfloor + 2^{MK} S; \\ & j = W + 2MK - 1; \; m = 2. \end{cases}$$

We were unable to get analytical solution for state probabilities P_{ij}, but using π_{in}^{jm} definition we can get state probabilities (P_{ij}) numerically and calculate $E[A]$ and $E[B]$ based on them using these formulas:

$$E[A] = \sum_{(i,j) \in SS \cup CA} i P_{ij} + \sum_{(i,j) \in FR} P_{ij} \frac{\sum_{k=1}^{i} (\lfloor (i + k - 1)/2 \rfloor + 1)}{i},$$

$$E[B] = \sum_{j=2}^{\lfloor W/2 \rfloor} P_{1j} + \sum_{(i,j) \in FR} P_{ij} \sum_{k=1}^{ML} k \frac{C_i^k F^{i-k}(1 - F)^k}{\sum_{n=1}^{ML} C_i^n F^{i-n}(1 - F)^n}.$$

As the protocol sends data in slow start, congestion avoidance and fast recovery modes and doesn't send anything awaiting timeout expiration, we use only CA, SS and FR states probabilities and congestion window sizes corresponding to them to calculate $E[A]$. TCP retransmits lost segment when it enters slow-start mode or fast recovery mode. That is why only FR states and initial states of SS mode are counted to calculate $E[B]$.

5   Reno Measurements and Traces

In order to verify modeling results, Reno TCP connection throughput measurements were made in environment with properties similar to modeling assumptions.

Fig. 1. The logical diagram of the network topology used for TCP throughput measurements.

Required round trip time and packet loss rate were emulated with LANforge Manager. Linux hosts were used as data sender and receiver. Linux sender was configured to use Reno congestion control procedure. The logical diagram of network topology used for measurements is shown on the Fig. 1.

Measurements were made for each combination of receiver's advertised window (W), success rate (F) and round trip time (RTT) parameters. To get average Reno throughput number we used five separate 5-minute long measurements of single TCP connection. The following section provides comparison of modeling results and Reno TCP measurements.

6   Results Comparison and Analysis

Numerical results of proposed model, measurements of real TCP Reno traces and results of approximation provided in [1] are shown in Table 2. We used advertised window size of 64 KB which in modeling terms means $W = 45$, i.e. the maximum window size is 45 segments.

Comparison of numerical results for cumulative acknowledgments with captured traces and approximation suggested in [1] is provided on Fig. 2[1]: RTT is 200 ms, advertised window size is 64 KB (W is 45 segments).

According to provided results proposed model produces more accurate throughput estimation than approximation suggested in [1], and closely matches average results of real TCP traces measurements. We noticed that the model is 9 % more optimistic than real traces at average.

We compared model predictions for cumulative ($ML = 1$) and selective ($ML = 4$) acknowledgments. Comparison for environment with round trip time of 200 ms and advertised window size equal to 64 KB is shown on Fig. 3.

Model predicts that SACK provides potential for TCP throughput increase of up to 6–7.5 % for environments with packet loss rate not greater than 10 %. It can be concluded that networks with higher loss rate and lower round-trip time benefit from SACK more (see results in Table 2).

---

[1] Please note that all presented graphs show TCP throughput axis in logarithmic scale.

Table 2. Numerical results of proposed model in comparison with Reno TCP traces and approximation proposed in [1].

Parameters		TCP Reno Throughput, Bps			
RTT ms	F	Reno model	Reno model, SACK	Approximation result from [1]	Reno TCP traces
50	0.9	25783.6	27681.6	10045	23360
	0.92	38865.2	41931.2	14928	36709
	0.94	63072	68182	23435	55455
	0.95	83074	89614.8	30340	75683
	0.96	111952.8	120158	40653	90522
	0.97	154468	164074.8	57424	153452
	0.98	219759.2	229804	88799	201193
	0.99	344706	353612	167075	355237
	0.995	506182	513774	281329	523507
	0.999	995398.8	997851.6	740499	904240
	0.9999	1276682.4	1276740.8	1310700	1105182
100	0.9	20075	21403.6	8780	15172
	0.92	28353.2	30309.6	12572	24472
	0.94	41945.8	44792.8	18759	36065
	0.95	52092.8	55436.2	23491	46614
	0.96	65743.8	69598.2	30208	58480
	0.97	84826	89045.4	40479	87106
	0.98	114026	118376.8	58300	116073
	0.99	173345.8	177419.2	98921	172485
	0.995	253310	256945.4	154899	238719
	0.999	497714	498925.8	378397	438822
	0.9999	638341.2	638370.4	655350	563017
200	0.9	13913.8	14731.4	7013	13756
	0.92	18403.3	19498.3	9556	17696
	0.93	21403.6	22659.2	11264	20495
	0.94	25126.6	26557.4	13407	23020
	0.95	29835.1	31448.4	16185	28472
	0.96	36018.2	37792.1	19955	35321
	0.97	44603	46515.6	25456	40105
	0.98	58115.3	60100.9	34560	53187
	0.99	86921.1	88870.2	54477	78248
	0.995	126713.4	128487.3	81577	125746
	0.999	248857	249462.9	191303	227786
	0.9999	319170.6	319185.2	327675	266427

In environments with high loss rate TCP sender has no chance to increase congestion window to significant values. As a result a transport connection shows poor performance. It can be concluded that in such conditions TCP can't gain benefit from high receiver's advertised window size. The proposed model confirms this assumption. Comparison of TCP Reno throughput predicted by the model for different values of advertised window size in case of 200 ms round-trip time is shown Table 3.

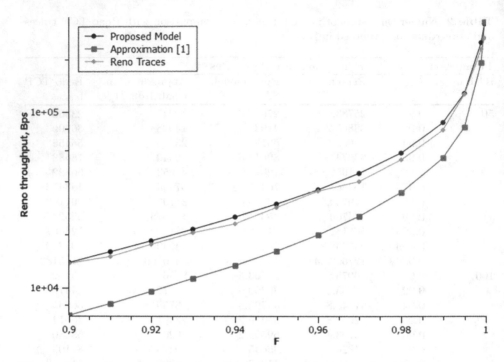

Fig. 2. Comparison of numerical results with captured traces and well-known approximation.

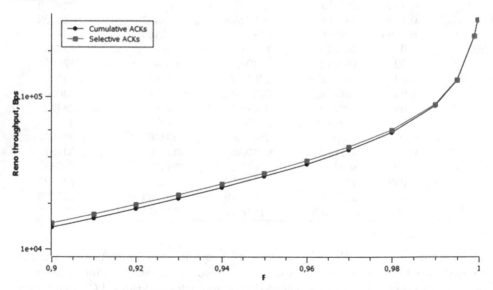

Fig. 3. Comparison of model predictions for cumulative and selective acknowledgments (200 ms RTT, 64 KB advertised window).

Table 3. Comparison of throughput predictions for different values of advertised window size in 200 ms network environment.

Parameters		TCP Reno Throughput, Bps		
RTT ms	F	32 KB, $W = 23$	64 KB, $W = 45$	128 KB, $W = 90$
200	0.9	13913.8	13913.8	13913.8
	0.91	15943.2	15943.2	15943.2
	0.92	18403.3	18403.3	18403.3
	0.93	21403.6	21403.6	21403.6
	0.94	25126.6	25126.6	25126.6
	0.95	29835.1	29835.1	29835.1
	0.96	36018.2	36018.2	36018.2
	0.97	44603	44603	44603
	0.98	58064.2	58115.3	58115.3
	0.99	85264	86921.1	86921.1
	0.995	113996.8	126713.4	126852.1
	0.999	154365.8	248857	293659.8
	0.9999	166476.5	319170.6	590445.9

When the protocol faces situations in which losses can't be detected by duplicate acknowledgments it enters timeout phase and simply waits timer expiration. For example if the number of successfully transmitted segments during RTT cycle is less than four fast retransmission procedure can't be used for loss recovery and timeout happens. Timeout leads to retransmission and slow start re-initiation. At the same time timeout exponential back-off algorithm adjust current timeout value. MK parameter is used to limit exponential back-off behavior of Karn scheme. We analyzed the influence of MK value on time spent by Reno congestion control procedure awaiting timeout. It is obvious that for environments with high packet loss rate the protocol spends more time in TO phase and the influence of MK parameter is more noticeable. At the same time, as it was shown above, for environments with high packet loss connection throughput can't be improved by large receiver's advertised window, because due to losses data sender has no chance to increase congestion window size significantly. Thus we compared integral part of time (T) spent by Reno algorithm in TO phase for different values of MK parameter for a receiver with advertised window size of 32 KB (see Table 4). Formula (3) shows how to calculate T. Based on results presented in Table 4 it can be concluded that for environments with packet loss rate less than or equal to 0.1 it is sufficient to use MK value of 3 to gather numerical results for the proposed model. Despite the fact that modern TCP implementations use back-off limit of 5 or 6, the usage of MK value of 3 will produce an error less than 0.3 % in numerical results but this will significantly reduce the Markov chain size.

$$T = \sum_{(i,j)\in TO} P_{ij}, \qquad (3)$$

Table 4. Integral part of time spent by Reno in TO phase, T.

Parameters		Integral part of time spent by Reno in TO phase, T						
RTT ms	F	$MK = 0$	$MK = 1$	$MK = 2$	$MK = 3$	$MK = 4$	$MK = 5$	$MK = 6$
200	0.9	0.38574	0.41052	0.41567	0.41678	0.41702	0.41707	0.41708
	0.91	0.35192	0.37339	0.37741	0.37819	0.37834	0.37837	0.37837
	0.92	0.31474	0.33277	0.33578	0.33629	0.33638	0.33639	0.33639
	0.93	0.27410	0.28864	0.29076	0.29107	0.29112	0.29113	0.29113
	0.94	0.23018	0.24126	0.24264	0.24281	0.24283	0.24284	0.24284
	0.95	0.18357	0.19135	0.19216	0.19224	0.19225	0.19225	0.19225
	0.96	0.13560	0.14044	0.14084	0.14087	0.14087	0.14087	0.14087
	0.97	0.08869	0.09118	0.09133	0.09134	0.09134	0.09134	0.09134
	0.98	0.04679	0.04770	0.04774	0.04774	0.04774	0.04774	0.04774
	0.99	0.0152	0.01541	0.01541	0.01541	0.01541	0.01541	0.01541
	0.995	0.00499	0.00501	0.00501	0.00501	0.00501	0.00501	0.00501
	0.999	0.00030	0.00030	0.00030	0.00030	0.00030	0.00030	0.00030

7   Conclusions

In this paper we proposed the TCP Reno analytical model based on Discrete-Time Markov Chain. The model allows estimating throughput of bulk TCP transfer as a function of loss rate and round-trip time. The model provides a way to compare SACK performance against cumulative acknowledgments. It was shown that SACK option is more beneficial for high-loss low delay networks. We compared numerical results gather from our model with real TCP traces measured in conditions close to modeling assumptions. It was shown that our model closely matches real measurements. Therefore the model can be safely used to estimate TCP Reno throughput over different network environments with uncorrelated losses such as router links with Active Queue Management algorithms (RED, Blue, etc.) or fading wireless channels. In addition the model allows estimating the influence of receiver's advertised window size on connection throughput, and the influence of Karn exponential back-off limit on time spent awaiting timeout event.

References

1. Padhey, J., Firoiu, V., Towsley, D., Kurose, J.: Modeling TCP throughput: a simple model and its empirical validation. Technical report TR98-008, UMASS CMPSI, University of Massachusetts, February 1998
2. Postel, J.: Transmission control protocol, Internet RFC 0793/STD 0007, September 1981
3. Fall, K.: TCP/IP illustrated. The Protocols, 2nd edn. Addison-Wesley Professional Computing Series, Boston (2012)
4. Jacobson, V., Karels, M.: Congestion avoidance and control. In: SIGCOMM'88, November 1988
5. Allman, M., Paxson, V., Blanton, E.: TCP congestion control, Internet RFC 5681, September 2009

 6. Lakshman, T.V., Madhow, U.: The performance of TCP/IP for networks with high bandwidth-delay products and random loss. ACM/IEEE Trans. Netw. **5**, 336–350 (1997)
 7. Kumar, A.: Comparative performance analysis of versions of TCP in a local network with a lossy link. ACM/IEEE Trans. Netw. **6**, 485–498 (1998)
 8. Ewald, N.L., Kemp, A.H.: Analytical model of TCP NewReno through a CTMC. In: Bradley, J.T. (ed.) EPEW 2009. LNCS, vol. 5652, pp. 183–196. Springer, Heidelberg (2009)
 9. Padhey, J., Firoiu, V., Towsley, D.: A stochastic model of TCP reno congestion avoidance and control. Technical report UMASS-CS-TR-1999-02, University of Massachusetts, 1999
10. Wierman, A., Osogami, T., Olsen, J.: A unified framework for modeling TCP-Vegas, TCP-SACK, and TCP-Reno. In: Proceedings of the 11th IEEE/ACM International Symposium on Modeling, Analysis and Simulation of Computer Telecommunications Systems (MASCOTS03), pp. 1526–7539, March 2003
11. Casetti, C., Meo, M.: An analytical framework for the performance evaluation of TCP Reno connections. Comput. Netw. **37**, 669–682 (2001)
12. Mathis, M., Mahdavi, J., Floyd, S., Romanow, A.: TCP selective acknowledgement options, Internet RFC 2018, October 1996
13. Floyd, S., Mahdavi, J., Mathis, M., Podolsky, M.: An extension to the selective acknowledgement (SACK) option for TCP, Internet RFC 2883, July 2000
14. Karn, P., Partridge, C.: Improving round-trip time estimates in reliable transport protocols. In: SIGCOMM'87, 1987
15. Fall, K., Floyd, S.: Simulation-based comparison of Tahoe, Reno, and SACK TCP. Comput. Commun. Rev. **26**(3), July 1996
16. Paxson, V.: Automated packet trace analysis of TCP implementations. In: Proceedings of SIGCOMM'97, 1997
17. Callegari, C., Giordano, S., Pagano, M., Pepe, T.: A survey of congestion control mechanisms in linux TCP. In: Proceedings of DCCN-2013 International Conference, October 2013
18. Floyd, S., Henderson, T., Gurtov, A., Nishida, Y.: The NewReno modification to TCPs fast recovery algorithm, RFC6582, April 2012
19. Paxson, V., Allman, M., Chu, J., Sargent, M.: Computing TCPs retransmission timer, RFC 6298, June 2011
20. Blanton, E., Allman, M., Fall, K., Wang, K.: A conservative selective acknowledgment (SACK)-based loss recovery algorithm for TCP, RFC 3517, April 2003
21. Allman, M.: TCP congestion control with appropriate byte counting (ABC), RFC 3465, February 2003
22. Braden, R.: Requirements for internet hosts communication layers, RFC 1122, October 1989
23. Lee, D.J., Carpenter, B.E., Brownlee, N.: Media streaming observations: trends in UDP to TCP ratio. Int. J. Adv. Syst. Measur. **3**, 3–4 (2010)

Asymptotic Method of Traffic Simulations

Ivanna Dronjuk[1](✉), Maria Nazarkevych[2], and Olga Fedevych[1]

[1] Automated Control Systems Department, Institute of Computer Science,
Lviv National Polytechnic University, Lviv Oblast, Ukraine
[2] Publishing Information Technology Department, Institute of Computer Science,
Lviv National Polytechnic University, Lviv Oblast, Ukraine
{Ivanna.Droniuk,OlhaFedevych}@gmail.com,
Nazarkevich@mail.ru
http://www.lp.edu.ua

Abstract. A method of simulation and predicting the behavior of daily traffic in computer or mobile communication networks based on differential equations of nonlinear oscillations with one degree of freedom and with a small parameter was proposed. Small perturbations are modelled using the sum of delta functions. The asymptotic method was used for solving the differential equations. Some numerical experiments are presented.

Keywords: Asymptotic method · Traffic simulation · Ateb-functions · Small perturbations

1 Introduction

There is a tendency to integrate communication networks and computer networks. For providing qualified customers cellular mobile communications service it is necessary to implement continuous users access to the Internet [1]. Modern communication networks, as well as computer networks, as their structure do not optimally use their functionality. One reason is the complexity of behavior of the network traffic. It is necessary to ensure optimum between used resources and characteristics of the network traffic in order to fully use network potential and ensure its successful operation. Thus, the task of predicting the network traffic behavior arises.

In this paper we aim to describe the daily periodicity of network traffic taking into account small pertubation. The mathematical model is based on the nonlinear differential equation, which describes oscillation motion. We obtained first approximation given by small parameter asymptotic method. As a result of computer simulation we discuss some numerical experiments. Obtained results were compared with real traffic data according to a maximum correlation criterion.

The paper is organized as follows: Sect. 2 contains review of some related works, Sect. 3 introduces the proposed mathematical model, Sect. 4 shows the traffic simulation experiment results and conclusion is presented in Sect. 5.

V. Vishnevsky et al. (Eds.): DCCN 2013, CCIS 279, pp. 136–144, 2014.
DOI: 10.1007/978-3-319-05209-0_12, © Springer International Publishing Switzerland 2014

2   Related Work

In this section we make a brief review of the known methods of mathematical modeling of traffic properties in computer or telecommunication networks, based on the articles [2–8]. The approximate method for computer system models was proposed in [2].

In the article [3] classical queuing theory is used to predict behavior of the traffic network with packet switching. The BMAP/PH/N/R queuing system operating in a finite state space Markovian random environment is presented. Disciplines of partial admission, complete rejection and complete admission are analyzed. The stationary distribution of the system states is calculated. The loss probability and other main performance measures of the system are derived. The LaplaceStieltjes transform of the sojourn time distribution of accepted customers is obtained. There are shown the effect of an admission strategy, a correlation in an arrival process, a variation of a service process. Poor quality of the loss probability approximation by means of simpler models is illustrated. The similar way in simulation tandem queuing system consisting of R multi-server network without buffers is analyzed in [4].

A method for calculating blocking probabilities for models involving both unicast and multicast traffic is suggested in the paper [5].

Another way for describing behavior of the network traffic is using the diffusion equation. The model of TCP/UDP connection with Active Queue Management in an intermediate IP router is studied using the fluid flow or diffusion approximation technique to model the interactions between the set of TCP/UDP flows and AQM. There is a model of three approaches: Markovian queues solved numerically, the diffusion approximation and fluid-flow approximation. The obtained results were compared [6]. There are some articles [7, 8] where self-similarity character of traffic network nature were discussed.

3   Simulation Traffic with Periodic Perturbation

On the basis of this analysis we propose our method. The main supposition is that the traffic network nature has the periodic character. It is known that network traffic depends from human reason periodicity [9]. Traffic fluctuations have daily, weekly, and monthly periods. This investigation focuses on simulation of daily cycle of the periodicity.

In previous articles we considered the problem of traffic modeling based on the differential equation for nonlinear oscillating system [10, 11]. We modelled main traffic trend. As it is shown on Figs. 1, 2 calculating simulation curve doesn't take into account sharp traffic fluctuations. Our article aims to eliminate this disadvantage.

In this paper it is continued our investigation in this field and suggested to use equations for modeling and predicting the behavior of traffic in mobile or computer network, which describe nonlinear oscillatory systems with one degree of freedom and with small perturbations. It is assumed that the main periodic

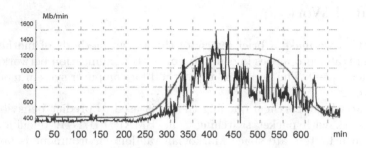

Fig. 1. Simulation traffic with periodic perturbation example 1

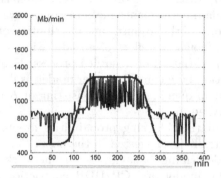

Fig. 2. Simulation traffic with periodic perturbation example 2

component simulates the daily fluctuations of traffic in the network, while the traffic deviation from the main trend is modelled using delta function with a small parameter. The Krylov - Bogoliubov - Mitropolsky asymptotic method for solving of the modeling differential equations was used [12].

4   Mathematical Model of Computer Network Traffic

Let us consider the changing traffic in the network over time, as a nonlinear oscillating system with one degree of freedom and with small perturbation. Modeling behavior of the network traffic $x(t)$ is generated by an ordinary differential equation with a small parameter ε in the form

$$\ddot{x} + \alpha^2 x^n = \varepsilon f(t, x, \dot{x}), \tag{1}$$

where $x(t)$ – is the number of packets in the network at time t; α – a constant that determines size of the period of traffic oscillation; $f(t, x, \dot{x})$ – any analytical function that is used to simulate small deviations from the main component of the traffic fluctuations, n – a number that determines the degree of nonlinearity of the equation that affects the period of the main component of fluctuations. In the performance of such conditions on α and n : $\alpha \neq 0, n = \frac{2k_1+1}{2k_2+1}, k_1, k_2 = 0, 1, 2 \ldots$

it is proved [13], that the analytical solution of Eq. (1) is represented as Ateb - functions.

For problems of forecasting behavior of the traffic in computer network or mobile network it is important to choose the type of function f, as these terms take into account the specific features of particular network. Previously, the authors examined a small perturbation in the form of periodic functions [14]. This approach is appropriate for modeling a network with smooth traffic changing. In the present work the perturbation as a sum of delta functions was proposed. This description corresponds to a network with abrupt changes of traffic. Consider the function f as

$$f(t, x, \dot{x}) = \sum_{i=1}^{N} a_i(t_i) f_i(x, \dot{x}), \tag{2}$$

where N is the number of disturbances over the period $[0, T]$, a_i - amplitude of perturbation, $-A \leq a_i \leq A$, A - the maximum amplitude of perturbation (randomly generated in simulation), f_i - perturbation function, t_i - time in which comes off the i-th perturbation, which is generated randomly. Here T - simulation time. Perturbation function f_i - in general case depends from x- number of packets and $\dot{x}$ - speed of packets changing.

To build the solution first consider the Eq. (1) without perturbation function

$$\ddot{x} + \alpha^2 x^n = 0. \tag{3}$$

Taking into account the change of variables $y = \dot{x}$, a second order differential Eq. (1) transforms to a following system of first order differential equations

$$\begin{cases} \frac{dx}{dt} - y = 0, \\ \frac{dy}{dt} + \alpha^2 x^n = 0. \end{cases} \tag{4}$$

The solution (4) is represented through periodic Ateb-functions [13] as follows

$$\begin{cases} x = aCa(n, 1, \phi), \\ y = a^{\frac{1+n}{2}} hSa(1, n, \phi), \end{cases} \tag{5}$$

where a is the amplitude of oscillation, $a = max(x)$ or $a = min(x)$, $Ca(n, 1, \phi)$, $Sa(1, n, \phi)$ are Ateb-cosine and Ateb-sine respectively, $h^2 = \frac{2\alpha^2}{1+n}$. Variable ϕ is associated with time t as follows

$$\phi = \frac{a^{\frac{n-1}{2}}}{L} t + \phi_0, \tag{6}$$

where L - is some constant, ϕ_0 - the initial phase of the oscillations, which are determined from the initial and periodical conditions for Eq. (3).

Periodical conditions are presented by expressions

$$\begin{cases} Ca(n, 1, \phi + 2\Pi) = Ca(n, 1, \phi), \\ Sa(1, n, \phi + 2\Pi) = Sa(1, n, \phi), \end{cases} \tag{7}$$

where Π is a half period of Ateb-function. When expressions (5) and (6) substitute in the Eq. (4) and taking into account periodical conditions (7), we obtain such expression for calculating constant L in formula (6)

$$L = \frac{2B(0.5, \frac{1}{1+n})}{\pi(1+n)h}.$$
(8)

In formula (8) denomination $B(x, y)$ means complete Beta-function with arguments $x = 0.5$, $y = \frac{1}{1+n}$. Taking into account formula (7) and identity $Ca(n, 1, \phi)^{m+1} + Sa(1, n, \phi)^2 = 1$ we result following formula for a half period of Ateb-function

$$\Pi(n, 1) = B(0.5, \frac{1}{1+n}).$$
(9)

Let us consider initial conditions for system of differential Eq. (4). In paper [15] such initial conditions for differential Eq. (1) are considered

$$x(0) = 0, \dot{x}(0) = const,$$
(10)

assuming that $n = \frac{1}{2k+1}$ and $k \rightarrow \infty$. Such initial condition is not useful for simulation traffic. We suppose that traffic value and changing traffic value are constantly defined in time $t = 0$. According to these assumptions we define initial conditions as

$$x(0) = c_1, \dot{x}(0) = c_2,$$
(11)

where c_1 defines initial traffic and $c_1 \neq 0$, c_2 defines changing initial traffic and it can equal to zero in initial time $t = 0$. Taking into account that $Ca(n, 1, 0) = 1$ and $Sa(1, n, 0) = 0$ and Eq. (5) we obtain from initial conditions (11) that $c_1 = a$ and $c2 = 0$.

We use the asymptotic method to create a solution of Eq. (1) based on (5). Asymptotic method creates the solution as series in the small parameter ε like following

$$x(t) = \sum_{i=1}^{\infty} \varepsilon^i x_i(t).$$
(12)

For numerical simulation we have to discard terms, which have ε in order greater than M. We obtain solution with accuracy of the order ε^{M+1}. We will find the solution as series in the small parameter ε like the following

$$x(t) = \sum_{i=1}^{M} \varepsilon^i x_i(t).$$
(13)

We substitute series (13) into left part of (1) and after that the terms with the same order of small parameter ε were equated each other. Now consider (1) which is transformed to a system of differential equations by changing variables (the same way as (3) into (4)) in the form

$$\begin{cases} \frac{dx}{dt} - y = 0 \\ \frac{dy}{dt} + \alpha^2 x^n = \varepsilon f(t, x, y). \end{cases}$$
(14)

Solution approach is represented by degrees of the small parameter ε. For creation the first approximation in (13) according the first order of ε ($M = 1$ in formula (13)) we make the change of variables of the type

$$x = \xi + \varepsilon f(t, \xi, \zeta), \qquad y = \zeta = \dot{x}, \tag{15}$$

and equate the coefficients of equal degrees of ε, and also cast members of higher order with respect to ε. As shown in [12], variables ξ, ζ calculated according to solution without perturbation (5) can be taken as the first approximation of the solution (14), and (15) as the improved first approximation.

5 Traffic Simulation

For numerical traffic simulation let us consider the functions $f(t, \xi, \zeta)$ in more simpler form than (2). We suppose that traffic and changing traffic value do not affect the small perturbation. Thus, we construct the perturbation function in following form

$$f(t, \xi, \zeta) = \sum_{i=1}^{N} a_i \delta_i(t_i), \tag{16}$$

where δ_i - Dirac function. By substituting expressions (16) and (5) to (15), we obtain formulas for modeling traffic in the network based on the improved approximation of asymptotic method for differential equations of oscillatory movements with small perturbation in the form

$$\begin{cases} x = aCa(n, 1, \phi) + \varepsilon f(t, \xi, \zeta), \\ y = a^{\frac{1+n}{2}} hSa(1, n, \phi). \end{cases} \tag{17}$$

In order to simulate the daily fluctuations, $n = 7$ was selected. This choice stems from the fact, that Ateb-function of this parameter has large gaps of values, which are close to the maximum and minimum, which corresponds to the behavior of network traffic. Figure 3 shows a graph of the function $Sa(7, 1, t)$. The simulation was conducted using the appropriate zoom. Diurnal variations of network traffic were modeled. In this case, we consider that the period $\Pi(7, 1) = 9, 21$, function $Sa(7, 1, t)$ correspond to 24 hours. Table 1 shows the data, which were used for modeling:

Table 1. Data for simulation

Experiment N	T, s	N	A	n	s	Π
E1	86400	50000	500	7	70	9,21
E2	86400	50000	600	7	70	9,21
E3	86400	50000	500	7	70	9,21

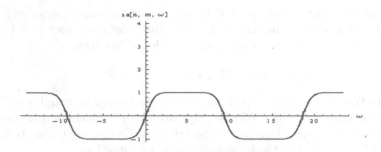

Fig. 3. Graph of the function $Sa(7, 1, \omega)$

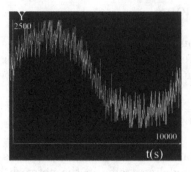

Fig. 4. The simulation results E1 based on Eq. (17)

Figures 4–6 show the results of modeling of network traffic based on formulas (17). The environment Simulink of MatLab package was used for simulation.

Traffic data of computer network Department of Automated Control System National University Lviv Polytechnic was used for testing simulation model. The simulation experiments E1, E2, E3 data were presented in Table 1 and numerical results - on Figs. 4–6. The comparison results was conducted by the criteria maximum correlation ρ, and calculated the ratio of the coefficient k standard deviation to the maximum. The calculated results of the comparison were presented in Table 2. Maximum amplitude perturbation value was change in experiment E2 and the number of perturbation - in experiment E3. The main difficulties in modeling are choosing the correct zoom of Ateb- function period with respect to the real time of day, and scaling large amplitude oscillation, because the amplitude of Ateb-function equals to 1 (see Fig. 3). We used the value c_1 from

Table 2. Comparison simulation results with real traffic data

Kriteria	E1	E2	E3
k	10,81 %	12,56 %	12,75 %
ρ	0,79	0,75	0,74

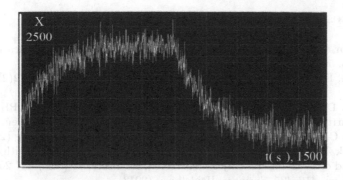

Fig. 5. The simulation results E2 based on Eq. (17)

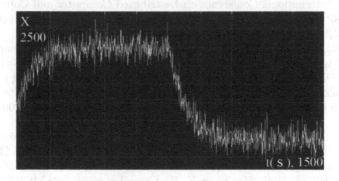

Fig. 6. The simulation results E3 based on Eq. (17)

initial condition (11), which was calculated from real traffic data based on averaging daily traffic amplitude during 1 month. The value c_1 determs the traffic amplitude in numerical simulation and scaling of Ateb-function amplitude.

6   Conclusion

Thus, the method of modeling and predicting the behavior of traffic in communication networks and computer networks was proposed. As a mathematical model, the differential equations of oscillatory movements with small perturbation were used. The solution of equations was based on the asymptotic method of Krylov - Bogoliubov - Mitropolsky. Improved first approximation was used in order to build the traffic model. It is shown, that the unperturbed solution is represented as an Ateb - functions. Small perturbations were modeled by using delta functions of random occurrence in time and with randomly selected amplitude perturbation. The environment Simulink of MatLab package was used for simulation. The data deviation is about 11–13 %. The received results can be used for simulation and prediction the behavior of traffic in computer networks and in mobile networks.

References

1. Goldstein, B.S., Ehriel, I.M., Ruhrle, R.D.-M.: Radio and communication. Intell. Netw., p. 500 (2000). (in Russian)
2. Gelembe, E.: On approximate computer systems models. J. ACM **22**(2), 261–269 (1975)
3. Kim, C., Dudin, A., Dudin, S., Dudina, O.: Queueing system MAP|PH|N|R with session arrivals operating in random environment. In: Gaj, P., Kwiecień, A., Stera, P. (eds.) CN 2013. CCIS, vol. 370, pp. 406–415. Springer, Heidelberg (2013)
4. Klimenok, V., Dudin, A., Vishnevsky, V.: Tandem queueing system with correlated input and cross-traffic. In: Gaj, P., Kwiecień, A., Stera, P. (eds.) CN 2013. CCIS, vol. 370, pp. 416–425. Springer, Heidelberg (2013)
5. Yu, G.: On mathematical modeling of P2P streaming networks. In: Proceedings of The 17th International Conference on Distributed Computer and Communication Networks: Control, Computation, Communications (DCCN-2013), pp. 188–191. JSC, TECHNOSPHERA, Moscow (2013)
6. Czachórski, T., Nycz, M., Nycz, T., Pekergin, F.: Analytical and numerical means to model transient states in computer networks. In: Kwiecień, A., Gaj, P., Stera, P. (eds.) CN 2013. CCIS, vol. 370, pp. 426–435. Springer, Heidelberg (2013)
7. Czachorski, T., Domanska, J., Sochan, A.: Samopodobny charakter natezenia ruchu w sieciach komputerowych. Studia Informatica **22**(1) (43), 93–108 (2001). (in Polish)
8. Leland, W.E., Taqqu, M.S., Willinger, W., Wilson, D.V.: On the self-similar nature of Ethernet (extended version). IEEE/ACM Trans. Netw. **2**(1), 1–15 (1994)
9. Xie, Y., Tang, S., Huang, X.: A periodic structural model for characterizing network traffic. In: Huang, D.-S., Ma, J., Jo, K.-H., Gromiha, M.M. (eds.) ICIC 2012. LNCS (LNAI), vol. 7390, pp. 545–552. Springer, Heidelberg (2012)
10. Dronjuk, I., Klymash, M., Oleksyn, M.: Mathematical modeling of passing connection synchronous telecommunications network in fuzzy conditions. In: Proceedings of International Conference on Modern Problems of Radio Engineering, Telecommunications and Computer Science, TCSET 2006, pp. 112–113 (2006)
11. Droniuk, I.: Modeling nonlinear oscillatory system under disturbance by means of Ateb-functions for the Internet. In: Droniuk, I.M., Nazarkevich, M.A. (eds.) Proceedings of 6th Working International Conference Het-Nets2010, Gliwice, pp. 325–335 (2009)
12. Bogolyubov, N.N., Mitropolsky, Y.A.: Asymptotic methods in the theory of nonlinear oscillations. M. Izd. Fiz-Mat. Lit., 407 (1963). (in Russian)
13. Sokol, B.I.: Solutions asymptotic approximations for a nonlinear nonautonomous equation. Ukr. Mat. J. **49**(11), 1580 (1997). (in Ukrainian)
14. Medykovsky, M., Droniuk, I., Nazarkevich, M., Fedevych, O.: Modelling the pertubation of traffic based on ateb-functions. In: Gaj, P., Kwiecień, A., Stera, P. (eds.) CN 2013. CCIS, vol. 370, pp. 38–44. Springer, Heidelberg (2013)
15. Awrejcewicz, J., Andrianov, I.V.: Oscillations of non-linear system with restoring force close to sign (X). J. Sound Vibr. **252**(5), 962–966 (2002)

New Generation of Safety Systems
for Automobile Traffic Control
Using RFID Technology and Broadband
Wireless Communication

Vladimir Vishnevsky[1], Dmitry Kozyrev[1,2]([✉]), and Vladimir Rykov[2,3]

[1] V.A. Trapeznikov Institute of Control Sciences of Russian Academy of Sciences,
65 Profsoyuznaya Street, Moscow 117997, Russia
[2] Peoples' Friendship University of Russia, 6 Miklukho-Maklaya Street,
Moscow 117198, Russia
[3] Gubkin Russian State University of Oil and Gas, 65, Leninsky prospekt,
Moscow 119991, Russia
vishn@inbox.ru, kozyrevdv@gmail.com, vladimir_rykov@mail.ru

Abstract. Broadband wireless data transmission network for providing of automobile transport system safety is considered. The network operates under IEEE802.11n-2012 protocol that guarantees high-speed transmission of multimedia information between automatic stationary and mobile systems of traffic control. The model of stochastic network with dependent service time for the problem solution is used. Based on product form representation of the steady state probabilities of such system the formulas and algorithms for calculating the main system characteristics such as: mean marginal queue length and mean sojourn times in the nodes, as well as mean packages delivery time in the whole network are proposed. Results of some numerical experiments are presented.

Keywords: Wireless broadband network · RFID · Stochastic model · Tandem type network

1 Introduction and Motivation

Road traffic crashes are one of the world's largest problems. According to the World Health Organization (WHO), about 1.2 million people are killed on the world?s roads each year (including 27000 in Russia). It looks like a war in piecetime. As a fighting tool against accidents on the roads the Automobile System Safety (ASS) is usually used. It is a system for automatic fixing of violations of traffic rules (TR) and for information transmission to the appropriate Police Center as it is shown in Fig. 1.

It consists of a fixing system located on Stationary or/and Mobile Modules as it is shown in Figs. 2, 3 and an appropriate transmission system.

The latter is a broadband wireless data transmission network deployed along the road.

V. Vishnevsky et al. (Eds.): DCCN 2013, CCIS 279, pp. 145–153, 2014.
DOI: 10.1007/978-3-319-05209-0_13, © Springer International Publishing Switzerland 2014

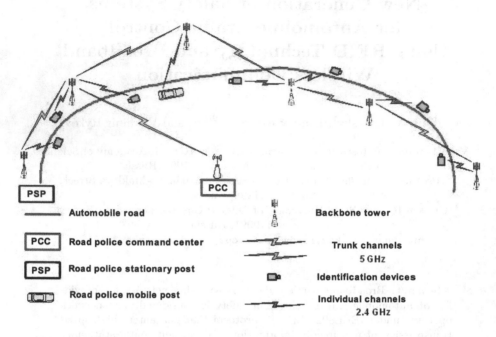

Fig. 1. A system for automatic fixing of TR violations and data transmission

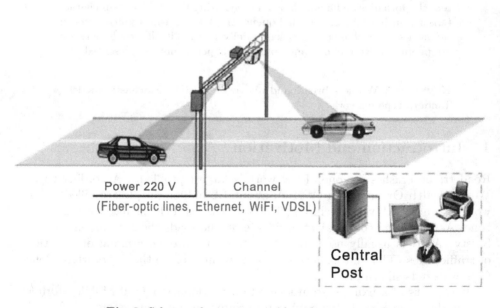

Fig. 2. Scheme of a stationary identification module

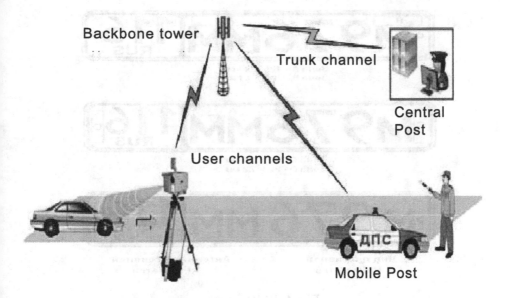

Fig. 3. Scheme of a mobile identification module

The system usually uses a radar too to measure the velocity and an optic camera for the vehicle number plate fixing.

However, this system has two defects:

- impossibility to recognize too dirty signs, and
- too long transmission of the information to the Central Police Office (CPO)

that decrease the system efficiency up to 40 %.

Another approach consists in using of the RFID-technology that makes use of a special Radio Fixing Identification Mark installed in the vehicle plate as it is shown in Fig. 4.

The Stochastic Networks have a wide spectrum of applications, including computer, data transmission and telecommunication networks. Nowadays telecommunication technologies give extremely wide possibilities for information interchange. In the paper a stochastic network with dependent service times is used for the problem description and solution.

It is necessary to take into account that in data transmission and telecommunication networks channels are usually considered as nodes (service stations), while stations (terminals) are usually considered as buffers.

Most of really applicable network characteristics are macro-state characteristics such as queue length in buffers, mean time message transmission etc. They are usually represented in terms of their steady-state probabilities (SSP). Therefore their calculation is one of the most important problems in this topic. Network decomposition and the product form representation of the SSP is a real way for the problem solution. There is a vast bibliography devoted to both precision and approximate methods of decomposition. For the bibliography and the

Fig. 4. RFID - marks

latest situation in the topic see an excellent book by R.Serfozo [1]. For applications of stochastic networks to data transmission systems see, for example, V.Vishnevsky, S.Portnoy and I.Shakhnovich [2].

However in most of all these publications it is assumed that the service times in different nodes are independent random variables. But this assumption is not adequate for real data transmission systems such as tele- and computer communication networks. Indeed, the same message (call) during its transmission through the network has the same (constant) size (work requirement, workload), but may be transmitted at different rates in different channels. That is, the service times in different nodes of the message route must be dependent.

Product form for networks with regenerative service mechanism and Kelly's networks, which service disciplines are determined by some collection of service rates in nodes, was obtained in [3]. Product form for equilibrium probabilities of the open hierarchial networks with dependent service times was obtained in [4]. In [5] another approach to decomposition of complex hierarchial stochastic networks was proposed. In [6] it was shown that the macro-state stationary probabilities for the network with Poisson input, infinite-servers nodes and processor sharing discipline have a product form. These results have been generalized for open and closed queueing networks with dependent service times in [7,8] that involve and generalize almost all previously considered models and where a vast bibliography has also been presented.

At the 9th International ISAAC Congress (ISAAC-2013) the model of the broadband wireless data transmission network deployed along the road and aimed at promotion of automobile transport system safety has been presented [9], and its theoretical study based on this approach was held. In this paper we continue and extend these investigations with some numerical examples. The paper is organized as follows. In the next section the mathematical model of the considered situation will be presented. In the third section the model will be investigated and the final section contains a numerical example.

2   The Model

It is necessary to take into account that in data transmission and telecommunication networks the role of nodes (service stations) is usually played by the links, while the stations (terminals) are usually considered as buffers.

The Data Transmission System (DTS) for Automobile Safety System (ASS) is modelled as a tandem type network with r nodes of infinite capacity servers, infinite buffers and a processor sharing discipline as it is presented at Fig. 5. This means that all calls are served simultaneously and transferred from k-th node to the $(k+1)$-th one, and after leaving the last node they go out of the system and arrive to the Road Police Command Center for elaboration and decision making.

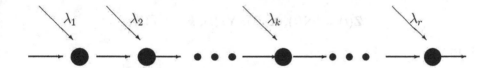

Fig. 5. The model of automobile safety systems

Throughout the paper the following assumptions and notations are used.

- $\mathbf{n} = (n_1, \ldots, n_k, \ldots n_r)$ is the system states vector, where n_k is the number of calls at the k-th node;
- The capacity of the k-th node (channel) is denoted by c_k, and the λ_k is the intensity of the input Poisson flow arriving to the k-th node from the outside and therefore the summary flow intensity to the k-th node is

$$\Lambda_k = \lambda_1 + \cdots + \lambda_k;$$

- The service discipline is supposed to be processor sharing discipline, which allows all calls (requests) to be served simultaneously at the rate $\frac{c_k}{n_k}$ for the k-th node.
- The sizes of all calls are independent random variables Y with the same cumulative distribution function (c.d.f.) $G(y) = \mathbf{P}\{Y \leq y\}$ and therefore the conditional c.d.f. of service times at each stage of service given the total requests volume y are

$$B_k(x|y) = \mathbf{P}\{X_k \leq x \,|\, Y = y\} = \Theta\left(x - \frac{yn_k}{c_k}\right) \quad (k = \overline{1,r}).$$

- This means that the unconditional c.d.f. of the service time is

$$B_k(x) = \mathbf{P}\{X_k \le x\} = \int_0^\infty \Theta\left(x - \frac{yn_k}{c_k}\right) dG(y) =$$

$$= \int_0^\infty \Theta\left(\frac{c_k x}{n_k} - y\right) dG(y) = G\left(\frac{c_k x}{n_k}\right) \qquad (k = \overline{1,r}), \qquad (1)$$

- and therefore the joint service times c.d.f. for the calls is

$$B(\mathbf{x}) = \mathbf{P}\{X_k \le x_k, \ k = \overline{1,r}\} = \prod_{1 \le k \le r} G\left(\frac{c_k x}{n_k}\right).$$

In the next section the main steps for the model investigation will be represented.

3   The Model Investigation

For the model investigation consider a stochastic process

$$\mathbf{Z}(t) = \{\mathbf{N}(t), \mathbf{X}_k(t), \mathbf{Y}_k(t); \ k = \overline{1,r}\}$$

where

- $\mathbf{N}(t) = (N_1, \dots N_k(t), \dots N_r(t))$ is the random vector of requests at each node;
- $\mathbf{X}_k(t) = (X_{k,1}, \dots, \dots, X_{k,n_k})$ is the vector of residual service times of requests at the k-th node $(k = \overline{1,r})$;
- $\mathbf{Y}_k(t) = (Y_{k,1}, \dots, \dots, Y_{k,n_k})$ is the vector of initial service times of requests at the k-th node $(k = \overline{1,r-1})$ arranged in the same order as vector $\mathbf{X}_k(t)$,

and its set of states for $t \ge 0$ is

$$\mathcal{E} = \{0 \le x_{k,i_k} \le y_{k_1} < \infty, \quad (i_k = \overline{1,n_k}, \ k = \overline{1,r})\}.$$

Denote by $\pi(t; \mathbf{z}) = \pi(t; \mathbf{n}, \mathbf{x}, \mathbf{y})$ the probability density function (p.d.f.) of the process $\mathbf{Z}(t)$. In order to write out the Kolmogorov equations for these p.d.f.s denote also by

- $\mathbf{e}_k = (0, \dots 0, 1, 0, \dots 0)$ the vector, which k-th component equals to 1, while others equal to 0,
- $\mathbf{x}_{k,i}^- = (x_{k,1}, \dots x_{k,i-1}, x_{k,i+1}, \dots x_{k,n})$ the vector, which i-th component is eliminated,
- $\mathbf{x}_i(k, y) = (x_{k,1}, \dots x_{k,i-1}, y, x_{k,i+1}, \dots x_{k,n})$ the vector, where additional value y is inserted in the i-th place, while others move to the next positions,

and introduce operators $A_{k,i}$, $T_{k,i,j}$, D_i in the system state space:

- $A_{k,i}$ — corresponds to the arrival of a request to the i-th position of the k-th node;

- $T_{k,i,j}$ — corresponds to the being of a request at the i-th position of the k-th node and transmitting it to the j-th position of the next node;
- D_i — corresponds to the service termination of a request at the i-th position of the last node and its leaving the system.

Formally these operators can be represented by the following relations:

•

$$A_{k,i}[\mathbf{z}](y) = (\mathbf{n} + \mathbf{e}_k, \ \mathbf{x}_{k,i}(y), \ \mathbf{y}_{k,i}(y));$$

•

$$T_{k,i,j}[\mathbf{z}] = (\mathbf{n} - \mathbf{e}_k + \mathbf{e}_{k+1}, \ \mathbf{x}_{k,i}^-, \ \mathbf{x}_{k+1,j}(y_{k,i}), \ \mathbf{y}_{k+1,i}^-(y_{k,i}));$$

•

$$D_i[\mathbf{z}] = (\mathbf{n} - \mathbf{e}_r, \ \mathbf{x}_{r,}^-).$$

The p.d.f. $\pi(t; \mathbf{z}) = \pi(t; \mathbf{n}, \mathbf{x}, \mathbf{y})$ for all $\mathbf{z} \in \mathcal{E}$ satisfies to the Kolmogorov system of equations,

$$\frac{\partial \pi(t, \mathbf{z})}{\partial t} - \sum_{\substack{1 \le k \le r, \\ 1 \le i_k \le n_k}} \frac{1}{n_k} \frac{\partial \pi(t, \mathbf{z})}{\partial x_{k, i_k}} + \Lambda \pi(t, \mathbf{z}) =$$

$$= \sum_{\substack{1 \le k \le r, \\ 1 \le i_k \le n_k}} \frac{\Lambda_k}{n_k} \pi(t; A_{k, i_k}^{-1} \mathbf{z}) + \sum_{\substack{1 \le k \le r-1, \\ 1 \le i_k \le n_k, \\ 1 \le i_{k+1} \le n_{k+1}+1}} \frac{1}{n_k (n_{k+1} + 1)} \pi(t, T_{k, i_k, i_{k+1}}^{-1} \mathbf{z}) +$$

$$+ \frac{1}{n_r + 1} \sum_{1 \le i_r \le n_r} \pi(t, D_{r, i}^{-1} \mathbf{z}) \tag{2}$$

while the initial conditions in terms of Dirac δ-function are

$$\pi_{\mathbf{j}}(0; 0) = \delta_{(\mathbf{j}, 0)}(t) \quad (\mathbf{j} \in E). \tag{3}$$

Appropriate equations for steady state p.d.f. can be represented as

$$\Lambda \pi(\mathbf{z}) - \frac{1}{n_r + 1} \sum_{1 \le i_r \le n_r} \pi(D_{r, i}^{-1} \mathbf{z}) +$$

$$+ \sum_{1 \le k \le r} \frac{\Lambda_k}{n_k} \sum_{1 \le i_k \le n_k} \pi(A_{k, i_k}^{-1} \mathbf{z}) + \sum_{\substack{1 \le k \le r-1, \\ 1 \le i_k \le n_k, \\ 1 \le i_{k+1} \le n_{k+1}+1}} \frac{1}{n_k (n_{k+1} + 1)} \pi(T_{k, i_k, i_{k+1}}^{-1} \mathbf{z})$$

$$\tag{4}$$

By a simple substitution it is possible to check that the last equations have the following solution

$$\pi(\mathbf{z}) = C \prod_{1 \le k \le r-1} \Lambda_k^{n_k} \prod_{1 \le i \le n_k} \Theta(y_{k, i} - x_{k, i}) b_k(y_{k, i}) \times \lambda_r^{n_r} \prod_{1 \le i \le n_r} (1 - B_r(x_{r, i}))$$

$$\tag{5}$$

At least integration with respect to all admissible continuous variables and calculation of the constant C gives in terms of calls size the final distribution of
macro-states probabilities

$$\pi(\mathbf{n}) = \prod_{1 \leq k \leq r} (1 - \rho_k)\rho_k^{n_k} \tag{6}$$

with

$$\rho_k = \frac{\Lambda_k m}{c_k} \qquad \text{and} \qquad m = \mathbf{E}[V] = \int_0^\infty (1 - G(x))dx.$$

The load of the channel parameter ρ_k shows its real throughput and for the
normal channel performance it should be about 70–75 %.

4   Numerical Study

The results show that the mean number $\bar{n}_k$ of units at the k-th node and its
mean sojourn time $\bar{w}_k$ according to Little's formula equal

$$\bar{n}_k = \frac{\rho_k}{1 - \rho_k} \qquad \text{and} \qquad \bar{w}_k = \frac{1}{\Lambda_k}\bar{n}_k.$$

Suppose that the number of retransmittings $r = 10$, the automobile traffic
is homogeneous and heavy enough $\lambda_k = \lambda = 1\,[auto/\text{min}] = 1/60\,[auto/\text{s}]$, the
mean volume of transferred data $m = 100\,[\text{Mb}]$, and the channels are homogeneous with the same capacity $c_k = 1\,[\text{Gb}/\text{s}]$.
   In this case simple calculations show that

$$\Lambda_k = k\lambda = \frac{k}{100}\,[\text{s}^{-1}], \quad \rho_k = \frac{\Lambda_k m}{c_k} = \frac{k}{100},$$

and therefore

$$\bar{n}_k = \frac{\rho_k}{1 - \rho_k} = \frac{k}{100 - k}\,[units], \quad \bar{w}_k = \frac{\bar{n}_k}{\Lambda_k} = \frac{1}{100(100 - k)}\,[\text{s}].$$

This means that the overall delivery time equals

$$\bar{w} = \sum_{1 \leq k \leq r} \bar{w}_k = \frac{1}{100} \sum_{1 \leq k \leq r} \frac{1}{100 - k} \approx \frac{1}{950}\,[\text{s}] \approx 1 \times 10^{-3}\,[\text{s}],$$

that shows a good enough performance of the system. Moreover, the calculations according to the formula $\rho_r = 100^{-1}r < 0.7$ show that for the admissible
throughput the number of retransmittings could be increased up to $r^* = 70$
units.

5   Conclusion

The safety system for automobile traffic control is considered. The tandem type queueing model with dependent service times and processor sharing service discipline is used to study the system. The product form for the steady state probabilities distribution of the considered model has been established. Some simple numerical calculations show the admissible results for the practical use of the system.

Acknowledgments. This research is supported by the Russian Foundation for Basic Research (grant No. 13-07-00737) and by the Ministry of education and science of Russia (grant No. 14.514.11.4071).

References

1. Serfozo, R.: Introduction to Stochastic Networks, 300pp. Springer, New York (1999)
2. Vishnevsky, V., Portnoy, S., Shakhnovich, I.: Encyclopaedia WiMAX. a way to 4G, 472pp. Technosphera, Moscow (2009)
3. Tolmachev, A.L.: Queueing network with regenerative service mechanism. Probl. Inf. Transm. **22**(2), 59–68 (1986). (in Russian)
4. Rykov, V.V.: Two approaches to decomposition of the complex hierarchical stochastic systems. Continuously interacting subsystems. Autom. Remote Control **(10)**, 91–104 (1997)
5. Rykov, V.V.: Two approaches to decomposition of the complex hierarchical stochastic systems. Aggregate systems. Autom. Remote Control **(12)**, 140–149 (1997)
6. Rykov, V.V.: On decomposition of hierarchical computer communication network. Vestnik RUDN. Ser. Appl. Math. Inform. **(2)**, 110–125 (1996)
7. Pechinkin, A.V., Rykov, V.V.: On Product form for open queuing systems with dependent service times. In: Proceedings of the International Workshop, pp. 34–48. IPPI, Moscow (1998)
8. Pechinkin, A.V., Rykov, V.V.: On decomposition of closed networks with dependent service times. Autom. Remote Control **60**(11), 1568–1576 (1999)
9. Vishnevsky, V.M., Rykov, V.V.: On an application of quality of service characteristics of stochastic networks with dependent service times. In: Abstracts of the 9th International ISAAC Congress, 5–9 August 2013, Krakow, Poland, p. 180 (2013)

Parameters Estimator of the Probabilistic Model of Moving Batches Traffic Flow

Michael Fedotkin$^{(\boxtimes)}$ and Maria Rachinskaya

N. I. Lobachevsky State University of Nizhni Novgorod,
National Research University,
Gagarina ave., 23, Nizhni Novgorod 603950, Russia
fma5@rambler.ru

Abstract. A probabilistic model for time characteristic of a traffic flow moving on a motorway is proposed and investigated in this paper. The time intervals between consecutive cars are supposed to be dependent and have different distribution. Cars with the slow and fast movement are distinguished in the traffic flow. The mathematical model of such traffic flow is represented as a control cybernetic system of a certain class. The methods to derive estimate for the parameters of the control cybernetic system are proposed in order to select the adequate traffic flow model in the form of batches flow. These methods are approved processing the statistical data of the Bartlett traffic flow. Effectiveness of suggested methods for the parameters estimation and algorithms for splitting a real traffic flow into the batches is demonstrated.

Keywords: Control cybernetic system · Kolmogorov equations · Batches traffic flow · Bartlett flow · Ergodic distribution · Non-ordinary Poisson flow · Algorithm splitting flow · Parameters estimator · Wallis-Moore statistics

1 Introduction

Model construction for the control of the conflict flows with the heterogeneous cars on motorways intersection starts with describing mathematically the time characteristic of investigated flows. Probabilistic properties of the mentioned time characteristics are classically described in the form of random sequence $\{\tau'_j; j = 1, 2, \ldots\}$. In this sequence τ'_j denotes a moment when the car with number j crosses a transverse stop-line on the motorway. As a rule, independence and identical distribution of random variables $\tau'_{j+1} - \tau'_j, j = 1, 2, \ldots$ are postulated in this description method. However, weather conditions, bad road conditions and other factors often make overtaking difficult and their movement becomes dependent. In this case hypothesis about independence and identical distribution of the mentioned time intervals between the successive cars does not take place. This is why the classical method to describe the traffic flow is not effective since finding finite-dimensional distributions of sequence $\{\tau'_j; j = 1, 2, \ldots\}$ becomes very unlikely.

V. Vishnevsky et al. (Eds.): DCCN 2013, CCIS 279, pp. 154–168, 2014.
DOI: 10.1007/978-3-319-05209-0_14, © Springer International Publishing Switzerland 2014

In this paper traffic flow description is implemented in the form of vectorial random sequence $\{(\tau_{i+1} - \tau_i, \eta_i); i = 0, 1, \ldots\}$. Here $\{\tau_i; i = 0, 1, \ldots\}$ is an ever-increasing sequence and η_i defines a number of cars at interval $[\tau_i, \tau_{i+1})$ on time axis Ot. Variable η_i will be also called i-th batch size. Flow $\{\tau_j'; j = 1, 2, \ldots\}$ should be split with the help of sequence $\{\tau_i; i = 0, 1, \ldots\}$ so as independence and identical distribution of variables $\tau_{i+1} - \tau_i, i = 0, 1, \ldots$ and also independence and identical distribution of $\eta_i, i = 0, 1, \ldots$ would take place. The main problem here is to find the distribution of sequences $\{\tau_{i+1} - \tau_i; i = 0, 1, \ldots\}$ and $\{\eta_i; i = 0, 1, \ldots\}$ based on observed finite realization of sequence $\{\tau_j'; j = 1, 2, \ldots\}$ and specified traffic flow split. In case of point model for the traffic flow, when the size of each car is adopted to equal zero, such well-known test distributions may be used to determine the distribution of sequence $\{\tau_{i+1} - \tau_i; i = 0, 1, \ldots\}$ as exponential, hyperexponential, lognormal, uniform, gamma distribution, chi-square distribution, etc. If each car is not supposed to be a point on time axis Ot, it crosses the stop-line during time T. In this case the corresponding displaced distributions may be used to find the distribution for each interval in sequence $\{\tau_{i+1} - \tau_i; i = 0, 1, \ldots\}$, e.g. displaced exponential or displaced gamma distribution, etc. It is clear that displacement value h depends on value of T and on the distribution of $\{\eta_i; i = 0, 1, \ldots\}$ or (in the simplest case) on mean value of variable η_i. Unfortunately, determining and choosing the test distributions for sequence $\{\eta_i; i = 0, 1, \ldots\}$ cause bigger difficulties. Ideology of the method proposed in this paper for describing the traffic flow with complicated probabilistic structure is that one traces not every car separately, but only every batch (group) in the flow.

The process of formation of each group in traffic flow is considered in this paper as functioning of simple control service system [1–7]. A big variety of well-learned control service systems allows us to propose one system class or another to choose an adequate mechanism for i-th group formation for even the most complicated real traffic flow. The flow of cars on infinite or circular single-lane motorway is investigated. Such idealization makes it possible to examine a stationary process of each group movement. In this investigation the flows consisting of nonhomogeneous cars are studied: the slow cars are distinguished in the flow from the fast cars. Only fast cars can overtake cars with slow movement. The unusual representation of batch with number i as a certain control cybernetic service system allows us to understand mechanism of such traffic flows formation and, in final analysis, to determine the probabilistic distribution laws for the flow description in the form of sequence $\{(\tau_{i+1} - \tau_i, \eta_i); i = 0, 1, \ldots\}$.

2   Problem Statement

In this paper probability space $(\Omega, F, \mathbf{P}(\cdot))$ is considered. Here Ω is a sample space. The description of a certain elementary outcome which defines both process of cars movement on the motorway and process of their crossing the stop-line is designated as $\omega \in \Omega$. The set of all observed outcomes $A \subset \Omega$ in this experiment composes σ-algebra F on which probability measure function

$\mathbf{P}(A)\colon F \to [0,1]$ is specified. In some cases we will not fix symbol ω as an argument of functions or variables still keeping in mind that all the random events and the random elements are considered on mentioned probability space.

Let $\eta(t) = \eta(\omega; t)$ by $t \geq 0$ count a number of the cars of all types that crossed the stop-line during time interval $[0, t)$. By fixed $\omega \in \Omega$ function $\eta(t)$ is left continuous and equals zero at zero. Then random process $\{\eta(t)\colon t \geq 0\}$ gives us the description of the time characteristic for the flow. In order to describe mathematically the time characteristic of the traffic flow it is desirable to determine finite-dimensional distributions of random process $\{\eta(t)\colon t \geq 0\}$. This task was solved [8] in the simplest idealization for the free movement [9] of the null-length homogenous cars. In such case process $\{\eta(t)\colon t \geq 0\}$ is Poisson process. However, in other cases this problem is so difficult that no papers are known in which it is considered.

The model for the mechanism of each traffic batch formation is represented as evolution of control cybernetic service system [1] of certain specified class. To that end the following blocks should be assigned for such system: input pole, external memory, internal memory, apparatus handling external memory information and apparatus handling internal memory information. The detailed mathematical description for each mentioned block should be provided.

The input pole is the flow of the fast cars. Every fast car is a point on time axis Ot. The intensity of the fast cars flow equals $\lambda_0 > 0$. The speed of each fast car has the same distribution density. The fast cars move freely [8,9] on the motorway sections without the slow cars. Intensity $\lambda > 0$ of the flow of the slow cars is such that distance between any neighboring slow cars ensures restoration of the stationary movement of the fast cars after overtaking process. In this case the fast cars movement satisfies all the requirements from monographs [8,9] and thereby it is a Poisson process. Let $\eta_0(\omega; t, \Delta t)$ denote a random number of the fast cars that arrive to a queue for overtaking certain slow car during time interval $[t, t + \Delta t)$. A set of random variables in the form $\{\eta_0(t, \Delta t)\colon t, \Delta t \geq 0\}$ defines a mathematical model for the input pole. Let symbol $o(\Delta t)$ denote nonnegative infinitesimal function with a higher order of smallness with respect to $\Delta t > 0$ as $\Delta t \to 0$. The following formulas are well-known for the Poisson flow with parameter λ_0:

$$
\begin{aligned}
\mathbf{P}(\{\omega\colon \eta_0(\omega; t, \Delta t) = 0\}) &= 1 - \lambda_0 \Delta t + o(\Delta t) \ , \\
\mathbf{P}(\{\omega\colon \eta_0(\omega; t, \Delta t) = 1\}) &= \lambda_0 \Delta t - o(\Delta t) \ , \\
\mathbf{P}(\{\omega\colon \eta_0(\omega; t, \Delta t) \geq 2\}) &= o(\Delta t) \ .
\end{aligned}
\tag{1}
$$

The observation of the real traffic flow shows that each fast car arrives to certain batch (queue) catching up with the slow one. Such queue consists of one mandatory slow car and possibly some fast cars. If random variable $\kappa(\omega; t) \geq 1$ counts number of cars of all types in the batch at moment $t \geq 0$, random process $\{\kappa(t)\colon t \geq 0\}$ is a mathematical description for the external memory block. A situation often takes place when overtaking intensity exceeds considerably small enough intensity of fast cars arrival to the batch. In this case an idealization when number of cars in every group does not exceed predetermined variable N is admissible. Thus, in our problem it is supposed that $1 \leq \kappa(\omega; t) \leq N$.

The internal memory block is responsible for the process of overtaking the slow car by fast ones. Each slow car is interpreted as a server for the fast cars. Then random overtaking time is considered as service time. Since the times for overtaking slow car by the fast ones are dependent and have the different distribution laws, a cumulative distribution function is not specified for the systems of this class. Moreover, the distribution laws for the stated variables are still not found. Therefore, it is easier in such systems to specify so called saturation flow [1,2] in the form of set $\{\xi(t, \Delta t): t, \Delta t \geq 0\}$ of random variables instead of set of many-dimensional cumulative distribution functions for service times. Here $\xi(\omega; t, \Delta t)$ denotes a maximal number of fast cars which can overtake the slow one during time interval $[t, t + \Delta t)$. Then the set of random variables in the form $\{\xi(t, \Delta t): t, \Delta t \geq 0\}$ defines a mathematical model for internal memory block.

Since the cars in the flow cannot be lost, a fast car catching up with the batch consisting of N cars, joins the batch all the same. However, at the same time the fast car following right after the slow one, overtakes it mandatory. On the practice the distribution of service (overtaking) time considerably depends on the size of the fast cars queue. It is quite natural to suppose that by the small values of $\Delta t > 0$ the conditional probabilities of events generated by discrete random variable $\xi(\omega; t, \Delta t)$ are characterized by the following relations:

$$\mathbf{P}(\{\omega: \xi(\omega; t, \Delta t) = 0\} | \{\omega: \kappa(\omega; t) = 1, \eta_0(\omega; t, \Delta t) = 1\}) = 1 - o(\Delta t) \,,$$
$$\mathbf{P}(\{\omega: \xi(\omega; t, \Delta t) = 0\} | \{\omega: \kappa(\omega; t) = 2, \eta_0(\omega; t, \Delta t) = 0\}) = 1 - \mu_1 \Delta t + o(\Delta t) \,,$$
$$\mathbf{P}(\{\omega: \xi(\omega; t, \Delta t) = 1\} | \{\omega: \kappa(\omega; t) = 2, \eta_0(\omega; t, \Delta t) = 0\}) = \mu_1 \Delta t - o(\Delta t) \,,$$
$$\mathbf{P}(\{\omega: \xi(\omega; t, \Delta t) = 0\} | \{\omega: \kappa(\omega; t) = 3, \eta_0(\omega; t, \Delta t) = 0\}) = 1 - \mu_2 \Delta t + o(\Delta t) \,,$$
$$\mathbf{P}(\{\omega: \xi(\omega; t, \Delta t) = 1\} | \{\omega: \kappa(\omega; t) = 3, \eta_0(\omega; t, \Delta t) = 0\}) = \mu_2 \Delta t - o(\Delta t) \,,$$
$$\mathbf{P}(\{\omega: \xi(\omega; t, \Delta t) = 0\} | \{\omega: \kappa(\omega; t) = k, \eta_0(\omega; t, \Delta t) = 0\}) = 1 - \mu_3 \Delta t + o(\Delta t) \,,$$
$$\mathbf{P}(\{\omega: \xi(\omega; t, \Delta t) = 1\} | \{\omega: \kappa(\omega; t) = k, \eta_0(\omega; t, \Delta t) = 0\}) = \mu_3 \Delta t - o(\Delta t) \,,$$
$$4 \leq k \leq N \,,$$
$$\mathbf{P}(\{\omega: \xi(\omega; t, \Delta t) = 1\} | \{\omega: \kappa(\omega; t) = N, \eta_0(\omega; t, \Delta t) = 1\}) = 1 \,.$$

$$(2)$$

Parameters μ_1^{-1} and μ_2^{-1} from (2) specify the conditional mean overtaking times in cases when the number of cars of all types in the queue equals 2 and 3 correspondingly. Assume also that mean overtaking time μ_3^{-1} does not change when the queue consists of more than three cars. Parameters μ_1, μ_2 and μ_3 will be called conditional overtaking intensities. This is how the fact that overtaking time distribution depends on the number of cars of all types in the batch is modeled. Note also that the last relation in (2) models immediate overtaking process for the sake of preserving the limitation requirement mentioned above for the number of the cars in the batch when another fast car arrives to the batch with N cars. A simple statement results from (2): when the size of the cars queue is fixed, the conditional probability that at least two cars overtake the slow one during interval with length Δt (located anywhere on the time axis) is an infinitesimal function with a higher order of smallness with respect to Δt. The conditional probabilities in (2) set a mathematical description for the apparatus handling internal memory information. In other words, (2) defines the changes of the probabilistic overtaking mechanism for the fast cars.

Denote $\overline{\xi}(\omega; t, \Delta t)$ as a number of the fast cars which actually overtake the slow one during time interval $[t, t+\Delta t)$. The apparatus handling external memory information can be described mathematically with the help of simple functional relation

$$\kappa(\omega; t + \Delta t) = \kappa(\omega; t) + \eta_0(\omega; t, \Delta t) - \overline{\xi}(\omega; t, \Delta t) . \tag{3}$$

It is evident the $\overline{\xi}(\omega; t, \Delta t) \leq \min\{\kappa(\omega; t) + \eta_0(\omega; t, \Delta t) - 1, \xi(\omega; t, \Delta t)\}$. Since the fast cars strive to leave the batch as soon as possible, it is natural to suppose that the following relation takes place:

$$\overline{\xi}(\omega; t, \Delta t) = \min\{\kappa(\omega; t) + \eta_0(\omega; t, \Delta t) - 1, \xi(\omega; t, \Delta t)\} . \tag{4}$$

In other words, the extremal policy of the traffic batch formation [1] is used here. The apparatus handling external memory information implements the functional law (3) for selecting the fast cars from the batch to overtake the slow car. The service (overtaking) process in the traffic batch should be according to FIFO discipline.

It is required to determine time characteristic description for the traffic flow in the form $\{(\tau_{i+1} - \tau_i, \eta_i); i = 0, 1, \ldots\}$ taking into account the mentioned assumptions concerning all the blocks in the control cybernetic service system modeling the process of the traffic batches formation. The traffic flow of nonhomogeneous cars should be considered as the flow of moving batches. Relations of types (1), (2), (3), (4) and control parameters λ_0, N, μ_1, μ_2, μ_3 is a convenient way to specify the class of mathematical models for the control waiting service systems with the limited queue, one server with variable structure, immediate service possibility and FIFO service discipline.

3   Properties of Time Characteristic of the Traffic Flow with Nonhomogeneous Cars

Let $Q(t, m)$ denote probability $\mathbf{P}(\{\omega: \kappa(\omega; t) = m\})$ for fixed $t \geq 0$ and for $m = 1, 2, \ldots, N$. Then set $\{Q(t, m); m = 1, 2, \ldots, N\}$ of the probabilities defines the distribution for the number of the cars of all types in the batch at moment $t \geq 0$. For random event $\{\omega: \kappa(\omega; t + \Delta t) = m\}$ by each fixed $m = 1, 2, \ldots, N$ the following equation in events is derived using (3):

$$\begin{aligned} &\{\omega: \kappa(t + \Delta t) = m\} \\ &= \bigcup_{k=1}^{N} \bigcup_{n=0}^{\infty} \{\omega: \kappa(t) = k, \eta_0(\Delta t) = n, \overline{\xi}(\Delta t) = k + n - m\} . \end{aligned} \tag{5}$$

Now the finite Kolmogorov linear differential equations system with Jacobi matrix for probabilities $Q(t, m)$, $t \geq 0$, $m = 1, 2, \ldots, N$, is derived with the help of (1), (2), (4), (5) and methodology from monograph [9]:

$$\begin{aligned} \frac{dQ(t,1)}{dt} &= -\lambda_0 Q(t, 1) + \mu_{1,0} Q(t, 2) , \\ \frac{dQ(t,2)}{dt} &= \lambda_0 Q(t, 1) - (\lambda_0 + \mu_{1,0}) Q(t, 2) + \mu_{2,0} Q(t, 3) , \\ \frac{dQ(t,3)}{dt} &= \lambda_0 Q(t, 2) - (\lambda_0 + \mu_{2,0}) Q(t, 3) + \mu_{3,0} Q(t, 4) , \\ \frac{dQ(t,m)}{dt} &= \lambda_0 Q(t, m - 1) - (\lambda_0 + \mu_{3,0}) Q(t, m) + \mu_{3,0} Q(t, m + 1) , \\ &\quad m = 4, 5, \ldots, N - 1 , \\ \frac{dQ(t,N)}{dt} &= \lambda_0 Q(t, N - 1) - \mu_{3,0} Q(t, N) . \end{aligned} \tag{6}$$

It is additionally supposed that at moment $t = 0$ the number of the cars of all types in the batch equals i. Then the dynamics of the distribution for the number of the cars in the batch is defined by the solution of differential equation system (6) with entry conditions $Q(0, i) = 1$, $Q(0, m) = 0$ by $1 \leq m \leq N$ and $m \neq i$.

Solving process for the differential equation system of type (6) is bulky enough [10]. Actually only properties of the distribution for the number of the cars in the batch (i.e. properties of the system (6) solution) by $t \to \infty$ is needed in our investigation. Limiting probabilities $Q(m) = \lim_{t \to \infty} Q(t, m)$, $m = 1, 2, \ldots, N$, can be derived solving the following finite linear algebraic equation system:

$$
\begin{aligned}
0 &= -\lambda_0 Q(1) + \mu_{1,0} Q(2) , \\
0 &= \lambda_0 Q(1) - (\lambda_0 + \mu_{1,0}) Q(2) + \mu_{2,0} Q(3) , \\
0 &= \lambda_0 Q(2) - (\lambda_0 + \mu_{2,0}) Q(3) + \mu_{3,0} Q(4) , \\
0 &= \lambda_0 Q(m-1) - (\lambda_0 + \mu_{3,0}) Q(m) + \mu_{3,0} Q(m+1) , \\
& \qquad 4 \leq m \leq N - 1 , \\
0 &= \lambda_0 Q(N-1) - (\lambda_0 + \mu_{3,0}) Q(N) + \lambda_0 Q(N) .
\end{aligned}
\tag{7}
$$

Limiting (ergodic) distribution $\{Q(m); m = 1, 2, \ldots, N\}$ characterizes so called stationary mode for number $\kappa(\omega)$ of the cars of all types in each batch in the traffic flow of nonhomogeneous moving cars. In paper [11] the solution of system (7) was derived in the form

$$
Q(1) = (1 + \nu_1 + \nu_1 \nu_2 \tfrac{\nu_3^{N-2}-1}{\nu_3-1})^{-1}, \quad Q(2) = \nu_1 Q(1) ,
$$
$$
Q(k) = \nu_1 \nu_2 \nu_3^{k-3} Q(1), \quad k = 3, 4, \ldots, N .
\tag{8}
$$

Here ν_1, ν_2, ν_3 denote the essential system parameters:

$$
\nu_1 = \lambda_0 \mu_1^{-1}, \quad \nu_2 = \lambda_0 \mu_2^{-1}, \quad \nu_3 = \lambda_0 \mu_3^{-1} .
\tag{9}
$$

In the real traffic flow it is often observed that the slow cars move along the motorway independently and without any obstacles from the rest of the cars. Suppose that the speed of each slow car has absolutely continuous distribution and does not depend on the car number. Hence it appears [8] that by the zero size of each car time characteristic $\{\tau_i; i = 0, 1, \ldots\}$ of the slow cars flow is Poisson with parameter λ. The following idealization should be additionally assumed. First, let the time intervals between the neighboring cars be big enough. Second, let the density of the fast cars be small enough and their mean speed value be large enough. The limitations on the mentioned variables are specified based on the road conditions and the standards established by the transport workers. Under these two suppositions for the stationary mode it can be considered that all the cars in each group arrive to the stop-line at the same time. In this case the real traffic flow of more sophisticated probabilistic structure is approximated by the non-ordinary Poisson flow [12] of the random points on time axis Ot. In our approximation at the moment when every slow car arrives to the stop-line m cars actually arrive with probability $Q(m)$. Recall that random variable $\eta(t)$

introduced earlier by $t \geq 0$ counts the number of the cars of all types that crossed the stop-line during interval $[0, t)$. This designation remains for the non-ordinary Poisson flow.

For simplicity we will research a traffic flow with $N = 3$. Then derive from (8) that at every calling instance the batch with one car arrives to the stop-line with probability p, with two cars – with probability q, with three cars – with probability s, where

$$p = \frac{1}{1 + \nu_1 + \nu_1\nu_2}, \quad q = \frac{\nu_1}{1 + \nu_1 + \nu_1\nu_2}, \quad s = \frac{\nu_1\nu_2}{1 + \nu_1 + \nu_1\nu_2} . \tag{10}$$

Prove a theorem for probabilities $P(t, k) = \mathbf{P}(\{\omega \colon \eta(\omega; t) = k\})$, $k = 0, 1, \ldots$ for non-ordinary Poisson flow $\{\eta(t) \colon t \geq 0\}$.

Theorem 1. *If $[m]$ denotes integer part of number m, then probability*

$$P(t, k) = e^{-\lambda t} \sum_{i=0}^{[\frac{k}{2}]} \sum_{j=0}^{[\frac{k-2i}{3}]} \frac{(k-i-2j)!}{i!j!(k-2i-3j)!} p^{k-2i-3j} q^i s^j \frac{(\lambda t)^{k-i-2j}}{(k-i-2j)!} . \tag{11}$$

Proof. The generating function for variable $\eta(\omega; t)$ distribution is derived using well-known method from [12]:

$$\begin{aligned} \Psi(t, z) = \sum_{k=0}^{\infty} P(t, k) z^k &= \exp\{\lambda t (sz^3 + qz^2 + pz - 1)\} \\ &= e^{-\lambda t} \exp\{\lambda t s z^3\} \exp\{\lambda t q z^2\} \exp\{\lambda t p z\} . \end{aligned} \tag{12}$$

Based on (12) decompose function $\Psi(t, z)$ into power series

$$\Psi(t, z) = e^{-\lambda t} \sum_{l=0}^{\infty} \frac{(\lambda t s z^3)^l}{l!} \sum_{i=0}^{\infty} \frac{(\lambda t q z^2)^i}{i!} \sum_{n=0}^{\infty} \frac{(\lambda t p z)^n}{n!} . \tag{13}$$

Now derive the following equation based on (13) using the diagonal method for collecting terms in series multiplication in Cauchy form:

$$\Psi(t, z) =$$

$$= e^{-\lambda t} \sum_{k=0}^{\infty} z^k \left\{ \sum_{i=0}^{[\frac{k}{2}]} \sum_{j=0}^{[\frac{k-2i}{3}]} \frac{(k-i-2j)!}{i!j!(k-2i-3j)!} p^{k-2i-3j} q^i s^j \frac{(\lambda t)^{k-i-2j}}{(k-i-2j)!} \right\} . \tag{14}$$

Ascertain formula (11) comparing the generating function definition given in (12) and relation (14). The theorem is proved.

Now derive the numerical characteristics (mathematical expectation, variance, skewness, kurtosis) for variable $\eta(t)$ using formula (10.1) from [9] and relation (12):

$$\begin{aligned} \mathbf{E}(\eta(t)) &= \lambda t (1 + q + 2s), \quad \mathbf{Var}(\eta(t)) = \lambda t (1 + 3q + 8s) , \\ \mathbf{Skew}(\eta(t)) &= \frac{\mathbf{E}(\eta(t) - \mathbf{E}\eta(t))^3}{(\mathbf{E}(\eta(t) - \mathbf{E}\eta(t))^2)^{1/2}} = \frac{1 + 7q + 26s}{\sqrt{\lambda t}(1 + 3q + 8s)^{3/2}} , \\ \mathbf{Kurt}(\eta(t)) &= \frac{\mathbf{E}(\eta(t) - \mathbf{E}\eta(t))^4}{(\mathbf{E}(\eta(t) - \mathbf{E}\eta(t))^4} - 3 = \frac{1 + 15q + 80s}{\lambda t (1 + 3q + 8s)^2} . \end{aligned} \tag{15}$$

4 Analyzing Traffic Batches Based on Observations

On the practice, the results of the traffic flow observation are written, as a rule, in the form of the realization of random sequence $\{\tau'_j; j = 1, 2, \ldots\}$ consisting of moments when the cars of all types arrive to the stop-line. Random intervals $\tau'_{j+1} - \tau'_j, j \geq 1$, between neighboring cars often occur to be dependent and to have different distribution functions. In such case it is impossible to detect the adequate finite-dimensional distributions for process $\{\eta(t): t \geq 0\}$ based on the only one realization. Another difficulty consists in fact that there is no opportunity to detect the consecutive moments of the slow cars arrivals. The different algorithms to determine sequence $\{\tau_{i+1} - \tau_i; i = 0, 1, \ldots\}$ consisting of independent and identically distributed random variables based on information on sequence $\{\tau'_j; j = 1, 2, \ldots\}$ realization are presented in [1,13], where they are also successfully approved using different experimental data from [14,15]. A simple algorithm for determining consecutive moments τ_i, $i = 0, 1, \ldots$ of the slow cars arrivals based on consecutive moments τ'_j, $j = 1, 2, \ldots$ of all cars arrivals is presented in this paper.

According to the algorithm source traffic flow $\{\tau'_j; j = 1, 2, \ldots\}$ is split into groups in successive steps. On the step number $m = 0, 1, \ldots$ sequence $\{(\tau_i^m, \eta_i^m); i \geq 0\}$ is derived. For fixed $m \geq 0$ moments τ_i^m, $i \geq 0$, coincide with certain moments $\tau'_j, j = 1, 2, \ldots$ In other words, we have $\tau_i^m = \tau'_{k_{m,i}}$, $k_{m,i} \geq 1$. Sequence $\{(\tau_i^m, \eta_i^m); i \geq 0\}$ is called sequence (virtual flow) of m-th level. The number of cars in i-th group in virtual flow of m-th level is defined as follows: $\eta_i^m = k_{m,i+1} - k_{m,i}$. Variable $\delta_i^m = \tau'_{k_{m,i+1}} - \tau'_{k_{m,i+1}-1}$ specifies the interval between i-th and $(i+1)$-th groups when the source flow is described with the help of sequence $\{(\tau_i^m, \eta_i^m); i \geq 0\}$ of m-th level. The recurrent formulas for moments τ_i^m, $m \geq 0$, $i \geq 0$, splitting the source flow are of the form

$$k_{0,0} = 1, \quad k_{0,i+1} = \inf\{k : k > k_{0,i}, \ \tau'_k - \tau'_{k-1} \geq h_0\},$$
$$s_m = \min\{\ \inf\{k : k \geq 0, \ \eta_k^m \leq d, \ \eta_{k+1}^m = d+1, \ \delta_k^m < h_1\},$$
$$\inf\{k : k \geq 0, \ \eta_k^m \leq d, \ \eta_{k+1}^m \leq d, \ \delta_k^m < h_2\}\ \},$$
$$\tau_i^{m+1} = \tau_i^m, \text{ if } i \leq s_m; \quad \tau_i^{m+1} = \tau_{i+1}^m, \text{ if } i > s_m. \tag{16}$$

Here we assume $\eta_{-1}^m = 0$ for $m \geq 0$. The algorithm parameters are real number d and constant variables h_0, h_1, h_2, where $0 < h_0 < h_1 < h_2$. Let us explain this heuristic algorithm. Initially the quantity of h_0 is set as follows. If the big enough groups can be formed in the flow, then $h_0 \gg \min\{x_1, x_2, \ldots, x_n\}$. Here $\{x_1, x_2, \ldots, x_n\}$ is a realization of random sequence $\{\tau'_{j+1} - \tau'_j; j = 1, 2, \ldots, n\}$ of size n. In case the big groups are not observed in the flow, then value is chosen according to relation $h_0 \geq \min\{x_1, x_2, \ldots, x_n\}$. Finally, if the number of the cars in the batch does not exceed two, then $h_0 \ll \min\{x_1, x_2, \ldots, x_n\}$. It is clear that conception of big sizes for batches is specified by the physical characteristics of the real traffic flow. As a result of applying the first line in (16) sequence $\{(\tau_i^0, \eta_i^0); i \geq 0\}$ of zero level is derived. The intervals between any neighboring cars in each i-th batch of virtual flow $\{(\tau_i^0, \eta_i^0); i \geq 0\}$ of zero level is smaller then h_0. It means the cars in source flow $\tau'_i, i \geq 0$, are associated into groups according to principle of time closeness of their arrivals to the stop-line. After that, the following procedure is applied to sequence $\{(\tau_i^0, \eta_i^0); i \geq 0\}$ of zero level. Successively, starting with zero batch η_0^0 we associate a couple of neighboring batches in two cases: (1) the first batch in this couple contains at the most d cars while the second one contains $d + 1$ cars, the distance between batches is less then h_1; (2) each batch in the couple contains at the most d cars and the interval between batches is less than h_2. It allows us to specify sequence $\{(\tau_i^1, \eta_i^1); i \geq 0\}$ of first level. Then the same procedure is applied to sequence $\{(\tau_i^1, \eta_i^1); i \geq 0\}$ of first level to derive sequence $\{(\tau_i^2, \eta_i^2); i \geq 0\}$ of second level, etc. Note that $\{\omega \colon \lim_{m \to \infty} \tau_i^m \; \exists\} = \Omega$. Therefore, specifying moment $\tau_i = \tau_{k_i}' = \lim_{m \to \infty} \tau_i^m$ and random variable $\eta_i = k_{i+1} - k_i$ by every $i \geq 0$ we derive the description of source flow $\{\eta(t) \colon t \geq 0\}$ of the form $\{(\tau_i, \eta_i); i \geq 0\}$. This algorithm analyses the traffic batches based on the observations, i.e. determines moments τ_i of slow cars arrivals to the stop-line.

Demonstrate the efficiency of this algorithm by example of describing time characteristic $\{\tau_{j+1}' - \tau_j'; j = 1, 2, \ldots\}$ of the certain traffic flow of the moving cars. The experimental data of such flow was received by Bartlett [14] who observed the process of crossing the certain motorway stop-line by the cars next to London during $T_0 = 2024\,\text{s}$. Realization $\{x_1, x_2, \ldots, x_n\}$ of size $n = 128$ for the flow of the form $\{\tau_{j+1}' - \tau_j'; j = 1, 2, \ldots, n\}$ is presented in Table 1, where the values (in seconds) for the intervals between the consecutive crossings should be read along the rows.

Physical road characteristics, types of cars, traffic regulations and weather influence the probabilistic structure of the traffic flow. Note that weather changes stochastically. Under the good weather conditions the movement of the cars along the motorway is Poisson, whereas under the bad conditions the overtaking times increases considerably that leads to the batches formation. Based on Table 1 data, note that in general the intervals values are relatively small. However, there are some cars that follow in the flow with the considerable delays. Such data illustrate the cars accumulation in small batches. Up to current moment no appropriate theoretical distribution for the intervals in Table 1 is known. Moreover, applying Wallis-Moore phase-frequency criteria [16] about independency and identical distribution for the intervals between neighboring cars to the statistical data in Table 1 we derive that statistics $\hat{z}$ value by $n = 128$ equals 3.167.

Table 1. Values x_j of interval $\tau_{j+1}' - \tau_j'$ by $j = 1, 2, \ldots, 128$; $\tau_1' = 0$

2.8	3.4	1.4	14.5	1.9	2.8	2.3	15.3	1.8	9.5	2.5	9.4	1.1	88.6	1.6	1.9
1.5	33.7	2.6	12.9	16.2	1.9	20.3	36.8	40.1	70.5	2.0	8.0	2.1	3.2	1.7	56.5
23.7	2.4	21.4	5.1	7.9	20.1	14.9	5.6	51.7	87.1	1.2	2.7	1.0	1.5	1.3	24.7
2.6	119.8	1.2	6.9	3.9	1.6	3.0	1.8	44.8	5.0	3.9	125.3	22.8	1.9	15.9	6.0
20.6	12.9	3.9	13.0	6.9	2.5	12.3	5.7	11.3	2.5	1.6	7.6	2.3	6.1	2.1	34.7
15.4	4.6	55.7	2.2	6.0	1.8	1.9	1.8	42.0	9.3	91.7	2.4	30.6	1.2	8.8	6.6
49.8	58.1	1.9	2.9	0.5	1.2	31.0	11.9	0.8	1.2	0.8	4.7	8.3	7.3	8.8	1.8
3.1	0.8	34.1	3.0	2.6	3.7	41.3	29.7	17.6	1.9	13.8	40.2	10.1	11.9	11.0	0.2

A threshold value on a significance level of 0.05 equals 1.96. Since the statistics value meets condition 3.167 > 1.96, according to the phase-frequency criteria the proposed hypothesis about independency and identical distribution for the intervals between neighboring cars should be rejected.

Now, split the given traffic flow into the batches according to proposed algorithm (16). Setting different values for parameters h_0, h_1, h_2, d the initial Bartlett flow can be split into the groups of the cars. For example, when values $d = 2$, $h_0 = 0.5 > \min\{x_1, x_2, \ldots, x_{128}\}, h_1 = 7$ and $h_2 = 9.6$ are used, we derive 66 values $y_0, y_1, \ldots, y_{65}$ for intervals τ_1-τ_0, τ_2-$\tau_1, \ldots, \tau_{66}$-$\tau_{65}$ between the slow cars and the sequence consisting of 67 values $r_0, r_1, \ldots, r_{66}$ for sizes $\eta_0, \eta_1, \ldots, \eta_{66}$ of the traffic batches. The processed data is presented in Table 2 which should be read along the rows. The last element in this table contains symbol "#". This means that the moment of 67-th batch arrival is not known. It should be noted that each traffic batch in Table 2 contains at the most three cars. Therefore, the class of the control service systems with control parameter $N = 3$ should be chosen for explaining the process of the batches formation in the Bartlett flow.

Table 2. Value (y_i, r_i) of vector $(\tau_{i+1} - \tau_i, \eta_i)$ by $i = 0, 1, \ldots, 65$; $\tau_0 = 0$

(7, 6; 3)	(14, 5; 1)	(7; 3)	(15, 3; 1)	(13, 8; 3)	(99, 1; 3)	(5; 3)	(33, 7; 1)
(15, 5; 2)	(16, 2; 1)	(22, 2; 2)	(36, 8; 1)	(40,1; 1)	(70, 5; 1)	(12, 1; 3)	(61, 4; 3)
(23, 7; 1)	(23, 8; 2)	(33, 1; 3)	(14, 9; 1)	(57, 3; 2)	(87, 1; 1)	(4, 9; 3)	(27, 5; 3)
(72, 6; 1)	(119, 8; 1)	(12; 3)	(6, 4; 3)	(44, 8; 1)	(134, 2; 3)	(22, 8; 1)	(17, 8; 2)
(26, 6; 2)	(12, 9; 1)	(16, 9; 2)	(21, 7; 3)	(17; 2)	(11, 7; 3)	(10, 5; 3)	(34, 7; 1)
(15, 4; 1)	(60, 3; 2)	(10; 3)	(45, 7; 3)	(101; 2)	(33; 2)	(16, 6; 3)	(49, 8; 1)
(58, 1; 1)	(5, 3; 3)	(32, 2; 2)	(11, 9; 1)	(2, 8; 3)	(20, 3; 3)	(13, 7; 3)	(34, 9; 2)
(9, 3; 3)	(41, 3; 1)	(29, 7; 1)	(17, 6; 1)	(15, 7; 2)	(40, 2; 1)	(10, 1; 1)	(11, 9; 1)
(11; 1)	(#; 2)						

The statistical analysis of Table 2 data ascertains the following. Both the hypothesis about independence and identical distribution for the intervals between neighboring cars and the same hypothesis for the number of cars in the traffic batches are not rejected on a significance level of 0.05. For example, the value of Wallis-Moore statistics $\hat{z}$ for realization $\{y_i; i = 0, 1, \ldots, 65\}$ of sequence $\{\tau_{i+1} - \tau_i; i = 0, 1, \ldots, 65\}$ equals 0.298 < 1.96. At the same time, the statistics $\hat{z}$ value calculated for realization $\{r_i; i = 0, 1, \ldots, 66\}$ of sequence $\{\eta_i; i = 0, 1, \ldots, 66\}$ equals 0.888 < 1.96.

5   Estimator for the Distribution Parameters

In Sect. 4 the source traffic flow is transformed into the batches flow where each batch contains at the most $N = 3$ cars. Recall that for this case distribution (10) for number η_i of the cars in i-th batch depends on two essential parameters ν_1 and ν_2. Consider now a method estimating the values of these parameters based on the real traffic flow observation.

Let the traffic flow be observed during time T_0. As a result of splitting the flow into the batches with the help of algorithm (16) we have l traffic batches. Let variables n_1, n_2, n_3 count a number of the batches containing one, two and three cars correspondingly. Here $n_1 + n_2 + n_3 = l$. Note that l specifies the number of all slow cars as well and $n_2 + 2n_3$ specifies the number of fast cars. Therefore, the fast cars intensity can be estimated by relation $\lambda_0^* = \frac{n_2 + 2n_3}{T_0}$ while the slow cars intensity – by relation $\lambda^* = \frac{l}{T_0}$. Let β_1 be the intensity of the batches containing one car. Then $\beta_1^* = \frac{n_1}{T_0}$ is an estimator for this variable. It was noticed on the practice that overtaking intensity μ_1 increases when the intensity of the flow of the batches containing one car increases. As a rule, such dependency can be described with the help of linear function $\mu_1 = K\beta_1$, where coefficient K value is defined by the road conditions. Subject to this dependency the estimator for μ_1 takes the form $\mu_1^* = K\beta_1^* = K\frac{n_1}{T_0}$, so the coefficient ν_1 value can be estimated as follows:

$$\nu_1^* = \lambda_0^*(\mu_1^*)^{-1} = (n_2 + 2n_3)(Kn_1)^{-1} . \tag{17}$$

Parameter ν_2 value may be estimated with the help of the modified minimum chi-square method stated in [17]. Since variable η_i may be equal to $1, 2$ or 3 with the corresponding probabilities (10), estimated value ν_2^* is derived from the solution of the equation formed according to formula (30.3.3a) in [17]:

$$\frac{n_1}{p(\nu_1, \nu_2)} \frac{\partial p(\nu_1, \nu_2)}{\partial \nu_2} + \frac{n_2}{q(\nu_1, \nu_2)} \frac{\partial q(\nu_1, \nu_2)}{\partial \nu_2} + \frac{n_3}{s(\nu_1, \nu_2)} \frac{\partial s(\nu_1, \nu_2)}{\partial \nu_2} = 0 . \tag{18}$$

Taking into account (10), this equation is transformed into

$$\frac{-n_1\nu_1}{(1 + \nu_1 + \nu_1\nu_2)} + \frac{-n_2\nu_1}{(1 + \nu_1 + \nu_1\nu_2)} + \frac{n_3(1 + \nu_1)}{\nu_2(1 + \nu_1 + \nu_1\nu_2)} = 0 , \tag{19}$$

and solving it we derive the estimated value for the parameter:

$$\nu_2^* = \frac{(1 + \nu_1^*)n_3}{\nu_1^*(n_1 + n_2)} = \frac{(Kn_1 + n_2 + 2n_3)n_3}{(n_2 + 2n_3)(n_1 + n_2)} . \tag{20}$$

Substitute the derived estimates (17) and (20) in (10). Now the estimate for the distribution parameters of the number of cars in the arbitrary batch in case $N = 3$ is obtained:

$$p^* = \left(1 + \frac{n_2 + 2n_3}{Kn_1} + \frac{(Kn_1 + n_2 + 2n_3)n_3}{Kn_1(n_1 + n_2)}\right)^{-1} ,$$
$$q^* = p^* \frac{n_2 + 2n_3}{Kn_1}, \quad s^* = p^* \frac{(Kn_1 + n_2 + 2n_3)n_3}{Kn_1(n_1 + n_2)} . \tag{21}$$

As has been mentioned in the introduction, the displaced exponential distribution with parameters $h \geq 0$ and $\sigma > 0$ can be used as the distribution for intervals $\tau_{i+1} - \tau_i, i \geq 0$, between the neighboring cars:

$$\mathbf{P}(\{\omega : \tau_{i+1} - \tau_i < t\}) = F(t, \sigma, h) = 1 - \exp\{-\frac{t - h}{\sigma}\} \text{ if } t > h \text{ and } 0 \text{ if } t \leq h . \tag{22}$$

It was proposed in [13] to estimate mentioned parameters with the help of modified minimum chi-square method. Let range $G = [0, \infty)$ of values for the intervals between the slow cars be split into v disjoint parts of the following form: $G_1 = [0, a), G_2 = [a, a + b), G_3 = [a + b, a + 2b), \ldots, G_v = [a + (v - 2)b, \infty)$. Let $m_c, c \in \{1, 2, \ldots, v\}$, denote a number of the elements in the sample for the intervals between slow cars that happen to be in G_c. The size of the sample equals $w = \sum_{c=1}^{v} m_c$. Then, according to the results of [13], the parameters estimators are of the following form:

$$\sigma^* = b \left(\ln(\sum_{c=2}^{v}(c-1)m_c - m_v) - \ln(\sum_{c=2}^{v}(c-2)m_c) \right)^{-1},$$
$$h^* = a - b \ln(\tfrac{w}{w-m_1}) \left(\ln(\sum_{c=2}^{v}(c-1)m_c - m_v) - \ln(\sum_{c=2}^{v}(c-2)m_c) \right)^{-1}.$$
(23)

6   Numerical Results

Now, estimate the distribution parameters for the number of the cars in the arbitrary batch for the Bartlett flow investigated above according to derived formulas (21). Based on Table 2 data values $n_1 = 27, n_2 = 15, n_3 = 24$ are calculated. Assume $K = 4.5$. The estimated values are obtained:

$$p^* = 0.419, \quad q^* = 0.217, \quad s^* = 0.364 . \tag{24}$$

In order to determine whether the processed data of the observed flow fits the distribution (10) by values (24) or not, calculate statistics χ^2 value according to formula (30.3.1) from [17]:

$$\chi^2 = \frac{(n_1 - lp^*)^2}{lp^*} + \frac{(n_2 - lq^*)^2}{lq^*} + \frac{(n_3 - ls^*)^2}{ls^*} = 0.046 . \tag{25}$$

The obtained value is less than threshold value 3.841 for the chi-square variable χ_1^2 with one degree of freedom on a significance level of 0.05. Therefore, according to criterion stated in [17], it may be considered that table data fits well distribution (10) with probabilities values (24).

Note that parameter K is a practical parameter. The values of the similar parameters are usually specified before the model construction. In our case parameter K value was not given. Therefore, we chose it so that distribution (10) with the probabilities values calculated with the specific value of K would describe experimental data well (in terms of χ^2 statistics). For example, by value $K = 4.3$ statistics value equals $\chi^2 = 0.005$. This result moves us away from the critical area limitations. Assume $K = 4.2$ and derive value $\chi^2 = 6.778 \times 10^{-6} < 0.005$, while by $K = 4.1$ value $\chi^2 = 0.006 > 6.778 \times 10^{-6}$ is calculated. Based on these reasonings we conclude that the described model fits in the best way the flow with the data from Table 1 if its physical parameter K value is equal to 4.2. Indeed, we would consider the Table 2 data have the distribution (10) with probabilities values

$$p^* = 0.409, \quad q^* = 0.227, \quad s^* = 0.364 . \tag{26}$$

Now we count the frequency of arrivals of the batches containing one, two or three cars according to Table 2 data and derive $\frac{27}{66}$, $\frac{15}{66}$ and $\frac{24}{66}$ correspondingly. Note that these frequencies coincide with the probabilities (26) with accuracy of 0.001. It's known that an event frequency is a good estimator for the event probability. We propose the other estimator for the probability and have the close result that points the good fitness the developed model to the real data.

Calculate now the estimates for the distribution parameters of the intervals between neighboring slow cars. According to (23) with values $v = 6$, $a = 3$ and $b = 26.24$ the following estimates are derived:

$$\sigma^* = 29.392, \quad h^* = 0.449 . \tag{27}$$

Hypothesize the fact that values $y_0, y_1, \ldots, y_{65}$ for the intervals from Table 2 is a sample collected from distribution (22) with parameters values (27). Statistics χ^2 is calculated:

$$\chi^2 = \sum_{c=1}^{6} \frac{(m_c - wp_c)^2}{wp_c} = 4.17 . \tag{28}$$

Here $p_c = \mathbf{P}(\{\omega : \tau_{i+1} - \tau_i \in G_c | F_{\tau_{i+1} - \tau_i}(t) = F(t, \sigma^*, h^*)\})$. The derived value (28) does not exceed the threshold value 7.815 for distribution of χ^2 with $(v - 3) = 3$ degrees of freedom on a significance level of 0.05. According to this result the hypothesis about the distribution of the form (22) with parameters values (23) for the intervals between slow cars from Table 2 is not rejected.

It should be noted that for the model parameters relation $\sigma = 1/\lambda$ takes place. According to estimator for the intensity of the slow cars mentioned above we have $\lambda^* = \frac{l}{T_0} = 0.033$. This value coincides with value $0.034 = 1/29.392$ derived as a result of chi-square estimator with accuracy of 0.001. This illustrates again the successful choice in estimator method for the system parameters.

7   Conclusion

The papers [11, 18–20] deal with the research of the traffic batches flow moving along the motorway. Development and the investigation of this model is urgent since the problem of the optimal regulation of the traffic flows moving on the intersections is increased. Researching the traffic flow we derive the formalization of the important constituent element of more extensive model that is a model of the traffic intersection as the queueing system. In this paper the investigation of the properties of the probabilistic model for the bathes flow movement under the conditions of small flow density has been completed. The easy computational estimators for the examined model parameters have been derived for the first time.

References

1. Fedotkin, M.A.: Service processes and control systems. Math. Ques. Cybern. (collected articles) **6**, 51–70 (1996). Science-Physmathlit, Moscow

2. Lyapunov, A.A., Yablonskiy, S.V.: Theoretical problems of cybernetics. Cybern. Probl. (collected scientific articles) **9**, 5–22 (1963). Physmathgiz, Moscow

3. Fedotkin, M.A.: Optimal control for conflict flows and market point processes with selected discrete component. I. Lietuvos Matematikos Rinkinys **28**(4), 783–794 (1988)

4. Fedotkin, M.A.: Optimal control for conflict flows and market point processes with selected discrete component. II. Lietuvos Matematikos Rinkinys **29**(1), 148–159 (1989)

5. Fedotkin, M.A., Zorine, A.V.: Optimization of control of doubly stochastic nonordinary flows in time-sharing systems. Autom. Remote Control **66**(7), 1115–1124 (2005)

6. Proidakova, E.V., Fedotkin, M.A.: Control of output flows in the system with cyclic servicing and readjustments. Autom. Remote Control **69**(6), 993–1002 (2008)

7. Fedotkin, M.A., Fedotkin, A.M.: Analysis and optimization of output processes of conflicting Gnedenko-Kovalenko traffic streams under cyclic control. Autom. Remote Control **70**(12), 2024–2038 (2009)

8. Haight, F.A.: Mathematical Theories of Traffic Flow. Academic Press, New York (1963)

9. Fedotkin, M.A.: Models in Theory of Probability. Physmathlit, Moscow (2011)

10. Saaty, T.L.: Elements of Queueing Theory with Applications. McGraw Hill Book Company Inc., New York (1961)

11. Fedotkin, M.A., Kudryavsev, E.V., Rachinskaya, M.A.: About correctness of probabilistic models for dynamics of the traffic flows on the motorway. In: Proceedings of the International Workshop on Distributed Computer and Communication Networks (DCCN-2010), 26–28 October 2010. R&D Company Information and Networking Technologies, Moscow, Russia (2010)

12. Gnedenko, B.V., Kovalenko, I.N.: Introduction in Queuing Theory. LKI, Moscow (2007)

13. Fedotkin, A.M., Fedotkin, M.A.: Model for refusals of elements of a controlling system. In: Materials of the First French–Russian Conference on Longevity, Aging and Degradation Models in Reliability, Public Health, Medicine and Biology (LAD 2004), vol. 2, pp. 136–151. St. Petersburg State Polytechnical University, St. Petersburg (2004)

14. Bartlett, M.S.: The spectral analysis of point processes. J. Roy. Stat. Soc. Ser. B. **25**, 264–296 (1963)

15. Cox, D.R., Lewis, P.A.W.: The Statistical Analysis of Series of Events. Wiley, London (1966)

16. Sachs, L.: Statistische Auswertungsmethoden. Springer, Heidelberg (1972)

17. Cramer, H.: Mathematical Methods in Statistics. Mir, Moscow (1975)

18. Fedotkin, M.A., Rachinskaya, M.A.: Investigation of traffic flows characteristics in case of small density. queues: flows, systems, networks. In: Proceedings of the International Conference Modern Probabilistic Methods for Analysis and Optimization of Information and Telecommunication Networks, vol. 21, pp. 82–87. BSU-RIVH, Minsk (2011)

19. Fedotkin, M.A., Rachinskaya, M.A.: Investigation of mathematical model of traffic based on Lyapunov-Yablonskiy method. Problems of theoretical cybernetics . In: Proceedings of XVI International Conference, pp. 508–512, 20–25 June 2011. Izd-vo of State University of Nizhni Novgorod, Nizhni Novgorod (2011)
20. Fedotkin, M., Kudryavtcev, E., Rachinskaya, M.: Simulation and research of probabilistic regularities in motion of traffic flows. In: Proceedings of the International Workshop on Applied Methods of Statistical Analysis, Simulations and Statistical Inference (AMSA'2011), pp. 11–18, 20–22 September 2011. Publishing House of NSTU, Novosibirsk, Russia (2011)

Simulation of Wireless Broadband Network Traffic in the Electromagnetic Interference

Vladimir S. Zhdanov(✉) and Victor M. Churkov

Moscow State Institute of Electronics and Mathematics, Moscow, Russia
{zhdanovvs,wchur}@mail.ru

Abstract. The article describes the authors' simulation system and the results of heterogeneous network traffic modeling in the physical channel of the wireless network for different values of the signal/noise ratio in a range of different broadband noise and intensity.

Designing infrastructure coverage of wireless networks LTE, WiMax, WI-Fi, Zigbee requires including the planning, noise and electromagnetic interference on data traffic is transmitted over the network radio channels [1]. Known indicators CINR and RSSI measured quality of the radio equipment in the client. CINR (Carrier to Interference + Noise Ratio) - The ratio of the carrier signal to noise, is used in telecommunications for determining signal quality and connectivity, if the quality of the signal above a certain standard for standards. RSSI (Received Signal Strength Indication) - the power level of the received signal can be used to approximate the signal quality. The message shown in Table 1 with these parameters for network WiMax (LTE).

Table 1. Information message when you connect to the network Wimax

Signal (CINR/RSSI)	7/−68
Signal level (RSSI)	−68 dBm
Signal level (RSSI)	−68 dBm
Transmit power (TX Power)	32.7 dBm

Indicator CINR measured periodically in the channel and a radio receiver transfers the source for further analysis and decision-making about the connection. Thus, CINR is an integral indicator radio wave quality, taking into account:

- Clients distance from sources of radio (base station, router, router, bridge, adapter);
- Natural and artificial obstacles;
- External and internal noise, electromagnetic interference;
- Dynamics of all factors in space and time.

V. Vishnevsky et al. (Eds.): DCCN 2013, CCIS 279, pp. 169–177, 2014.
DOI: 10.1007/978-3-319-05209-0_15, © Springer International Publishing Switzerland 2014

Obviously, the use of an indicator CINR fundamentally required when working as a broadband network, and the qualitative design and development of its infrastructure, taking into account the relative position coordinates effective and minimizes the number of elements.

In [2], the authors have developed and examined the processes of modeling and optimization of the spatial model of the wireless signal in the form of a matrix based on the quality of the radio as a function of the above factors influence.

Formation of the matrix occurs on the basis of the pilot study parameters CINR radio network coverage. For making design decisions necessary to simulate and examine network traffic. External factors - the useful signal level, noise, electromagnetic disturbance to a large extent determine the bandwidth of the radio channel, which can significantly reduce the data transmission rate and, consequently, the quality of the traffic.

In the simulation bandwidth using developed in [2] unit is assumed that the channel is transmitted broadband traffic as a binary sequence with the current capacity of p_s in the presence of broadband additive noise with the current power p_n [3].

In accordance with the theorem of Shannon - Hartley [3], the capacity of the physical channel wireless network should not be less than the capacity of the source of information $C = f_s \log(1 + \frac{p_s}{p_n})$, bits/s.

Bandwidth can be changed by changing the width of the signal spectrum f_s, and the ratio "signal/noise" $10 \log \frac{p_s}{p_n}$.

The input data for modeling a wireless network are given or known parameters radio, telecommunications equipment, network infrastructure, electromagnetic interference traffic.

Mathematical model of the spatial organization of wireless coverage includes a plurality of functions, variables and constants where: U - universal set of elements of the system wireless network (coverage):

S – A lot of system coverage subsets (regions);
C – The set of centers of areas;
R – Radius of the area;
(x, y, z) – Spatial coordinates;
$u(x, y, z)$, $c(x, y, z)$, $s(x, y, z)$ – Elements of the set;
m, n – Cardinality of the set;
$\rho(x, y, z)$ – An indicator of the effectiveness of the radio signal coverage.

Let $U, C, S, R, (x, y, z), \rho(x, y, z)$ — variable mathematical model:

$U, R, (x, y, z), \rho(x, y, z)$ – Asked variables;
C, S – Expanded variables.

Let the sets U, C, S associated functional dependency.

We introduce additional features $f, \phi, \psi, |\nu_u|$ specifying constraints and performance indicators for coverage, operating:

f – Computation $C \subset U$ and subject to the limitations of the selected performance criteria (for example $|S_j| \to \max \to |U|$);
ϕ – Calculation of the elements subsets U_i;
ψ – Calculation of the subset $U_j \setminus U_{j+1}$;
$|\overrightarrow{\nu_u}|$ – Calculation of magnitude of the vector in the coordinates U;

Express $f, \phi, \psi, |\nu_u|$ through the original variables:

$$
\begin{cases}
f\left[\phi(u, |\overrightarrow{\nu_u(x, y, z)}|, R, \rho(x, y, z), \psi(U))\right]; \\
\phi\left(u, |\overrightarrow{\nu_u(x, y, z)}|, R, \rho(x, y, z), \psi(U)\right); \\
|\overrightarrow{\nu_u(x, y, z)}|; \\
\psi(U);
\end{cases}
$$

Then the desired sets C, S can be written as a function of variables $U, R, (x, y, z), \rho(x, y, z), f, \phi, \nu_u, \psi$; following system of equations.

$$
\begin{cases}
C = \left\{u : f\left[\phi(u, |\overrightarrow{\nu_u(x, y, z)}|, R, \rho(x, y, z), \psi(U))\right]\right\}; \\
S = \left\{u : \phi(c, |\overrightarrow{\nu_u(x, y, z)}|, R, \rho(x, y, z), \psi(U))\right\}.
\end{cases}
\tag{1}
$$

We introduce the function ξ convolution common to the mathematical description of the sets C, S. Function $\xi = (|\overrightarrow{\nu_u(x, y, z)}|, R, \rho(x, y, z), \psi(U))$ calculates the occurrence of an element u_i in the set $\psi(U)$ with parameters R for $\rho(x, y, z)$ all elements $\psi(U)$. Then the system (1) can be minimized.

$$
\begin{cases}
C = \left\{u : f[\phi(u, \xi)]\right\}; \\
S = \left\{u : \phi(c, \xi)\right\}.
\end{cases}
\tag{2}
$$

Thus, the system (3) is a generalized mathematical model of the spatial organization of wireless coverage. The model provides the calculation of coverage areas and centers for wireless network infrastructure based on defined performance criteria and constraints. Figures 1 and 2 are given examples of the mathematical model for the design of different coverage areas of a wireless computer network.

$$\left\{ \begin{array}{l} \text{variables of the model} \\ U, R, (x,y,z), \rho(x,y,z); \text{input} \\ C, S; \text{outputs} \end{array} \right\}$$

initial conditions

$$\left. \begin{array}{l} S_1 \subset U, S_2 \subset U, \ldots, S_j \subset U, \ldots, S_n \subset U; \\ U = \bigcup_{j=1}^{n} S_j; \ j = k = \overline{1,n}; \\ S_j \cap S_j = \varnothing; \\ \bigcup_{k=1}^{n} \bigcup_{j=1}^{n} S_j \cap S_k = \varnothing; \\ C \subset U, C_j \subset S_j; \\ u \in U, c \in C, s \in S, c \in U, s \in U, c_j \in S_j; \\ u_i, c_j, s_j; i = \overline{1,m}; \end{array} \right\}$$

efficiency criteria

$$|U| = const, |S| \to \max \to |U|, |C| \to \min \to 1$$

transformation function f, ϕ, ν, ψ;

$$\left. \begin{array}{l} [\phi(u, |\overrightarrow{\nu_u(x,y,z)}|, R, \rho(x,y,z), \psi(U))]; \\ \phi(u, |\overrightarrow{\nu_u(x,y,z)}|, R, \rho(x,y,z), \psi(U)); \\ |\overrightarrow{\nu_u(x,y,z)}|; \\ \psi(U); \end{array} \right.$$

folding function ξ

$$\xi = (|\overrightarrow{\nu_u(x,y,z)}|, R, \rho(x,y,z), \psi(U));$$

area coverage

$$\left\{ \begin{array}{l} C = \left\{ u : f[\phi(u, |\overrightarrow{\nu_u(x,y,z)}|, R, \rho(x,y,z), \psi(U))] \right\}; \\ S = \left\{ u : \phi(c, |\overrightarrow{\nu_u(x,y,z)}|, R, \rho(x,y,z), \psi(U)) \right\}; \end{array} \right.$$

minimized

$$\left\{ \begin{array}{l} C = \left\{ u : f[\phi(u, \xi)] \right\}; \\ S = \left\{ u : \phi(c, \xi) \right\}; \end{array} \right.$$

$$(3)$$

System simulation of heterogeneous network created under the supervision of the authors. In the database of topology, speed and volume of information processing in the network node, formed of modeling input data:

- The type of physical connection between the subnet and subnets;

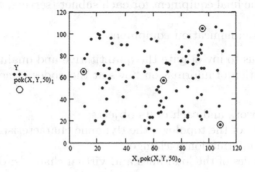

Fig. 1. Clustering of the two-dimensional zone wireless coverage

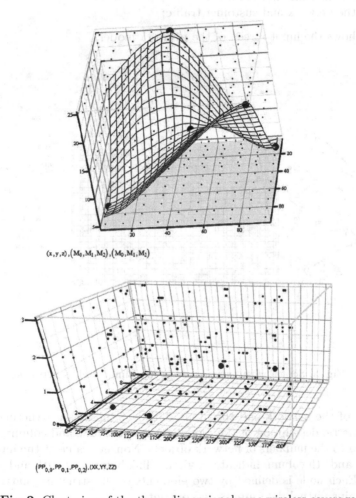

Fig. 2. Clustering of the three-dimensional zone wireless coverage

- Parameters of the final equipment for each subnet (servers, workstations, controllers, I/O);
- Parameters of communication equipment.

Simulation allows us to investigate the quantitative and qualitative characteristics of the distribution of information flows at the nodes of networks in a given time interval:

- monitor the network and each of its objects;
- assess the impact of the topology, the dynamic characteristics of the network on the speed of information exchange;
- assess the dynamics of the load lines and virtual channels: direct - from hosts to servers; reverse - from servers;
- simulate the network taking into account the dynamic change of input variables and their distribution laws;
- design the network and customer traffic;

Figure 3 shows the input screen of the network model.

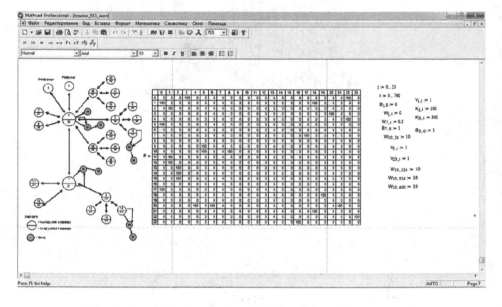

Fig. 3. Screen when entering data modeling wireless network

Graph of the network and the input data is entered as a structural matrix and parametric description of the modeling process. Row and column numbers correspond to the numbers of network objects. Nonzero value at the intersection of ith row and jth column indicates a virtual link between the ith and jth of the network. Each node is defined by two elements of the structure matrix. Model node contains virtual data inputs and outputs.

For each object set:

- Speed of information processing,
- The law of transformation and distribution of information,
- Parameters of the original boot.

The network model is given matrix $|P|$, the graph showing network connectivity. The network has the following parameters: the speed of information processing in the nodes (matrix $|V|$), the coefficient of transmission node (K), and the volume boot nodes (W), and the discrete data (A).

Figure 4 shows the results of traffic modeling in the physical channel of the wireless network, with different signal/noise ratio.

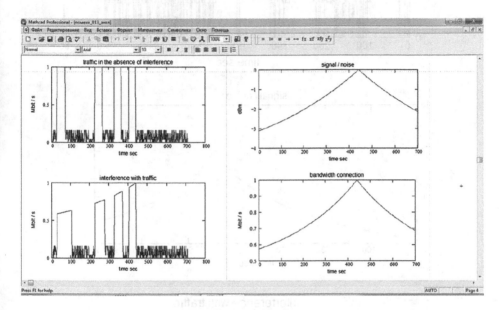

Fig. 4. Diagrams modeling wireless traffic when the signal/noise ratio

Analysis of simulation results clearly and with precision shows the character changes the transmission of information packets, depending on the signal/noise ratio in the channel wireless network.

In this case clearly shows that an increase in the actual power level of noise or interference in the broadband bandwidth traffic significantly increases the time a packet is broken due to the reduction of their synchronization bandwidth.

Modeling Methodology transmitting information in the computer network based on the simulation of a traffic channel, servers and clients.

Figure 5 shows the parameters of the noise and interference, the simulation results.

The results of using a computer network system simulation confirms its efficiency, accuracy of the mathematical model, sufficient functionality for designing networks of varying complexity.

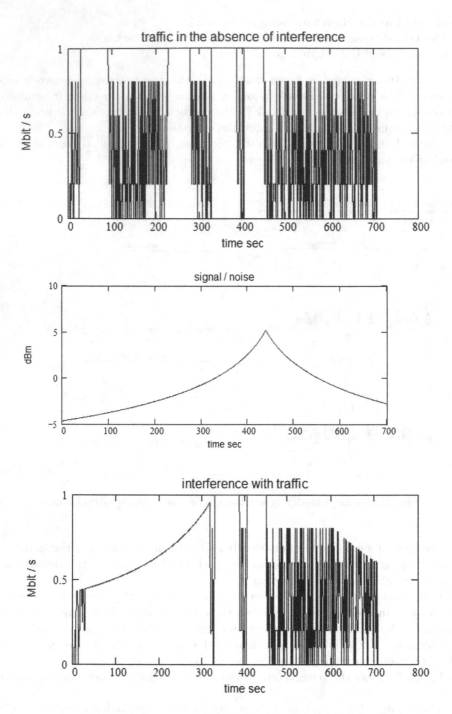

Fig. 5. Screen when modeling wireless traffic

The developed system is a heterogeneous network simulation to enhance the quality of design solutions through research in the operating parameters of the projected or existing broadband network infrastructure, information, and signal levels.

References

1. Vishnevsky, V.M.: Broadband Wireless Data Transmission Networks, 592 p. Technosphera, Moscow (2005)
2. Churkov V.M., Kosilov N.A., Zhdanov V.M. Optimization of spatial models wireless sensor networks. In: Proceedings of Seventh International Conference on Information and Telecommunication Technologies in Intelligent System, Schweiz, 3–9 July 2010, pp. 9–16 (2010)
3. Siebert, W.M.: Circuits, Signals, and Systems, 651 p. MIT Press, Cambridge (1986)

On the Effective Envelopes for Fluid Queues with Gaussian Input

Oleg Lukashenko[1,2], Evsey Morozov[1,2], and Michele Pagano[3]($\boxtimes$)

[1] Institute of Applied Mathematical Research, Karelian Research Center
RAS, Petrozavodsk, Russia
[2] Petrozavodsk State University, Petrozavodsk, Russia
[3] University of Pisa, Pisa, Italy
lukashenko-oleg@mail.ru, emorozov@karelia.ru,
m.pagano@iet.unipi.it

Abstract. Thanks to their flexibility and compact characterization, Gaussian processes have emerged as popular models to describe the traffic dynamics in a wide class of the modern telecommunication networks. A relatively new characterization of traffic flows is based on the effective envelopes, which represent a probabilistic generalization of the arrival curve of Network Calculus. In this paper, we analyse the effective envelopes for a general Gaussian process and use these results to derive non-asymptotic performance bounds for a fluid queuing system. To highlight the effectiveness of the proposed approach, numerical results are shown taking into account heterogeneous traffic flows as well as different correlation structures.

1 Introduction

Towards the end of the last century, the evolution of communication networks, in which new services such as Multimedia over IP, audio/video streaming and virtual private networks (VPNs) with Quality of Service (QoS) are being offered to end users, has led to the introduction of network architectures (e.g. Intserv [3] and DiffServ [2]) supporting QoS, usually defined in terms of loss rate, end-to-end delay and delay jitter.

As a consequence, realistic traffic modelling became a key issue in queueing performance evaluation and, at the same time, a great interest arose for novel traffic engineering frameworks, able to deal with the issues raised by the new network architectures. For instance, apart from traditional Best Effort, IntServ can offer two different types of service, one with deterministic guarantees (Guaranteed Service) and one (Controlled-Load Service), which emulates the behaviour of a unloaded network. In order to provide any kind of guarantee, the network should be able to limit the amount of traffic through *Call Admission Control* functionalities.

The most natural way to address this problem for the Controlled-Load traffic is through the concept of *effective bandwidth* (see, for instance [8]), which corresponds to the minimum bandwidth necessary to provide an expected service for a

V. Vishnevsky et al. (Eds.): DCCN 2013, CCIS 279, pp. 178–189, 2014.
DOI: 10.1007/978-3-319-05209-0_16, © Springer International Publishing Switzerland 2014

specific flow. On the other side, Guaranteed service asks for bounded end-to-end delays and an elegant framework for worst-case analysis of network performance is provided by the so-called *Network Calculus* [10]. Roughly speaking, the latter is based on the concepts of arrival curve $\alpha(t)$, a deterministic bound on the amount of traffic entering into the network in any interval of duration t (for instance, due to the presence of a token-bucket filter), and service curve $\beta(t)$, describing the minimum amount of work that can be guaranteed to the flow.

As far as traffic modelling is concerned, Gaussian processes are rather popular to describe the traffic dynamics in a wide class of the modern telecommunication networks. Indeed, these models naturally arise by virtue of central-limit-type arguments (superposition of a large number of independent sources) and are able to capture in a parsimonious and flexible way such specific properties as self-similarity and long-range dependence (LRD), which are inherent to network traffic flows [11, 19] and have a deep impact on queueing performance [6]. In a nutshell, self-similarity means invariance in the distribution under suitable scaling of time and space, while LRD implies a slow decay of the autocorrelation function; in particular, Fractional Brownian Motion (FBM) has become the *canonical* model with a LRD correlation structure [14].

These properties make difficult the probabilistic analysis of the related queueing systems and, as a consequence, obtaining the key characteristics in an explicit form. As already mentioned, one of the main performance parameters is the *overflow probability*, i.e. the probability that the workload process exceeds a given (finite) threshold. In Gaussian queueing systems with infinite buffer, the calculation of the overflow probability is reduced to the analysis of the extremes of Gaussian processes. Although this problem has been thoroughly studied in the past [15], in case of general Gaussian input there are no explicit expressions (some results have been derived for specific simple cases like standard Brownian motion thanks to its Markovian structure) and only asymptotics for the tail probability are available [4, 5, 7].

Unlike traditional results, typically developed in the framework of Large Deviation Theory (LDT), a relatively new approach, based on stochastic network calculus and sample path envelope [12, 16], can provide non-asymptotic performance bounds for Gaussian queues. Thanks to the special autocorrelation structure, analytical expressions of such bounds have been derived by Rizk and Fidler [16, 17] in case of FBM traffic.

The goal of this work is to enlarge their results to general Gaussian inputs, in order to take into account heterogeneous traffic flows (for instance the sum of independent FBMs with different values of the Hurst parameter) and different correlation structures. Unlike the FBM case, no closed-form results can be achieved and the performance bounds are calculated through point-wise numerical optimization.

The rest of the paper is organized as follows. Section 2 defines the reference queueing system (including the analysed traffic models and the underlying assumptions), while Sect. 3 introduces the notion of effective envelopes, highlighting how they can be calculated for a generic Gaussian process. Section 4 deals with queueing performance analysis starting from the previous definitions, and the corresponding numerical results are presented in Sect. 5. Finally, Sect 6 concludes the paper with some final remarks.

2   Problem Setting

We consider a *fluid queue* with a constant service rate C driven by the input process $A(t)$, which is defined as follows:

$$A(t) = mt + X(t), \tag{1}$$

where $m > 0$ is the mean input rate and the process X is a centred Gaussian process with stationary increments, i.e.

$$A(t) - A(s) \stackrel{(d)}{=} A(t-s) \quad s \leq t,$$

where $\stackrel{(d)}{=}$ denotes equality in distribution. Physically, $A(t)$ describes the amount of data (input traffic) arrived into a communication node within the time interval $[0, t]$, $t \geq 0$ and $X(t)$ takes into accounts its random fluctuations.

Let us denote the variance of $X(t)$ as $v(t)$; it is worth mentioning that the variance function $v(t)$ fully determines the correlation structure of the Gaussian source:

$$\Gamma(t, s) \stackrel{\Delta}{=} \mathbb{E}\left[X(t)\, X(s)\right] = \frac{1}{2}\left[v(t) + v(s) - v(|t-s|)\right].$$

In the following we only assume that $v(t)$ is regularly varying at infinity with index $V \in (0, 2)$, i.e.

$$\lim_{t \to \infty} \frac{v(\alpha t)}{v(t)} = \alpha^V \quad \text{for all } \alpha > 0. \tag{2}$$

We consider the following important cases of Gaussian inputs, which obey to the previous condition:

1. FBM: in this case $v(t) = t^{2H}$, with Hurst parameter $H \in (0, 1)$. In the teletraffic framework usually $H \in (0.5, 1)$ is considered, corresponding to traffic processes with LRD, while $H = 0.5$ identifies the traditional Brownian motion with independent increments. It has been shown in [18] that FBM arises as the scaled limit process when the cumulative workload is a superposition of on-off sources with mutually independent heavy-tailed on and/or off periods.
2. Sum of independent FBMs with $v(t) = \sum_{i=1}^{n} t^{2H_i}$. The use of this model is also motivated by the fundamental result in [18] in case of heterogeneous on-off sources, corresponding to traffic flows with different degree of LRD.
3. Integrated Ornstein-Uhlenbeck process (IOU) with $v(t) = t + e^{-t} - 1$. As can be seen from the structure of the variance function, IOU is a short-range dependent (SRD) process and it is known in teletraffic as the Gaussian counterpart of Anick-Mitra-Sondi fluid model [1,13] (its relevance is further motivated in [9]).

For a queueing system with infinite buffer size and deterministic service rate C (with $C < m$ to assure the stability of the queue), it is well-known that the stationary workload Q satisfies the Lindley recursion

$$Q \stackrel{(d)}{=} \sup_{t \geq 0}(A(t) - Ct). \tag{3}$$

As already mentioned in the Introduction, there are no explicit expressions for (3) in case of general Gaussian input, but only asymptotic results (namely, large-buffer and many sources asymptotics) for the tail probability have been derived [4,5,7].

3   Effective Bandwidth and Effective Envelopes

The effective envelope of an arrival process can be easily derived from the expression of its effective bandwidth [12]. Hence, in the following paragraphs we recall the main definitions and some specific results for Gaussian input flows, making use of the notation introduced in Sect. 2.

3.1   Effective Bandwidth

The Effective Bandwidth (EB) $\alpha(\theta, t)$ describes the minimum bandwidth required to provide an expected service for a given amount of traffic [8]. As can be expected, $\alpha(\theta, t)$ lies between the mean and peak of the arrival rate (measured over an interval of length t), more formally

$$\frac{\mathbb{E}A(t)}{t} \leq \alpha(\theta, t) \leq \frac{\bar{A}(t)}{t},$$

where

$$\bar{A}(t) \stackrel{\Delta}{=} \sup\Big\{x : \ \mathbb{P}\big(A(t) > x\big) > 0\Big\}.$$

In more detail, the EB of an arrival process $A(t)$ is defined as

$$\alpha(\theta, t) = \frac{1}{\theta t} \log E\left[e^{\theta A(t)}\right] \quad \text{for all } \ \theta, t \in (0, \infty), \tag{4}$$

where t is a *time parameter*, indicating the length of a time interval, and θ is a *space parameter*, containing information about the distribution of the arrivals: indeed, near $\theta = 0$, the EB is dominated by the mean rate, while for $\theta \to \infty$, it is primarily influenced by the peak rate.

EB expressions have been derived for many traffic models including those with LRD; in particular, for the Gaussian process defined by (1), the function $\alpha(\theta, t)$ is determined, for all θ and t, by the first two moments, expectation $m > 0$ and variance $v(t)$,

$$\alpha(\theta, t) = m + \frac{\theta}{2t}v(t). \tag{5}$$

3.2   Effective Envelopes

Effective envelopes $E(t)$ are statistical upper bounds for an arrival process and represent the stochastic generalization of the *Arrival Curve* $\alpha(\tau)$ of Network Calculus, which satisfies the following inequality:

$$A(\tau, t) \leq \alpha(t - \tau) \quad \text{for all } \tau \leq t,$$

where

$$A(\tau, t) \stackrel{\triangle}{=} A(t) - A(\tau)$$

denotes the cumulative arrivals over the interval $[\tau, t]$ and, for sake of simplicity, we use $A(t)$ to mean $A(0, t)$ since $A(0) = 0$. Hence, during any interval of length t the number of arrivals is bounded by $\alpha(t)$. The arrival curve concept leads to worst-case bounds for the delay and the backlog, which may be quite conservative. However, in many cases the users can tolerate some level of losses (depending on the specific application) and so it could be more suitable to provide probabilistic bounds on the arrival process, by defining a traffic profile that can be overtaken with a probability $\varepsilon_p > 0$. This leads in a natural way to the definition of the so-called *local (point-wise) effective envelope* of $A(t)$, a *deterministic* function $E(t)$ that satisfies

$$\mathbb{P}\Big[A(\tau, t) - E(t - \tau) > 0\Big] \leq \varepsilon_p \quad \text{for all } \tau \leq t \text{ and all } t. \qquad (6)$$

Making use of the Markov inequality, the duality between effective envelopes and EB [12] can be easily proved:

$$E(t) = \inf_{\theta > 0} \left\{ t\alpha(\theta, t) - \frac{\log \varepsilon_p}{\theta} \right\}$$

and, taking into account (5), for any Gaussian process with variance $v(t)$, the local effective envelope is given by

$$E(t) = mt + \sqrt{-2 \log \varepsilon_p \cdot v(t)}. \qquad (7)$$

As highlighted by (3), the overflow probability is related to the supremum of $A(t) - Ct$, so the point-wise information on the arrival process is not enough. A natural way to overcome this problem consists in introducing the so-called *global (sample path) effective envelope*:

$$\mathbb{P}\Big[\sup_{0 \leq \tau \leq t} \{A(\tau, t) - E(t - \tau)\} > 0\Big] \leq \varepsilon_s \quad \text{for all } \tau \geq 0 \text{ and all } t. \qquad (8)$$

In the next section, an adequate sample path effective envelope will be derived for a general Gaussian process, and it will be used to find a bound on the overflow probability.

4   Sample Path Envelopes and Overflow Probability

In order to define a sample path effective envelope for the arrival process $A(t)$, starting from (7), we consider a family of local envelopes

$$E(\tau) = m\tau + \sqrt{-2\log\eta \cdot f(\tau)\, v(\tau)} \tag{9}$$

where the parameter $\eta \in (0,1)$ and the function f should be chosen so that the following condition holds:

$$\limsup_{t\to\infty} \frac{E(t)}{t} < C, \tag{10}$$

being C the constant service rate of the considered queueing system.

Taking into account that $A(0) = E(0) = 0$, the left hand side of (8) can be upper bounded as follows

$$\mathbb{P}\left[\sup_{\tau\geq 0}\{A(\tau) - E(\tau)\} > 0\right] \leq \sum_{\tau=1}^{\infty} \mathbb{P}\left[A(\tau) > E(\tau)\right],$$

and, due to Markov's inequality, we have from (5) that

$$\mathbb{P}\left[A(\tau) > E(\tau)\right] \leq e^{-\theta E(\tau)}\, \mathbb{E}e^{\theta\, A(t)} = e^{-\theta E(\tau)}\, e^{\theta m\tau + \frac{\theta^2}{2}v(\tau)}. \tag{11}$$

Minimizing the right hand side of this inequality over $\theta > 0$ yields

$$\theta^* = \frac{E(\tau) - m\tau}{v(\tau)},$$

and, by the substitution of this value into inequality (11), we obtain after some algebra (and taking into account expression (9)) that

$$\mathbb{P}\left[A(\tau) > E(\tau)\right] \leq \eta^{f(\tau)}.$$

Finally, we have the following bound:

$$\mathbb{P}\left[\sup_{\tau\geq 0}\{A(\tau) - E(\tau)\} > 0\right] \leq \sum_{\tau=1}^{\infty} \mathbb{P}\left[A(\tau) > E(\tau)\right] = \sum_{\tau=1}^{\infty} \eta^{f(\tau)}$$

$$\leq \int_0^{\infty} \eta^{f(\tau)}\, d\tau \triangleq \varepsilon_s, \tag{12}$$

and hence,

$$E(\tau) = m\tau + \sqrt{-2\log\eta \cdot f(\tau)\, v(\tau)} \tag{13}$$

is a sample path envelope for $A(t)$ with the overflow probability ε_s.

In order to get a meaningful bound, it is reasonable to require that

$$\int_0^{\infty} \eta^{f(\tau)}\, d\tau < 1.$$

In case of regularly varying variance with some index $V \in (0,2)$, we can choose (as in [16,17]) a function $f(\tau) = \tau^\beta$ with $\beta \in (0, 2-V)$ (the latter condition guarantees that the limit (10) is verified), and hence the envelope (13) becomes

$$E(\tau) = m\tau + \sqrt{-2\log\eta \cdot \tau^\beta v(\tau)}. \tag{14}$$

If

$$E(t) \leq Ct + b \quad \text{for all } t \geq 0,$$

then the following inequality for the overflow probability holds:

$$\mathbb{P}\Big[Q > b\Big] = \mathbb{P}\Big[\sup_{t\geq 0}\{A(t) - (Ct + b)\} > 0\Big] \leq \mathbb{P}\Big[\sup_{t\geq 0}\{A(t) - E(t)\} > 0\Big] \leq \varepsilon_s.$$

In other words, the overflow probability of the analysed queueing system will be bounded by the overflow probability of the sample path envelope of the input traffic, i.e.

$$\mathbb{P}\Big[Q > b\Big] \leq \varepsilon_s \triangleq \int_0^\infty \eta^{\tau^\beta} \, d\tau = \varepsilon_s(\eta, \beta),$$

where, as shown in [16,17],

$$\varepsilon_s(\eta, \beta) = \frac{\Gamma\left(\frac{1}{\beta}\right)}{\beta(-\log\eta)^{\frac{1}{\beta}}},$$

and Γ denotes the gamma function. Hence, given the buffer size b and the service rate C, it is possible to find the largest envelope that satisfies the condition $E(t) \leq b + Ct$ by solving the following system of non-linear equations

$$\begin{cases} E'(t) = C, \\ E(t) = b + Ct. \end{cases} \tag{15}$$

Closed-form expressions can be found only in case of a single FBM flow (in [16,17] the corresponding bounds are determined), while in the general case the system of Eq. (15) can be solved by numerical methods and, as a result, the values of η can be found, using β as a free parameter, in order to optimize the bound.

5   Numerical Results

The goal of this section is to highlight the goodness of our bounds for different correlation structures and check how the optimal value of β depends on the queue parameters.

At first we verified the accuracy of the upper bound for different values of the service rate in comparison with simulation results. In more detail, Fig. 1 refers to the sum of two independent FBMs (here and in the following, unless otherwise stated, the following values of parameters have been used: $H_1 = 0.75$, $H_2 = 0.6$, $m = 1$ and $b = 0.3$), while IOU is considered in Fig. 2 (here the value

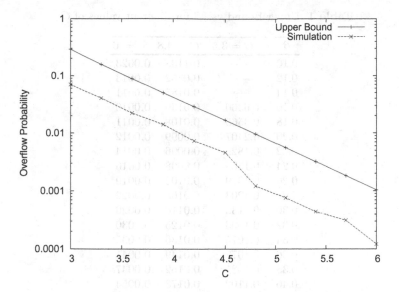

Fig. 1. Simulation vs. Upper Bound: sum of FBMs

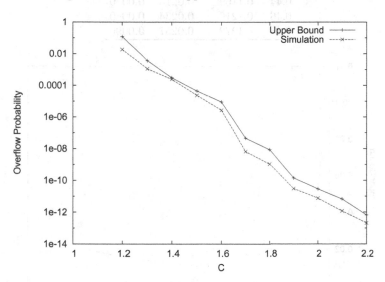

Fig. 2. Simulation vs. Upper Bound: IOU

of buffer size was $b = 10$). In both scenarios (corresponding to LRD and SRD traffic processes, respectively) simulation results are quite close to our bound (the difference is always less than one order of the magnitude) over a wide range of values. Unlike classical LDT–based asymptotics, which capture only the decay-rate, our estimations are always conservative and so can be safely used in dimensioning network elements under QoS constraints.

Table 1. Upper Bound vs. β for different values of C

β	$C = 3.5$	$C = 4.8$	$C = 6$
0.10	—	0.1138	0.0023
0.12	—	0.0352	0.0013
0.14	—	0.0185	0.0011
0.16	0.6300	0.0130	0.0010
0.18	0.3565	0.0108	0.0011
0.20	0.2407	0.0099	0.0012
0.22	0.1829	0.0096	0.0014
0.24	0.1508	0.0098	0.0016
0.26	0.1319	0.0102	0.0019
0.28	0.1203	0.0108	0.0022
0.30	0.1133	0.0116	0.0026
0.32	0.1093	0.0125	0.0030
0.34	0.1075	0.0136	0.0035
0.36	0.1072	0.0149	0.0041
0.38	0.1081	0.0162	0.0047
0.40	0.1102	0.0178	0.0054
0.42	0.1131	0.0195	0.0062
0.44	0.1169	0.0213	0.0070
0.46	0.1216	0.0234	0.0080
0.48	0.1272	0.0257	0.0090

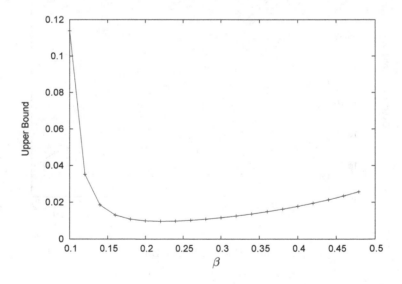

Fig. 3. Upper Bound vs. β for C=4.8

Since LRD is an essential feature of multimedia traffic, we investigated in more detail the first scenario, focusing on the value β^* of the parameter β for which the tightest bound is achieved. For sake of brevity, the results for three different values of the server capacity C are reported in Table 1, and only the case $C = 4.8$ is presented in Fig. 3: the upper bound is a convex function of β, and typically there is a relatively big neighbourhood of β^* in which similar estimates are obtained, highlighting the stability of the numerical procedure.

Finally, Figs. 4 and 5 show the dependence of β^* on the server capacity C and the buffer size b (with the fixed server capacity $C = 3.5$) respectively: the

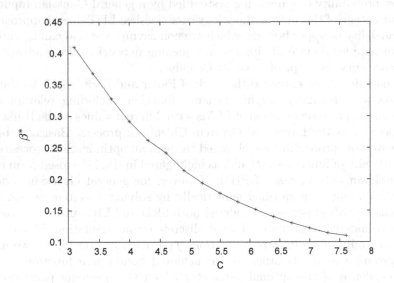

Fig. 4. β^* vs. C

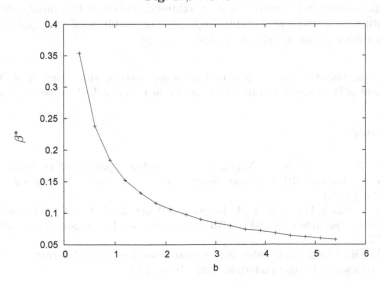

Fig. 5. β^* vs. b

decay is monotone and quite regular, so we can conclude that the optimal value of β decreases with the overflow probability.

6   Conclusions

In this paper, we analysed the applicability of effective envelopes to estimate the overflow probability of a queueing system fed by a general Gaussian input. The main advantage of this new methodology over classical LDT–based approaches is the availability of upper bounds, which are non asymptotic and can be extended to more complex service disciplines and queuing networks taking advantage of the service curve concept of Network Calculus.

In more detail, we extended the work of Fidler and Rizk [16,17] to Gaussian processes with regularly varying variance function, including relevant traffic models such as the combination of FBMs with different values of the Hurst parameter as well as the Integrated Ornstein-Uhlenbeck process. Basically, bounds for the overflow probability are obtained through an optimization process over a suitable family of functions $f(t)$ and, as highlighted in [16,17], closed-form results can be obtain only in case of FBM. However, for general Gaussian processes optimization can be performed numerically by solving a system of non–linear equations. In this paper, we considered both SRD and LRD processes, comparing the bounds with estimates through discrete-event simulation. Moreover, for a special family of *optimization functions* $f(t)$, namely $f(t) = t^\beta$, we numerically investigated the tightness of the achieved bound as a function of β and the dependence of the optimal parameter β^* on the queueing parameters. In particular, the convex relation between the upper bound and β (with a unique minimum and stable estimates in its neighbourhood) and the monotone decay of β^* with the overflow probability highlight the stability of the approach and its applicability to more complex traffic mixtures.

Acknowledgements. This work is partially supported by the Program of Strategy development of Petrozavodsk State University in the framework of the research activity.

References

1. Addie, R., Mannersalo, P., Norros, I.: Most probable paths and performance formulae for buffers with Gaussian input traffic. Eur. Trans. Telecommun. **13**(3), 183–196 (2002)
2. Blake, S., Black, D., Carlson, M., Davies, E., Wang, Z., Weiss, W.: An Architecture for Differentiated Service. RFC 2475 (Informational), December 1998. Updated by RFC 3260
3. Braden, R., Clark, D., Shenker, S.: Integrated services in the internet architecture: an overview. RFC 1633 (Informational), June 1994
4. Debicki, K.: A note on LDP for supremum of Gaussian processes over infinite horizon. Stat. Probab. Lett. **44**(3), 211–219 (1999)

5. Duffield, N.G., O'connell, N.: Large deviations and overflow probabilities for the general single-server queue, with applications. Math. Proc. Cambridge Philos. Soc. **118**, 363–374 (1995)
6. Erramilli, A., Narayan, O., Willinger, W.: Experimental queueing analysis with long-range dependent packet traffic. IEEE/ACM Trans. Netw. **4**(2), 209–223 (1996)
7. Lewis, J.T., Duffy, K., Sullivan, W.G.: Logarithmic asymptotics for the supremum of a stochastic process. Ann. Appl. Probab. **13**(2), 430–445 (2003)
8. Kelly, F.P.: Notes on effective bandwidths. In: Kelly, F.P., Zachary, S., Ziedins, I. (eds.) Stochastic Networks: Theory and Applications, Royal Statistical Society Lecture Notes Series, pp. 141–168. Oxford University Press, Oxford (1996)
9. Kulkarni, V., Rolski, T.: Fluid model driven by an Ornstein-Uhlenbeck process. Probab. Eng. Inf. Sci. **8**, 403–417 (1994)
10. Le Boudec, J.-Y., Thiran, P.: Network Calculus: A Theory of Deterministic Queuing Systems for the Internet. Springer, Heidelberg (2001)
11. Leland, W.E., Taqqu, M.S., Willinger, W., Wilson, D.V.: On the self-similar nature of Ethernet traffic (extended version). IEEE/ACM Trans. Netw. **2**(1), 1–15 (1994)
12. Li, Ch., Burchard, A., Liebeherr, J.: A network calculus with effective bandwidth. IEEE/ACM Trans. Netw. **15**(6), 1442–1453 (2007)
13. Mandjes, M.: Large Deviations for Gaussian Queues: Modelling Communication Networks. Wiley, New York (2007)
14. Norros, I.: On the use of fractional brownian motion in the theory of connectionless networks. IEEE J. Sel. A. Commun. **13**(6), 953–962 (2006)
15. Piterbarg, V.I.: Asymptotic Methods in the Theory of Gaussian Processes and Fields. Memoirs of the American Mathematical Society, Providence (1996)
16. Rizk, A., Fidler, M.: Sample path bounds for long memory FBM traffic. In: Proceedings of the 29th Conference on Information Communications, INFOCOM'10, pp. 61–65. IEEE Press, Piscataway (2010)
17. Rizk, A., Fidler, M.: Non-asymptotic end-to-end performance bounds for networks with long range dependent FBM cross traffic. Comput. Netw. **56**(1), 127–141 (2012)
18. Taqqu, M.S., Willinger, W., Sherman, R.: Proof of a fundamental result in self-similar traffic modeling. SIGCOMM Comput. Commun. Rev. **27**(2), 5–23 (1997)
19. Willinger, W., Taqqu, M.S., Leland, W.E., Wilson, D.V.: Self-similarity in high-speed packet traffic: analysis and modeling of Ethernet traffic measurements. Stat. Sci. **10**(1), 67–85 (1995)

Algorithmic Analysis of Dual Tandem Queue with Batch Markovian Arrival Process and Repeated Attempts at the First Station

Valentina Klimenok and Alexander Dudin[(✉)]

Department of Applied Mathematics and Computer Science,
Belarusian State University, 220030 Minsk, Belarus
{klimenok,dudin}@bsu.by

Abstract. A dual tandem queueing system with a Batch Markovian Arrival Process ($BMAP$) is considered. The first station is represented by a retrial multi-server queueing system with infinite orbit. General form of dependence of the total retrial intensity on the current number of customers in orbit is suggested. The second station is a multi-server queue with a finite buffer and impatient customers. The stationary distribution of the system states and the main performance measures of the tandem queue under consideration are calculated.

Keywords: Tandem queueing system · Retrials · Multi-server stations · Stationary state distribution

1 Introduction

Queueing networks and tandem queues as their important particular case are widely used in capacity planning, performance evaluation and optimization of computer and communication systems, contact centers, distributed data bases, manufacturing and logistic systems, etc. The theory of tandem queues is well developed, for more references see, e.g. [1,3,7,19].

Usually in literature, the first station of a tandem is represented by a queue with a finite or an infinite buffer. Because the retrial phenomenon is typical for many real life systems, wireless communication systems in particular, it is important to extend the theory of retrial queues to the case of retrial tandem queues. Current state of the art in algorithmic analysis of retrial queues is presented in the paper [5].

To the best of our knowledge, papers [4,10,11,14,17], were the only ones devoted to investigation of tandem retrial queues which do not have waiting space for arriving customers and a customer who does not find available servers upon arrival makes repeated attempts to get service in random times. The paper [17] considers the model with the stationary Poisson arrival process, general service time distribution at both stations and the constant retrial rate. The paper [4] considers the $MAP/PH/1 \rightarrow \cdot/PH/1/K + 1$ tandem retrial queue.

V. Vishnevsky et al. (Eds.): DCCN 2013, CCIS 279, pp. 190–203, 2014.
DOI: 10.1007/978-3-319-05209-0_17, © Springer International Publishing Switzerland 2014

Recently, more general compared to [4] tandem retrial queues where considered in literature. The model considered in [11] differs from the one analyzed in [4] by the following features.

- the $BMAP$ (Batch Markovian Arrival Process is considered as the descriptor of the input to the system. It allows to investigate the real life systems where the customers are allowed to arrive in batches of arbitrary size while only the single arrivals are suggested in [4];
- the service time distribution at the first station is assumed be general;
- different types of service for primary customers, which enter the service without visiting the orbit, and repeated customers are allowed;
- more general form of the total retrial intensity dependence on the number of customers in the orbit is allowed.

In the paper [10], a similar retrial tandem queue is dealt with, but, in contrast to all mentioned papers, the second station of a tandem is represented by the multi-server queue, there are two types of customers and there is a special mechanism for reservation of the servers at the second station for priority customers. In [14], the retrial tandem queueing system with two multi-server stations was investigated. That queueing model is the most closer to the one under consideration. The essential difference is that in [14] the input is described as a stationary Poisson flow while in this paper we consider correlated and bursty input using the model of $BMAP$.

The importance of taking into account bursty and correlated behavior of the input in queueing models is mentioned, e.g., in [8,15]. The $BMAP$ can be used to model such a behavior which appears in real flows in modern telecommunication networks and distributed data bases.

The $BMAP$ was introduced as versatile Markovian point process ($VMPP$) by M. Neuts in [18]. Further, his notations were simplified greatly by [16] and ever since these processes bear the name Batch Markovian Arrival Process ($BMAP$). The class of $BMAP$s includes many input flows considered previously, such as stationary Poisson (M), Erlangian (E_k), Hyper-Markovian (HM), Phase-Type (PH), Interrupted Poisson Process (IPP), Switched Poisson Process (SPP), Markov Modulated Poisson Process ($MMPP$). In opposite to recurrent (GI) flows and the PH renewal flow in particular, the $BMAP$ flow can be correlated.

The rest of the paper is organized as follows. In Sect. 2, the model under consideration is described. In Sect. 3, the process of the system states is defined as the level dependent multi-dimensional continuous time Markov chain. Its generator is specified and it is established that the chain pertains to the class of asymptotically quasi-Toeplitz Markov chains. Short Sect. 4 contains the results concerning the existence of the stationary distribution of the Markov chain and computation of this distribution. Section 5 contains formulae for computing the main performance measures of the tandem queue under study. In particular, formulae for probability of immediate access of an arbitrary customer at the first station upon arrival and probability of a customer loss due to the second station overflow. In Sect. 6, distribution of waiting time and probability that an arbitrary customer will not be lost at the second station due to obsolescence are derived. Finally, Sect. 7 concludes the paper, some possible generalizations of the model are discussed.

2 Model Description

We consider a tandem queueing system consisting of two stations in series. The first station is represented by the retrial N-server queue without a buffer. The primary customers arrive to this station according to a $BMAP$. Arrivals in the $BMAP$ is directed by an irreducible continuous time Markov chain ν_t, $t \geq 0$, with the finite state space $\{0, ..., W\}$. Sojourn time of the Markov chain ν_t, $t \geq 0$, in the state ν has exponential distribution with parameter $\lambda_\nu, \nu = \overline{0, W}$. After this sojourn time expires, with probability $p_k(\nu, \nu')$, the process $\nu_t, t \geq 0$, transits to the state ν', and a batch consisting of k customers, $k \geq 0$, arrives into the system. The intensities of jumps from one state into another, which are accompanied by an arrival of a batch consisting of k customers, are combined into the matrices D_k, $k \geq 0$, of size $(W + 1) \times (W + 1)$.

As follows from [16], the behavior of the $BMAP$ is completely characterized by the matrix generating function $D(z) = \sum\limits_{k=0}^{\infty} D_k z^k$, $|z| < 1$. The matrix $D(1)$ represents the generator of the process $\nu_t, t \geq 0$. The average arrival rate λ is defined as

$$\lambda = \boldsymbol{\theta} D'(1)\mathbf{e}$$

where $\boldsymbol{\theta}$ is the invariant vector of the stationary distribution of the process $\nu_t, t \geq 0$. The row vector $\boldsymbol{\theta}$ is the unique solution to the system

$$\boldsymbol{\theta} D(1) = \mathbf{0}, \ \boldsymbol{\theta}\mathbf{e} = 1.$$

Here $\mathbf{e}$ is the column-vector of appropriate size consisting of units and $\mathbf{0}$ is the row-vector of appropriate size consisting of zeroes. Intensity λ_g of group arrivals is defined as

$$\lambda_g = \boldsymbol{\theta}(-D_0)\mathbf{e}.$$

The variance v of intervals between group arrivals is calculated as

$$v = 2\lambda_g^{-1}\boldsymbol{\theta}(-D_0)^{-1}\mathbf{e} - \lambda_g^{-2},$$

while the coefficient of correlation c_{cor} of intervals between successive group arrivals is given by

$$c_{cor} = (\lambda_g^{-1}\boldsymbol{\theta}(-D_0)^{-1}(D(1) - D_0)(-D_0)^{-1}\mathbf{e} - \lambda_g^{-2})/v.$$

For more information about the history, properties, special cases of the $BMAP$ input and related research of queues see [16].

All N servers of the first station are identical and independent of each other. The service time of a customer by a server is exponentially distributed with parameter μ_1.

If the arriving batch of the primary customers meets several servers being idle, the primary customers occupy the corresponding number of the servers. If the number of the idle servers is insufficient (or all servers are busy) the rest of the batch (or all the batch) goes to some virtual place called the orbit. Each customer from the orbit repeats its attempts to reach a server in a random time intervals which have exponentially distributed duration. If the repeated attempt fails because all servers are busy, the customer comes back to the orbit and tries its attempts later.

The total flow of retrials from the orbit is assumed to be such that the probability of generating the retrial attempt in the small interval $(t, t + \Delta t)$ is equal to $\alpha_i \Delta t + o(\Delta t)$ when the orbit size (the number of customers in the orbit) is equal to $i, \alpha_0 = 0$.

We assume that α_i is a nondecreasing function for $i \geq i_*$ where i_* is some fixed finite integer and $\lim_{i \to \infty} \alpha_i = \infty$. This assumption holds good for, e.g., the classic retrial strategy ($\alpha_i = i\alpha$ where α is intensity of individual retrials) and the linear retrial strategy ($\alpha_i = i\alpha + c$) as special cases.

After completion of the service at the first station, a customer has to proceed to the second station. This station is represented by R-server queue with a buffer of a finite capacity M. Service time of a customer at any server of the second station is exponentially distributed with parameter μ_2. If a customer proceeding from the first station does not find a free space at the second station it leaves the system permanently. Otherwise, it occupies an idle server, if any, or takes a place in the buffer. Customers from the buffer are served according the FIFO (First In - First Out) discipline.

For each customer placed into the buffer, the waiting time is bounded. It is assumed that available waiting time of a customer in the buffer (obsolescence or impatience time) is defined as a random variable with exponential distribution with parameter γ. If the obsolescence time expires before a customer is picked-up from the buffer to the server, it is assumed that this customer leaves the buffer immediately and is lost. The obsolescence times of different customers are independent of each other. The customers, which succeed to reach the server, become patient and leave the system only after the service completion.

Our aim is to compute the steady state distribution of the system states and the main performance measures of the system.

3   Process of the System States

The process of the system states can be described in terms of the irreducible multi-dimensional continuous-time Markov chain

$$\xi_t = \{i_t, n_t, \nu_t, r_t\}, t \geq 0,$$

where i_t is the number of customers in the orbit, n_t is the number of busy servers at the first station, r_t is the number of customers at the second station, ν_t is the state of the $BMAP$ underlying process at time t, $i_t \geq 0$, $n_t \in \{0, \ldots, N\}$, $r_t \in \{0, \ldots, R + M\}$, $\nu_t \in \{0, \ldots, W\}$.

For further use in the sequel, we introduce the following notation:

- I is an identity matrix of appropriate dimension. When needed the dimension of the matrix is identified with a suffix;
- $diag\{a_l,\ l = \overline{1,L}\}$ is a diagonal matrix with diagonal entries or blocks $a_l,\ l = \overline{1,L}$;
- $\mathcal{N} = diag\{n,\ n = \overline{0,N}\},\ \bar{I} = diag\{1,1,\ldots,1,0\}$;
- $\otimes$ and $\oplus$ are symbols of the Kronecker product and sum of matrices, respectively, see, e.g., [6];
- $\bar{W} = W + 1;\ a = \bar{W}(R + M + 1)$;
- $I^- = \begin{pmatrix} 0\,0\,0\ldots0\,0 \\ 1\,0\,0\ldots0\,0 \\ \vdots\ \vdots\ \vdots\ \ddots\ \vdots\ \vdots \\ 0\,0\,0\ldots1\,0 \end{pmatrix}$ $\quad I^+ = \begin{pmatrix} 0\,1\,0\ldots0\,0 \\ 0\,0\,1\ldots0\,0 \\ \vdots\ \vdots\ \vdots\ \ddots\ \vdots\ \vdots \\ 0\,0\,0\ldots0\,1 \\ 0\,0\,0\ldots0\,0 \end{pmatrix}.$

Let us enumerate the states of the Markov chain $\xi_t,\ t \geq 0$, in lexicographic order. The matrices $Q_{i,l},\ i,l \geq 0$, contain the transition rates from the states having the value i of the component i_t (level i) to the states having the value l of this component (level l).

Lemma 1. *Infinitesimal generator Q of the Markov chain $\xi_t,\ t \geq 0$, has the following block structure:*

$$Q = \begin{pmatrix} Q_{0,0} & Q_{0,1} & Q_{0,2} & Q_{0,3} & \cdots \\ Q_{1,0} & Q_{1,1} & Q_{1,2} & Q_{1,3} & \cdots \\ O & Q_{2,1} & Q_{2,2} & Q_{2,3} & \cdots \\ O & O & Q_{3,2} & Q_{3,3} & \cdots \\ \vdots & \vdots & \vdots & \vdots & \ddots \end{pmatrix}$$

where the blocks of dimension $(N + 1)a$ are defined by

$$Q_{i,i-1} = \alpha_i I^+_{N+1} \otimes I_a,\ i \geq 1,$$

$$Q_{i,i+k} = \begin{pmatrix} O & \cdots & O & D_{k+N} \\ O & \cdots & O & D_{k+N-1} \\ \vdots & \ddots & & \vdots \\ O & \cdots & O & D_k \end{pmatrix} \otimes I_{R+M+1},\ i \geq 0, k \geq 1,$$

$$Q_{i,i} = \begin{pmatrix} D_0 - \alpha_i I_{\bar{W}} & D_1 & \cdots & D_{N-1} & D_N \\ (\mu_1 + \alpha_i)I_{\bar{W}} & D_0 - (\mu_1 + \alpha_i)I_{\bar{W}} & \cdots & D_{N-2} & D_{N-1} \\ O & (2\mu_1 + \alpha_i)I_{\bar{W}} & \cdots & D_{N-3} & D_{N-2} \\ \vdots & \vdots & \ddots & \vdots & \vdots \\ O & O & \cdots & D_0 - ((N-1)\mu_1 + \alpha_i)I_{\bar{W}} & D_1 \\ O & O & \cdots & N\mu_1 I_{\bar{W}} & D_0 - N\mu_1 I_{\bar{W}} \end{pmatrix}$$

$$\otimes I_{R+M+1} - I_{(N+1)\bar{W}} \otimes \mathcal{K} + I_{(N+1)\bar{W}} \otimes \mathcal{K}I^-_{R+M+1},\ i \geq 0.$$

Here

$$\mathcal{K} = \mu_2 \, diag \, \{r, \ r = \overline{0, R}; \ R, \ldots, R\} + \gamma \, diag \, \{0, \ldots, 0; \ r, \ r = \overline{1, M}\}.$$

Proof of the Lemma follows from analysis of transitions of the Markov chain $\xi_t, t \geq 0$, during an infinitesimal interval.

Corollary 1. *The Markov chain $\xi_t, t \geq 0$, belongs to the class of continuous time asymptotically quasi-Toeplitz Markov chains (AQTMC).*

Proof. According to the definition given in [12], the chain ξ_t, $t \geq 0$, belongs to the class of continuous time $AQTMC$ if there exist the limits

$$Y_k = \lim_{i \to \infty} R_i^{-1} Q_{i,i+k-1}, k = 0, 2, 3, \ldots; \ Y_1 = \lim_{i \to \infty} R_i^{-1} Q_{i,i} + I,$$

and the matrix $\sum\limits_{k=0}^{\infty} Y_k$ is a stochastic one.

Here R_i is a diagonal matrix defined by modules of diagonal entries of the matrix $Q_{i,i}$.

It is easy to see that the diagonal entries of the matrix R_i, $i \geq 1$, corresponding to the value $n_t = n, n = \overline{0, N-1}$, of the second component of the chain ξ_t, $t \geq 0$, include the term α_i while the rest of diagonal does not depend on i.

Taking into account dependence (or not dependence) of the blocks of the matrices $Q_{i,i+k}$, $k \geq -1$, on α_i we calculate the limits defining the matrices Y_k as follows:

$$Y_0 = I_{N+1}^+ \otimes I_{\bar{W}(R+M+1)}, \quad Y_k = \begin{pmatrix} O \cdots O & O \\ \vdots \ddots \vdots & \vdots \\ O \cdots O & O \\ O \cdots O & R^{-1}D_k \end{pmatrix} \otimes I_{R+M+1}, \ k > 1,$$

$$Y_1 =$$

$$\begin{pmatrix} O \cdots & O & & O \\ \vdots \ddots & \vdots & & \vdots \\ O \cdots & O & & O \\ O \cdots & R^{-1}N\mu_1 I_a & R^{-1}(D_0 \otimes I_{R+M+1} - N\mu_1 I_a - I_{\bar{W}} \otimes \mathcal{K} + I_{\bar{W}} \otimes \mathcal{K} I_{R+M+1}^-) + I \end{pmatrix},$$

where R is a diagonal matrix defined by modules of diagonal entries of the matrix

$$D_0 \otimes I_{R+M+1} - N\mu_1 I_a - I_{\bar{W}} \otimes \mathcal{K} + I_{\bar{W}} \otimes \mathcal{K} I_{R+M+1}^-.$$

It is easy to see that the sum of these matrices is a stochastic matrix. Thus, the chain ξ_t, $t \geq 0$, is asymptotically quasi-Toeplitz Markov chain.

The matrices like Y_k, $k \geq 0$, play an important role in steady state analysis of $AQTMC$. The use of such matrices in mathematical analysis of asymptotically quasi-Toeplits Markov chains allowed (see [12]) to handle formally with asymptotic properties of these chains and to derive the ergodicity condition and elaborate the stable algorism for calculation the stationary distribution of general form $AQTMC$.

In the following we will use the results of [12] to investigate the stationary distribution of the chain under consideration.

Let $Y(z)$ be the generating function of the matrices Y_k, $k \geq 0$.

Corollary 2. *The matrix generating function* $Y(z) = \sum_{k=0}^{\infty} Y_k z^k$, $|z| \leq 1$, *has the following form:*

$$Y(z) =$$

$$\begin{pmatrix} O_a\ I_a \cdots & O & & O \\ O\ O \cdots & O & & O \\ \vdots\ \vdots\ \ddots & \vdots & & \vdots \\ O\ O \cdots & O & & I_a \\ O\ O \cdots R^{-1}N\mu_1 I_a z & R^{-1}[D(z) \otimes I - N\mu_1 I - I \otimes \mathcal{K} + I \otimes \mathcal{K} I^-_{R+M+1}]z + zI \end{pmatrix}.$$

4   Stationary Distribution

It is easy to see that the sufficient condition for existence or non-existence of the ergodic (stationary) distribution of the queue under consideration coincides with the corresponding condition for the first station. This station is represented by the retrial $BMAP/M/N$ queueing system. The sufficient ergodicity (non-ergodicity) condition for such a queue has been proven in [2].

Using that results, we immediately obtain the following statement.

Theorem 1. *The Markov chain* ξ_t, $t \geq 0$, *is ergodic if the following inequality holds*

$$\lambda/N\mu_1 < 1. \tag{1}$$

and non-ergodic if inequality (1) has an opposite sign.

In the following we will assume that inequality (1) holds good. Introduce notation for the steady state probabilities of the chain under consideration:

$$p(i, n, \nu, r) = \lim_{t \to \infty} P\{i_t = i, n_t = n, \nu_t = t, r_t = r\},$$

$$i \geq 0,\ n \in \{0, \ldots, N\},\ \nu \in \{0, \ldots, W\},\ r \in \{0, \ldots, R\}.$$

Let us enumerate the probabilities corresponding the value i of the first component in lexicographic order and form from these probabilities the row vectors $\mathbf{p}_i$, $i \geq 0$.

To find the steady state vectors $\mathbf{p}_i$, $i \geq 0$, of a level-dependent Markov chain under consideration we use the algorithm for computing the stationary distribution of $AQTMC$ developed in [12] based on the notion of so called censored Markov chains [9]. Taking into account the specific features of the chain ξ_t, $t \geq 0$, we obtain the algorithm consisting of the following steps.

- Calculate the matrix G as the minimal nonnegative solution of the matrix equation

$$G = Y(G).$$

- For preassigned sufficiently large integer i_0 calculate the matrices G_{i_0-1}, $G_{i_0-2}, \ldots, G_0$ using the equation of the backward recursion

$$Q_{i+1,i} + \sum_{n=i+1}^{\infty} Q_{i+1,n} G_{n-1} G_{n-2} \cdots G_i = O,$$

$i = i_0 - 1, i_0 - 2, \ldots, 0$, with the boundary condition $G_i = G$, $i \geq i_0$.
- Calculate the matrices

$$\bar{Q}_{i,l} = Q_{i,l} + \sum_{n=l+1}^{\infty} Q_{i,n} G_{n-1} G_{n-2} \cdots G_l, \, l \geq i, \, i \geq 0,$$

where $G_i = G$, $i \geq i_0$.
- Calculate the matrices Φ_l using the recurrent formulae

$$\Phi_l = (\bar{Q}_{0,l} + \sum_{i=1}^{l-1} \Phi_i \bar{Q}_{i,l})(-\bar{Q}_{l,l})^{-1}, \, l \geq 1.$$

- Calculate the vector $\boldsymbol{p}_0$ as the unique solution of the system

$$\boldsymbol{p}_0 \bar{Q}_{0,0} = \boldsymbol{0}, \quad \boldsymbol{p}_0(\boldsymbol{e} + \sum_{l=1}^{\infty} \Phi_l \boldsymbol{e}) = 1.$$

- Calculate the vectors $\boldsymbol{p}_l$ by formula: $\boldsymbol{p}_l = \boldsymbol{p}_0 \Phi_l$, $l \geq 1$. Note that all inverse matrices appearing in the above formulas exist and are non-negative because the inverted matrices are all sub-generators. This makes the proposed algorithm numerically stable.

5   Performance Measures

As soon as the probability vectors $\boldsymbol{p}_i$, $i \geq 0$, have been computed, we are able to calculate different performance measures of the system. Below we present some of them. Some nontrivial performance measures are given with brief explanations.

- Stationary distribution of the number of customers in the orbit

$$p_i = \boldsymbol{p}_i \boldsymbol{e}, \, i \geq 0.$$

- Mean number of customers in the orbit

$$N_{orb} = \sum_{i=1}^{\infty} i \boldsymbol{p}_i \boldsymbol{e}.$$

- Joint distribution of the number of customers at the first station (not including the customers in the orbit) and at the second station is given by the vector

$$\mathbf{p}_{busy}^{(1,2)} = \sum_{i=0}^{\infty} \mathbf{p}_i (I_{N+1} \otimes \mathbf{e}_{\bar{W}} \otimes I_{R+M+1}).$$

- Stationary distribution of the number of busy servers at the first station is given by the vector

$$\mathbf{p}_{busy}^{(1)} = \mathbf{p}_{busy}^{(1,2)} (I_{N+1} \otimes \mathbf{e}_{R+M+1}).$$

- Mean number of busy servers at the first station

$$N_{busy}^{(1)} = \mathbf{p}_{busy}^{(1)} \mathcal{N} \mathbf{e}.$$

- Stationary distribution of the number of customers at the second station is given by the vector

$$\mathbf{p}_{busy}^{(2)} = \mathbf{p}_{busy}^{(1,2)} (\mathbf{e}_{N+1} \otimes I_{R+M+1}).$$

- Mean number of customers at the second station

$$L_{busy}^{(2)} = \mathbf{p}_{busy}^{(2)} diag\{0, 1, \ldots, R+M\} \mathbf{e}.$$

- Mean number of busy servers at the second station

$$N_{busy}^{(2)} = \mathbf{p}_{busy}^{(2)} diag\{0, 1, \ldots, R, \ldots, R\} \mathbf{e}.$$

- **Assertion 1.** *Probability that an arbitrary primary customer reaches a server of the first station immediately upon arrival*

$$P_{imm} = \lambda^{-1} \sum_{i=0}^{\infty} \mathbf{p}_i (I_{(N+1)\bar{W}} \otimes \mathbf{e}_{R+M+1}) \sum_{k=0}^{N} (k\mathcal{N}_k - NI + \mathcal{N}) \otimes D_k \mathbf{e} \quad (2)$$

where $\mathcal{N}_k = diag\{\min\{1, \frac{N-n}{k}\}, n = \overline{0,N}\}$.
Proof. The brief derivation of formula (2) can be done as follows. Under the natural assumption that position of an arbitrary customer in an arriving batch is uniformly distributed, the rate of primary customers, which had a luck to occupy a server immediately upon arrival, is calculated as

$$\sum_{i=0}^{\infty} \mathbf{p}_i (I_{(N+1)\bar{W}} \otimes \mathbf{e}_{R+M+1}) \sum_{k=1}^{\infty} k(\mathcal{N}_k \otimes D_k).$$

Dividing this rate by the fundamental rate λ of the $BMAP$, we get the following expression for probability P_{imm}:

$$P_{imm} = \lambda^{-1} \sum_{i=0}^{\infty} \mathbf{p}_i (I_{(N+1)\bar{W}} \otimes \mathbf{e}_{R+M+1}) \sum_{k=1}^{\infty} k(\mathcal{N}_k \otimes D_k) \mathbf{e}. \quad (3)$$

The sum $\sum\limits_{k=1}^{\infty} k(\mathcal{N}_k \otimes D_k)\mathbf{e}$ can be represented in the form $\sum\limits_{k=1}^{\infty} k(\mathcal{N}_k \otimes D_k)\mathbf{e} =$

$\sum\limits_{k=0}^{N} k(\mathcal{N}_k \otimes D_k)\mathbf{e} + diag\{N - n, \ n = \overline{0,N}\} \otimes \sum\limits_{k=N+1}^{\infty} D_k\mathbf{e}.$

Taking into account the relation

$$\sum_{k=N+1}^{\infty} D_k\mathbf{e} = -\sum_{k=0}^{N} D_k\mathbf{e}$$

we easily get (2) from (3).

- **Assertion 2.** *Probability that an arbitrary customer arriving at the second station will be lost*

$$P_{loss} = \frac{\mathbf{p}_{busy}^{(1,2)}(\mathcal{N} \otimes (I - I_{R+M+1}))\mathbf{e}}{\kappa} \qquad (4)$$

where

$$\kappa = \mathbf{p}_{busy}^{(1,2)}(I_{N+1} \otimes \mathbf{e}_{R+M+1})\mathcal{N}\mathbf{e}. \qquad (5)$$

Proof. Formula (4) becomes clear if we note that the numerator multiplied by μ_1 is the rate of lost customers and the denominator multiplied by μ_1 is the rate of all customers proceeding from the first station to the second station. It is worth to note that, because there is no customers loss at the first station, the following equality holds:

$$\lambda = \mu_1\kappa.$$

This equality is useful for control of accuracy of computer implementation of the presented algorithm.

6   Distribution of Waiting Time and Probability That an Arbitrary Customer Will Not Be Lost at the Second Station Due to Obsolescence

Let $W^{(2)}(t)$ be the probability that an arbitrary customer arriving at the second station will not be lost due to the obsolescence (impatience) and its waiting time is less than t.

Let also $W_r^{(2)}(t)$ be the probability that an arbitrary customer arriving at the second station will not be lost due to the obsolescence and its waiting time is less than t conditional that r customers are at the station at the arriving epoch, $r = \overline{0, R + M}$.

Denote

$$w_r^{(2)}(s) = \int\limits_0^{\infty} e^{-st}dW_r^{(2)}(t), \quad w^{(2)}(s) = \int\limits_0^{\infty} e^{-st}dW^{(2)}(t), \ Re \ s \geq 0,$$

the Laplace-Stieltjes transforms (*LST*s) of the functions $W_r^{(2)}(t)$ and $W^{(2)}(t)$ respectively.

Assertion 3. *LSTs $w_r^{(2)}(s)$ are calculated as follows:*

$$w_r^{(2)}(s) = 1, \ r = \overline{0, R-1}, \tag{6}$$

$$w_{R+k}^{(2)}(s) = \prod_{l=0}^{k} \frac{R\mu_2 + (k-l)\gamma}{s + R\mu_2 + (k-l+1)\gamma}, \ k = \overline{0, M-1}, \tag{7}$$

$$w_{R+M}^{(2)}(s) = 0. \tag{8}$$

Proof. Formulas (6) and (8) are obvious. Expression (7) for $w_{R+k}^{(2)}(s)$, $k = \overline{1, M-1}$, becomes clear if we take into account the following recurrent relations:

$$w_R^{(2)}(s) = \int_0^\infty e^{-st} e^{-\gamma t} d(1 - e^{-R\mu_2 t}),$$

$$w_{R+1}^{(2)}(s) = \int_0^\infty e^{-st} e^{-\gamma t} d(1 - e^{-(R\mu_2+\gamma)t}) w_R^{(2)}(s),$$

$$w_{R+k}^{(2)}(s) = \int_0^\infty e^{-st} e^{-\gamma t} d(1 - e^{-(R\mu_2+k\gamma)t}) w_{R+k-1}^{(2)}(s), \ k = \overline{2, M-1}.$$

Calculating integrals, we easy get formula (7).

Assertion 4. *LST $w^{(2)}(s)$ is calculated as follows:*

$$w^{(2)}(s) = \frac{\mathbf{p}_{busy}^{(1,2)}(\mathcal{N} \otimes \mathcal{W}(s))\mathbf{e}}{\kappa} \tag{9}$$

where $\mathcal{W}(s) = diag\{w_m^{(2)}(s), \ m = \overline{0, R+M}\}$, *and* κ *is defined by (5).*

Proof. Let $I^{(r)}$ be a matrix of size $R + M + 1$ whose rth diagonal entry is equal to 1 and the rest entries are zeroes.

It is easy to see that the expression

$$\frac{\mathbf{p}_{busy}^{(1,2)}\mu_1(\mathcal{N} \otimes I^{(r)})\mathbf{e}}{\mathbf{p}_{busy}^{(1,2)}(I_{N+1} \otimes \mathbf{e}_{R+M+1})\mu_1\mathcal{N}\mathbf{e}} \tag{10}$$

defines the probability that an arbitrary customer served at the first station sees r customers at the second station (for more detail explanation see proof of formula (4)).

Multiplying (10) by $w_r^{(2)}(s)$ we obtain *LST $w^{(r)}(s)$* of the distribution of the waiting time at the second station of a customer which sees r busy servers at the second station after service at the first station and will not be lost. Then expression (9), that is calculated as the sum of *LSTs $w^{(r)}(s)$* over $r = \overline{0, R+M}$, defines the *LST* of the distribution of the waiting time at the second station of a customer that will not be lost due to the obsolescence.

Corollary 3. *Probability that an arbitrary customer will not be lost at the second station due to obsolescence is computed by*

$$P_{success}^{(2)} = \frac{\mathbf{p}_{busy}^{(1,2)}(\mathcal{N} \otimes \mathcal{W}(0))\mathbf{e}}{\kappa}$$

where the matrix $\mathcal{W}(0)$ *is given by*

$$\mathcal{W}(0) = diag\{1, 1, \ldots, 1, \frac{N\mu + \gamma}{N\mu + (k+1)\gamma}, \, k = \overline{1, M-1}, \, 0\}.$$

Proof follows from relation $P_{success}^{(2)} = w^{(2)}(0)$ evident due to the probabilistic meaning of the *LST*.

7   Conclusion

Tandem queueing model considered in this paper extends possibility of adequate mathematical modeling of many real life systems where an arriving customer should get two sequential steps of the service into the following three directions. Firstly, multi-server stations are considered while single server queues are considered in the overwhelming majority of the existing papers. Secondly, possibility of customers retrials (not queueing) at the first station and customers impatience at the second station is taken into account. Thirdly, the more general Batch Markov Arrival Process of primary customers is assumed instead of the stationary Poisson arrival process.

The results can be used, e.g., for performance evaluation and capacity planning of wireless contact centers or distributed data bases. The the first station models a remote wireless access of customers to resources of a contact center or distributed data base while the second station describes the process of utilization of these resources (servers, agents, operators, tables, indices, etc.) by the customers requests. Using the presented results, different optimization problems can be solved, e.g., planning the number of servers at both stations sufficient for providing desired value of probability of immediate access to the system and required level of probability of successful service at both stations of the system or the optimal matching the number of servers at both stations to minimize maintenance cost of servers and, at the same time, to avoid overflow or starvation of some stations.

Presented results have algorithmic nature. So, they can be more or less easily extended in different aspects. E.g., service time distribution at one or at both stations can be assumed having phase type distribution, which is much more general than exponential distribution supposed for simplicity of presentation in this paper. Tool of Kronecker products and sums of matrices can be successfully applied to this end, as it was done, e.g., in [2,13]. The structure of the generator does not change. Only the blocks of generator become more complicated due to the necessity to keep track of phases of a service at all servers of stations. The case of the constant retrial rate also can be easily considered based on the results presented above and the related results in [2]. Ergodicity condition coincides with condition for the corresponding retrial model in [2]. Algorithm for computation of the stationary distribution follows from the presented above just by the formal replacement of the total retrial rate α_i by a constant. Models with a finite orbit and finite or infinite buffer also can be handled by analogy with the presented results. Non-persistence of the orbiting customers can be taken into account by analogy with [13]. Cross traffic (additional flow of customers arriving directly to the second station) or partial departure of customers after the service at the first station can be taken into account easily as well.

References

1. Balsamo, S., Persone, V.D.N., Inverardi, P.: A review on queueing network models with finite capacity queues for software architectures performance prediction. Perform. Eval. **51**, 269–288 (2003)
2. Breuer, L., Dudin, A.N., Klimenok, V.I.: A retrial $BMAP/PN/N$ system. Queueing Syst. **40**, 433–457 (2002)
3. Gnedenko, B.W., Konig, D.: Handbuch der Bedienungstheorie. Akademie Verlag, Berlin (1983)
4. Gomez-Corral, A.: A matrix-geometric approximations for tandem queues with blocking and repeated attempts. Oper. Res. Lett. **30**, 360–374 (2002)
5. Gomez-Corral, A.: A bibliographical guide to the analysis of retrial queues through matrix analytic techniques. Ann. Oper. Res. **141**, 163–191 (2006)
6. Graham, A.: Kronecker Products and Matrix Calculus with Applications. Ellis Horwood, Cichester (1981)
7. Hall, N.G., Sriskandarajah, C.: A survey of machine scheduling problems with blocking and no-wait in process. Oper. Res. **44**, 510–525 (1996)
8. Heindl, A.: Decomposition of general queue networks with $MMPP$ inputs and customer losses. Perform. Eval. **1**, 117–136 (2003)
9. Kemeni, J.G., Snell, J.L., Knapp, A.W.: Denumerable Markov Chains. Van Nostrand, New York (1966)
10. Kim, C.S., Klimenok, V., Taramin, O.: A tandem retrial queueing system with two Markovian flows and reservation of channels. Comput. Oper. Res. **37**, 1238–1246 (2010)
11. Kim, C.S., Park, S.H., Dudin, A., Klimenok, V., Tsarenkov, G.: Investigaton of the $BMAP/G/1 \rightarrow ./PH/1/M$ tandem queue with retrials and losses. Appl. Math. Model. **34**, 2926–2940 (2010)
12. Klimenok, V.I., Dudin, A.N.: Multi-dimensional asymptotically quasi-Toeplitz Markov chains and their application in queueing theory. Queueing Syst. **54**, 245–259 (2006)
13. Klimenok, V.I., Orlovsky, D.S., Dudin, A.N.: A $BMAP/PH/N$ system with impatient repeated calls. Asia Pac. J. Oper. Res. **24**, 293–312 (2007)
14. Klimenok, V., Savko, R.: A retrial tandem queue with two types of customers and reservation of channels. Commun. Comput. Inform. Sci. **356**, 105–114 (2013)
15. Klemm, A., Lindermann, C., Lohmann, M.: Modelling IP traffic using the batch Markovian arrival process. Perform. Eval. **54**, 149–173 (2003)
16. Lucantoni, D.M.: New results on the single server queue with a batch Markovian arrival process. Commun. Stat.-Stoch. Models **7**, 1–46 (1991)

17. Moutzoukis, E., Langaris, C.: Two queues in tandem with retrial customers. Probab. Eng. Inf. Sci. **15**, 311–325 (2001)
18. Neuts, M.F.: A versatile Markovian point process. J. Appl. Probab. **16**, 764–779 (1979)
19. Perros, H.G.: A bibliography of papers on queueing networks with finite capacity queues. Perform. Eval. **10**, 255–260 (1989)

Performance Characteristics for DD Priority Discipline with Semi-Markov Switching

Gheorghe Mishkoy[✉] and Lilia Mitev

Institute of Mathematics and Computer Sciences
of the Academy of Sciences of Moldova, Free International University
of Moldova, 5 Academiei street, 2028 Chisinau, Republic of Moldova
gmiscoi@asm.md, liliausate@yahoo.com

Abstract. Priority Discretionary Discipline (DD) with an arbitrary number of priority classes and with switching losses at switching process of service from one class of priority to another one is considered. Such characteristics as distribution of busy periods, queue lengths, probability of states, as well as various supplementary distributions are presented. Symmetric priority discipline named Relative DD discipline also was analyzed. Even through, the main characteristics are obtained in the terms of Laplace, Laplace-Stieltjes transforms and recurrent functional equations, the obtained relations are convenient for algorithmization.

Keywords: Priority · Busy period · Queue length · Probability of state · Laplace-Stieltjes transform · Generating function

1 Introduction: The Model Description: Formalization of the Interruption Scheme

Priority Discretionary Discipline, it seems, has its origin in the Jaiswal's monograph [1], where it is studied regarding problem of service with one server and two flows of requests. This discipline is more flexible than classical disciplines of preemptive (absolute) and head-of-the-line (relative) priority disciplines, which are characterized by a high level of conservatives. For example, in case of the preemptive priority discipline, the arrival of a request with a higher priority interrupts inevitably the serving of the request with a lower priority although the service of this request it is almost finished. And vice versa, for the relative priority discipline, a request with a higher priority has to wait the service completion of the request with a lower priority, although the serving of this request has just begun. For two flows of requests, the DD discipline, following [1], it is described as follows: if the service time of a request is less than set value θ, then it achieved the absolute priority, otherwise - relative.

Among the works devoted to this topic we mention the works [2] and [3], as well [4–6], where the discipline DD is analyzed more generally. Namely, it is supposed that the number of priority classes is arbitrary, and secondly, it is

V. Vishnevsky et al. (Eds.): DCCN 2013, CCIS 279, pp. 204–218, 2014.
DOI: 10.1007/978-3-319-05209-0_18, © Springer International Publishing Switzerland 2014

assumed that service process of switching from one class of requests to another requires to spend some time for switching. The duration of switching is a random variable with an arbitrary distribution function. The last assumption is important for applications as it allows to model and analyze different waiting times that objectively takes place in real systems. Using the analyzed models' results, for example, in the practice of engineering design, it has an important value as it allows the possibility to obtain more accurate evaluation about functioning of the real systems.

Let's consider a queueing system $M_r|G_r|1|\infty$ with priority DD : if the service time of a_k–request is less than set value $\theta_k, (k = 2, \ldots, r)$, then the arrived request with priority higher k (σ_{k-1}–request) achieves absolute priority, otherwise - relative. The durations of service a_k–requests are independent random variables B_k with distribution function $B_k(x), (k = 1, \ldots, r)$.

The switching takes place only at service's interruption and at returning to the interrupted service. If the service of a_j-request is interrupted by arriving a_i-request, $i < j$, then, at once switching to L_i ($\rightarrow i$) flow begins. When the system will be free from requests of priority higher j, the switching ($\rightarrow j$) begins, and only then the server is ready to serve the interrupted request. The durations of switching ($\rightarrow i$) are random variables C_i with distribution function $C_i(x), (i = 1, \ldots, r)$. The random variables B_k and C_i are independent. An arbitrary switching ($\rightarrow k$) also may be interrupted by arriving σ_{k-1}-request.

Let's introduce the following scheme's classification. We'll denote every scheme by two indexes IJ. The first index will show the future state of interrupted switching, the second - the future state of interrupted service. We'll consider that at $I = 1$ the interrupted switching begins anew (preemptive repeat-different discipline), $I = 2$ the interrupted switching continues from the interrupted point (preemptive resume discipline) and at $I = 3$ identically begins anew (preemptive repeat identically discipline). At $J = 1$ the interrupted request is served again (preemptive repeat-different discipline) and at $J = 2$ the interrupted request is served from the interrupted point (preemptive resume discipline).

In some practical situations, the interruption accepted on service a_i–request is supposed only through "quantum" of a service time. That's why we'll analyze and "symmetric" discipline for the mentioned above discipline. Namely: to realize a relative priority if the service time is less θ_k and absolute - otherwise.

For this "symmetric" discipline we will use the term Relative DD discipline.

2   Busy Period Distribution and Auxiliary Characteristics

We'll introduce the following notations. We'll denote by $\Pi(x), \overline{\Pi}_k(x), \overline{\Pi}_{kk}(x)$, $H_k(x), \overline{\Pi}_{kk}^{(n)}(x), N_k(x), \Pi_k(x), \Pi_{kk}(x)$ the distribution function of busy period, k–period, kk–period, k–service cycle, kkn–period, k–cycle of switching, Π_k–period, Π_{kk}–period and by $\pi(s) \ldots \pi_{kk}(s)$ - their Laplace-Stieltjes transforms (the definition of these see [6]). Let's consider also $\sigma_k = a_1 + \cdots + a_k$, where a_k–the parameter of Poisson flow of k–th priority.

2.1   Absolute DD Discipline

Lemma 1. For schemes I.1

$$h_k(s) = \int_0^{\theta_k} e^{-(s+\sigma_{k-1}[1-\pi_{k-1}(s)\nu_k(s)])x} dB_k(x)$$

$$+ e^{-(\sigma_{k-1}[\overline{\pi}_{k-1}(s)-\pi_{k-1}(s)\nu_k(s)])\theta_k} \int_{\theta_k}^{\infty} e^{-(s+\sigma_{k-1}[1-\overline{\pi}_{k-1}(s)])x} dB_k(x). \tag{1}$$

Lemma 2. For schemes I.2

$$h_k(s) = \left\{ \int_0^{\theta_k} e^{-(s+\sigma_{k-1})x} dB_k(x) + e^{-\sigma_{k-1}\overline{\pi}_{k-1}(s)\theta_k} \int_0^{\infty} e^{-(s+\sigma_{k-1}[1-\pi_{k-1}(s)])x} dB_k(x) \right\}$$

$$\times \left\{ 1 - \sigma_{k-1}\pi_{k-1}(s)\nu_k(s) \int_0^{\theta_k} e^{-(s+\sigma_{k-1})x}[1 - B_k(x)]dx \right\}^{-1}, \tag{2}$$

where $\nu_k(s)$ is expressed:
for schemes 1.J

$$\nu_k(s) = c_k(s + \sigma_{k-1}) \left\{ 1 - \frac{\sigma_{k-1}}{s + \sigma_{k-1}}[1 - c_k(s + \sigma_{k-1})]\pi_{k-1}(s) \right\}^{-1}, \tag{3}$$

for schemes 2.J

$$\nu_k(s) = c_k(s + \sigma_{k-1}[1 - \pi_{k-1}(s)]), \tag{4}$$

for schemes 3.J

$$\nu_k(s) = (s + \sigma_{k-1}) \int_0^{\infty} e^{-(s+\sigma_{k-1})\tau} \{s + \sigma_{k-1}[1 - \pi_{k-1}(s)$$

$$\times 1 - e^{-(s+\sigma_{k-1})\tau}]\}^{-1} dC_k(\tau), \tag{5}$$

the mentioned above $\overline{\pi}_k(s)$ and $\pi_{k-1}(s)$ are uniquely determined from recurrent relations of Theorem 1.

Theorem 1. For all schemes

$$\sigma_k\overline{\pi}_k(s) = \sigma_{k-1}\overline{\pi}_{k-1}(s + a_k - a_k\overline{\pi}_{kk}(s)) + a_k\pi_{kk}(s), \tag{6}$$

$$\overline{\pi}_{kk}(s) = h_k(s + a_k - a_k\overline{\pi}_{kk}(s)), \tag{7}$$

$$\sigma_k \pi_k(s) = \sigma_{k-1} \pi_{k-1}(s + a_k) + \sigma_{k-1}\{\pi_{k-1}(s + a_k[1 - \overline{\pi}_{kk}(s)])$$
$$- \pi_{k-1}(s + a_k)\}\nu_k(s + a_k[1 - \overline{\pi}_{kk}(s)]) + a_k \pi_{kk}(s), \qquad (8)$$

$$\pi_{kk}(s) = \nu_k(s + a_k[1 - \overline{\pi}_{kk}(s)])\overline{\pi}_{kk}(s), \qquad (9)$$

where $h_k(s + a_k - a_k \overline{\pi}_{kk}(s))$ and $\nu_k(s + a_k - a_k \overline{\pi}_{kk}(s))$, for each of the schemes I.J, are determined from (1)–(5) respectively, for $s = s + a_k - a_k \overline{\pi}_{kk}(s)$.

Remark 1. Lemma 1, lemma 2 and theorem 1 are demonstrated applying method of "catastrophes" [6]. Bellow we'll present only the sketch of proof for relations (1) and (2). According to the structure of k cycle of service we have: for schemes I.1

$$h_k(s) = \int_0^{\theta_k} e^{-(s+\sigma_{k-1}[1-\pi_{k-1}(s)\nu_k(s)])x} dB_k(x) + e^{-(\sigma_{k-1}[1-\pi_{k-1}(s)\nu_k(s)])\theta_k}[1 - B_k(\theta_k)]$$

$$\times \sum_{n \geq 0} \int_0^{\infty} e^{-sx} \frac{(\sigma_{k-1}x)^n}{n!} e^{-\sigma_{k-1}x} \frac{dB_k(\theta_k + x)}{1 - B_k(\theta_k)} [\overline{\pi}_{k-1}(s)]^n.$$

for schemes I.2.

$$h_k(s) = \int_0^{\theta_k} e^{-(s+\sigma_{k-1})x} dB_k(x) + \int_0^{\theta_k} e^{-sx}[1 - B_k(x)]e^{-\sigma_{k-1}x}\sigma_{k-1}dx$$

$$\times \pi_{k-1}(s)\nu_k(s)h_k(s) + e^{-(s+\sigma_{k-1})e^{-(s+\sigma_{k-1})x}}[1 - B_k(\theta_k)]$$

$$\times \sum_{n \geq 0} \int_0^{\infty} e^{-sx} \frac{(\sigma_{k-1}x)^n}{n!} e^{-\sigma_{k-1}x} \frac{dB_k(\theta_k + x)}{1 - B_k(\theta_k)} [\overline{\pi}_{k-1}(s)]^n.$$

After summation and integration, we receive (1) and (2). From mentioned relations can be obtained numerical characteristics of service. Let's consider

$$\rho_k = \sum_{i=1}^{k} a_i b_i, \quad b_1 = \frac{\beta_{11} + c_{11}}{1 + a_1 c_{11}},$$

$$b_i = \beta_{i1} + [\theta_i - \int_0^{\theta_i} B_i(x)dx][\Phi_{i-1} \ldots \Phi_2(1 + a_1 c_{11})q_i - 1],$$

$$\Phi_1 = 1, \Phi_i = 1 + \frac{\sigma_{i-1}\pi_{i-1}(a_i)}{\sigma_{i-1}}[q_i - 1], i = 2, \ldots, r.$$

where:

for scheme 1.1: $q_i = \frac{1}{c_i(\sigma_{i-1})}$,

for scheme 1.2: $q_i = 1 + \sigma_i c_{i1}$,

for scheme 1.3: $q_i = \frac{1}{c_i(-\sigma_{i-1})}$.

If $\rho_k < 1$, then the first moments of k - cycle of switching, k - cycle of service, $\overline{\Pi}_{kk}$ - period, $\overline{\Pi}_k$ - period, Π_{kk} - period and Π_k - period are equal, respectively:

$$\nu_{k1} = \frac{1}{\sigma_{k-1}}[q_k - 1]\frac{\Phi_{k-1}\cdots\Phi_2(1 + a_1c_{11})}{1 - \rho_{k-1}},$$

$$h_{k1} = \frac{b_k}{1 - \rho_{k-1}}, \quad \overline{\pi}_{kk1} = \frac{b_k}{1 - \rho_k}, \quad \sigma_k\overline{\pi}_{k1} = \frac{\rho_k}{1 - \rho_k},$$

$$\pi_{kk1} = [b_k + \Phi_{k-1}\cdots\Phi_2\frac{\rho_{k-1}}{\sigma_{k-1}}]\frac{1}{1 - \rho_k},$$

$$\sigma_k\pi_{k1} = \frac{\Phi_k\cdots\Phi_2(1 + a_1c_{11}) + \rho_k - 1}{1 - \rho_k}.$$

Remark 2. At formalization of the schemes for Absolute DD discipline the scheme with "loss" of the interrupted request has not been considered. We will denote this scheme by identifier I.3 and we'll consider its following modification: scheme I.3_0 - the interrupted request is lost at once as soon as there was an interruption in its service; scheme I.3_1 - the interrupted request "is lost" as soon as service of all requests of priority higher interrupted is finished; scheme I.3_2 - the "loss" of request occurs when the server is ready to start service of the next request of the same priority.

Below-mentioned lemmas give the relations for k–cycles of service distribution (10)–(12) for schemes I.3_0–I.3_2.

Lemma 3. For schemes I.3_0

$$h_k(s) = \int\limits_0^{\theta_k} e^{-(s+\sigma_{k-1})x}dB_k(x)$$

$$+e^{-\sigma_{k-1}\pi_{k-1}(s)\theta_k}\int\limits_{\theta_k}^{\infty} e^{-(s+\sigma_{k-1}-\sigma_{k-1}\overline{\pi}_{k-1}(s))u}dB_k(u). \tag{10}$$

Lemma 4. For schemes I.3_1

$$h_k(s) = \int\limits_0^{\theta_k} e^{-(s+\sigma_{k-1})x}dB_k(x) + \int\limits_0^{\theta_k} e^{-(s+\sigma_{k-1})x}[1 - B_k(x)]\sigma_{k-1}dx\pi_{k-1}(s)$$

$$+e^{-\sigma_{k-1}\overline{\pi}_{k-1}(s)\theta_k}\int\limits_{\theta_k}^{\infty} e^{-(s+\sigma_{k-1}-\sigma_{k-1}\overline{\pi}_{k-1}(s))u}dB_k(u). \tag{11}$$

Lemma 5. For schemes I.3$_2$

$$h_k(s) = \int_0^{\theta_k} e^{-(s+\sigma_{k-1})x} dB_k(x) + \int_0^{\theta_k} e^{-(s+\sigma_{k-1})x} [1 - B_k(x)] \sigma_{k-1} dx \pi_{k-1}(s) \nu_k(s)$$

$$+ e^{-\sigma_{k-1}\overline{\pi}_{k-1}(s)\theta_k} \int_{\theta_k}^{\infty} e^{-(s+\sigma_{k-1}-\sigma_{k-1}\overline{\pi}_{k-1}(s))u} dB_k(u). \tag{12}$$

2.2   Particular Cases

From mentioned above results follow the results for classical systems of relative and absolute priority. Really, if we'll consider $c_j = 0$, $j = 1, \ldots, r$, $\theta_k = 0$ then we will receive $M_r|G_r|1$ queueing system with relative priority. It is known [7], that distribution of busy period is defined (for $k = r$) from relations:

$$h_k(s) = \beta_k(s + \sigma_{k-1} - \sigma_{k-1}\overline{\pi}_{k-1}(s)), \tag{13}$$

$$\overline{\pi}_{kk}(s) = h_k(s + a_k - a_k\overline{\pi}_{kk}(s)), \tag{14}$$

$$\sigma_k \overline{\pi}_k(s) = \sigma_{k-1}\overline{\pi}_{k-1}(s + a_k - a_k\overline{\pi}_{kk}(s)) + a_k\overline{\pi}_{kk}(s)), \tag{15}$$

Here our notation for $\overline{\pi}_{k-1}(s)$ corresponds to $\pi_{k-1}(s)$ from [7]. We'll consider for definiteness the scheme I.2. From lemma 2, for $\theta_k = 0$, $c_j = 0$ follows $h_k(s) = \beta_k(s + \sigma_{k-1} - \sigma_{k-1}\overline{\pi}_{k-1}(s))$ that coincides with (13). Relations (6) and (7) from theorem 1 coincide with (14) and (15) as at demonstration of the theorem, the structure k - cycles of service has remained former. Let's consider now $c_j \neq 0$, $\theta_k = \infty$. We are in conditions of $M_r|G_r|1$ system with orientation and absolute priority [6]. On the other hand, from lemma 2 follows, that

$$h_k(s) = \beta_k(s + \sigma_{k-1}) + \frac{\sigma_{k-1}[1 - \beta_k(s + \sigma_{k-1})]}{s + \sigma_{k-1}} \pi_{k-1}(s)\nu_k(s)h_k(s),$$

or

$$h_k(s) = \beta_k(s + \sigma_{k-1}) \left\{ 1 - \frac{\sigma_{k-1}}{s + \sigma_{k-1}} [1 - \beta_k(s + \sigma_{k-1})] \pi_{k-1}(s)\nu_k(s) \right\}^{-1}.$$

In this way, for $\theta_k = \infty$ and $c_j \neq 0$ it is obtained the results for systems with absolute priority and orientation [6].

2.3   Relative DD Discipline

If the service time a_k - request is less than set value θ_k, $(k = 2, \ldots, r)$, then the arrived request of a priority higher k (σ_{k-1} - request) receives relative priority, otherwise - absolute.

All notations and definitions mentioned above are also kept. We'll consider additional:

$$\overline{\gamma}_k(s) = s + \sigma_k - \sigma_k \overline{\pi}_k(s), \quad \gamma_k(s) = s + \sigma_k - \sigma_k \pi_k(s)\nu_{k+1}(s),$$

$$R_k(s, \theta_k) = 1 - \sigma_{k-1}\pi_{k-1}(s)\nu_k(s)e^{\sigma_{k-1}\overline{\pi}_{k-1}(s)\theta_k} \int_{\theta_k}^{\infty} e^{-(s+\sigma_{k-1})x}[1 - B_k(x)]dx.$$

Lemma 6. For schemes I.1

$$h_k(s) = \{ \int_0^{\theta_k} e^{-\overline{\gamma}_{k-1}(s)x}dB_k(x)$$

$$+ e^{-\sigma_{k-1}\pi_{k-1}(s)\theta_k} \int_{\theta_k}^{\infty} e^{-(s+\sigma_{k-1})x}dB_k(x) \} / R_k(s, \theta_k). \tag{16}$$

Lemma 7. For schemes I.2

$$h_k(s) = \int_0^{\theta_k} e^{-\overline{\gamma}_{k-1}(s)x}dB_k(x) + e^{-(\overline{\gamma}_{k-1}(s)-\gamma_{k-1}(s))\theta_k} \int_{\theta_k}^{\infty} e^{-\gamma_{k-1}(s)x}dB_k(x), \tag{17}$$

where $\nu_k(s)$ for each of schemes correspondingly look like:
for schemes 1.J

$$\nu_k(s) = c_k(s + \sigma_{k-1})\left\{1 - \frac{\sigma_{k-1}}{s + \sigma_{k-1}}[1 - c_k(s + \sigma_{k-1})]\overline{\pi}_{k-1}(s)\right\}^{-1}, \tag{18}$$

for schemes 2.J

$$\nu_k(s) = c_k(\gamma_{k-1}(s)), \tag{19}$$

for schemes 3.J

$$\nu_k(s) = (s + \sigma_{k-1}) \int_0^{\infty} e^{-(s+\sigma_{k-1})x}dC_k(x)$$

$$\times (s + \sigma_{k-1} - \sigma_{k-1}[1 - e^{-(s+\sigma_{k-1})x}]\overline{\pi}_{k-1}(s)), \tag{20}$$

the mentioned above $\overline{\pi}_{k-1}(s)$ and $\pi_{k-1}(s)$ are unequivocally defined from recurrent relations of the theorem 2.

Theorem 2. For all schemes, Laplace-Stieltjes transform of the distribution function of busy period $\sigma\pi(s) = \sigma_r\pi_r(s)$ is defined from following recurrent relations:

$$\sigma_k\overline{\pi}_k(s) = \sigma_{k-1}\overline{\pi}_{k-1}(s + a_k - a_k\overline{\pi}_{kk}(s)) + a_k\pi_{kk}(s),$$

$$\overline{\pi}_{kk}(s) = h_k(s + a_k - a_k\overline{\pi}_{kk}(s)),$$

$$\sigma_k\pi_k(s) = \sigma_{k-1}\pi_{k-1}(s + a_k - a_k\overline{\pi}_{kk}(s)) + a_k\pi_{kk}(s),$$

$$\pi_{kk}(s) = \nu_k(s + a_k - a_k\overline{\pi}_{kk}(s))\overline{\pi}_{kk}(s),$$

where $h_k(s + a_k - a_k\overline{\pi}_{kk}(s))$ and $\nu_k(s + a_k - a_k\overline{\pi}_{kk}(s))$, for each of schemes, are expressed from (16)–(20) respectively for $s = s + a_k - a_k\overline{\pi}_{kk}(s)$.

Let's consider $\rho_k = \sum_{i=1}^{k} a_i b_i$, where $b_1 = \beta_{11}$ and b_i, $i = 2, \ldots, r$ are expressed:

for schemes I.1

$$b_i = \{\theta_i[1 - B_i(\theta_i)] + \int_0^{\theta_i} x dB_i(x) + (a_1 c_{11} + q_i + \sum_{j=2}^{i-1} \frac{a_j}{\sigma_{j-1}}(q_j - 1)$$

$$\times (\frac{1 - B_i(\theta_i)}{\sigma_{i-1}} - \frac{e^{\sigma_{i-1}\theta_i}}{\sigma_{i-1}} \int_{\theta_i}^{\infty} e^{-\sigma_{i-1}x} dB_i(x)\}\{B_i(\theta_i) + e^{\sigma_{i-1}\theta_i} \int_{\theta_i}^{\infty} e^{-\sigma_{i-1}x} dB_i(x)\}^{-1},$$

for schemes I.2

$$b_i = \theta_i[1 - B_i(\theta_i)] + \int_0^{\theta_i} x dB_i(x)$$

$$+ (a_1 c_{11} + q_i + \sum_{j=2}^{i-1} \frac{a_j}{\sigma_{j-1}}(q_j - 1)(\int_{\theta_i}^{\infty} x dB_i(x) - \theta_i[1 - B_i(\theta_i)]).$$

In these formulas q_i are equal:
for schemes 1.J: $q_i = \frac{1}{c_i(\sigma_{i-1})}$,
for schemes 2.J: $q_i = 1 + \sigma_{i-1}c_{i1}$,
for schemes 3.J: $q_i = \frac{1}{c_i(-\sigma_{i-1})}$.

If $\rho_k < 1$, then the first moments of k–cycle of orientation, k–cycle of service, $\overline{\Pi}_k$–period, $\overline{\Pi}_{kk}$–period, Π_{kk}–period, Π_k–period are equal, respectively:

$$\nu_{k1} = \frac{\rho_k - 1}{\sigma_{k-1}(1 - \rho_{k-1})}, \quad h_{k1} = \frac{b_k}{1 - \rho_{k-1}}, \quad \overline{\pi}_{kk1} = \frac{b_k}{1 - \rho_k},$$

$$\sigma_k\overline{\pi}_{k1} = \frac{\rho_k}{1 - \rho_k}, \quad \pi_{kk1} = (b_k + \frac{\rho_k - 1}{\sigma_{k-1}})/\rho_{k-1},$$

$$\sigma_k\pi_{k1} = (a_1 c_{11} + \sum_{i=2}^{k} \frac{a_i}{\sigma_{i-1}}(q_i - 1) + \rho_k)/(1 - \rho_k).$$

3 Queue Length Distribution

We'll denote by $P_m(t)$ the probability that at time t in the queue there are $m = (m_1, \ldots, m_r)$ requests, where m_i is the number of requests of i priority class. Let's consider $P(z,t) = \sum P_m(t)z^m$, $z^m = z_1^{m_1}, \ldots, z_r^{m_r}$, $0 \le z_i \le 1$ and $p(z,s) = \int\limits_0^\infty e^{-st} P(z,t)dt$.

3.1 Absolute DD Discipline

The distribution of queue length on separate k - cycle of service in terms of Laplace transform is given by following lemma.

Lemma 8. For schemes I.1

$$h_k(z,s) = z_k[1 + \sigma_{k-1}(\pi_{k-1}(z,s) + \pi_{k-1}(s + \omega_k)\nu_k(z,s))]$$

$$\times \int\limits_0^{\theta_k} [1 - B_k(x)]e^{-\eta_k(s,z)x}dx + z_k(1 + \sigma_{k-1}\overline{\pi}_{k-1}(z,s))$$

$$\times e^{-[\eta_k(s,z) - \xi_k(s,z)]\theta_k} \int\limits_{\theta_k}^{\infty} [1 - B_k(x)]e^{-\xi_k(s,z)}dx.$$

For schemes I.2

$$h_k(z,s) = \{z_k[1 + \sigma_{k-1}(\pi_{k-1}(z,s) + \pi_{k-1}(s + \omega_k)\nu_k(z,s))]$$

$$\times \int\limits_0^{\theta_k} [1 - B_k(x)]e^{-(s+\omega_k+\sigma_{k-1})x}dx + z_k(1 + \sigma_{k-1}\overline{\pi}_{k-1}(z,s))$$

$$\times e^{-[\eta_k(s,z) - \xi_k(s,z)]\theta_k} \int\limits_{\theta_k}^{\infty} [1 - B_k(x)]e^{-\xi_k(s,z)}dx\}$$

$$\times \{1 - \sigma_{k-1}\pi_{k-1}(s + \omega_k) \int\limits_0^{\theta_k} [1 - B_k(x)]e^{-(s+\omega_k+\sigma_{k-1})x}dx\}^{-1}.$$

where

$$\xi(s,z) = s + \omega_k + \sigma_{k-1}[1 - \overline{\pi}_{k-1}(s + \omega_k)],$$
$$\eta_k(s,z) = s + \omega_k + \sigma_{k-1}[1 - \pi_{k-1}(s + \omega_k)];$$

$\nu_k(z,s)$ for each of schemes are equal respectively:
for schemes 1.J

$$\nu_k(z,s) = \frac{1 - c_k(s + \omega_k) + \sigma_{k-1}[1 + \sigma_{k-1}\pi_{k-1}(z,s)]}{s + \omega_k + \sigma_{k-1}[1 - c_k(s + \omega_k + \sigma_{k-1})]\pi_{k-1}(s + \omega_k)}, \tag{21}$$

for schemes 2.J

$$\nu_k(z,s) = (1 + \sigma_{k-1}\pi_{k-1}(z,s))\frac{1 - c_k(s + \omega_k + \sigma_{k-1}[1 - \pi_{k-1}(s + \omega_k)])}{s + \omega_k + \sigma_{k-1}[1 - \pi_{k-1}(s + \omega_k)]}, \tag{22}$$

for schemes 3.J

$$\nu_k(z,s) = (1 + \sigma_{k-1}\pi_{k-1}(z,s))$$
$$\times \int_0^\infty \frac{\{1 - e^{-(s + \omega_k + \sigma_{k-1})x}\} dC_k(x)}{s + \omega_k + \sigma_{k-1} - \sigma_{k-1}\pi_{k-1}(s + \omega_k)\nu_k(s + \omega_k)[1 - e^{-(s + \omega_k + \sigma_{k-1})x}]}. \tag{23}$$

where $\nu_k(s + \omega_k)$, $\pi_{k-1}(s + \omega_k)$ and $\overline{\pi}_{k-1}(s + \omega_k)$ are defined from (3)–(5) and (6)–(9) for $s = s + \omega_k$; $\pi_{k-1}(z,s)$, $\overline{\pi}_{k-1}(z,s)$ - from formulas (24)–(26), recurrently.

Theorem 3.
$$p(z,s) = \frac{1 + \sigma\pi(z,s)}{s + \sigma - \sigma\pi(s)}, \tag{24}$$

$$\sigma_k\pi_k(z,s) = \sigma_{k-1}\pi_{k-1}(z,s) + \gamma_{k-1}(s,z)\nu_k(z,s) + \frac{h_k(z,s)}{z_k - h_k(s + \omega_k)}$$
$$\times [\gamma_{k-1}(s,z)\nu_k(s + \omega_k)\sigma_{k-1}\pi_{k-1}(s + a_k) - \sigma_k\pi_k(s)], \tag{25}$$

$$\sigma_k\overline{\pi}_k(z,s) = \sigma_{k-1}\overline{\pi}_{k-1}(z,s) + h_k(z,s)\frac{\xi_k(s,z) - \xi_{k-1}(s,z)}{z_k - h_k(s + \omega_k)}, \tag{26}$$

where $\gamma_{k-1}(s,z) = \sigma_{k-1}[\overline{\pi}_{k-1}(s + \omega_k) - \pi_{k-1}(s + a_k)] + a_k z_k$, and $\sigma\pi(s) = \sigma_r\pi_r(s)$ are defined from theorem 1.

Let denote through $P(z)$ the generating function of joint distribution of queue length in stationary state. If $\rho_r < 1$

$$P(z) = \frac{1 + \sigma\widehat{\pi}(z)}{1 + \sigma\pi_1}, \text{ where } \sigma\widehat{\pi}(z) = \sigma_r\pi_r(z,0), \quad \sigma\pi_1 = \sigma_r\pi_{r1}.$$

3.2   Relative DD Discipline

Let's consider

$$\omega_k = \nu_k = a_k(1 - z_k) + \cdots + a_r(1 - z_r), \quad u_k = (a_1 z_1 + \cdots + a_k z_k)/\sigma_k.$$

Lemma 9. For schemes I.1.

$$h_k(z,s) = \{z_k \int_0^{\theta_k} e^{-(s+v_1)x}[1 - B_k(x)]dx$$

$$+[1 + \sigma_{k-1}\pi_{k-1}(z,s)]z_k e^{\sigma_{k-1}u_{k-1}\theta_k} \int_{\theta_k}^{\infty} e^{-(s+v_k+\sigma_k)x}[1 - B_k(x)]dx$$

$$+v_k(z,s)\sigma_{k-1}\pi_{k-1}(s+v_k)z_k e^{\sigma_{k-1}\overline{\pi}_{k-1}(s+v_k)\theta_k} \int_{\theta_k}^{\infty} e^{-(s+v_k+\sigma_k)x}[1 - B_k(x)]dx$$

$$+\frac{\pi_{k-1}(z,s)}{u_{k-1} - \overline{\pi}_{k-1}(s+v_k)}[\int_0^{\theta_k} (e^{-(s+v_k)x} - e^{-\gamma_{k-1}(s)x})dB_k(x) + (e^{\sigma_{k-1}u_{k-1}\theta_k}$$

$$-e^{\sigma_{k-1}\overline{\pi}_{k-1}(s+v_k)\theta_k}) \int_{\theta_k}^{\infty} e^{-(s+v_k+\sigma_{k-1})x}dB_k(x) + \sigma_{k-1}\pi_{k-1}(s+v_k)z_k(e^{\sigma_{k-1}u_k\theta_k}$$

$$-e^{\sigma_{k-1}\overline{\pi}_{k-1}(s+v_k)\theta_k}) \int_{\theta_k}^{\infty} e^{-(s+v_k+\sigma_{k-1})x}[1 - B_k(x)]dx]\}/R_k(s+v_k,\theta_k).$$

For schemes I.2.

$$h_k(z,s) = \{z_k \int_0^{\theta_k} e^{-(s+v_1)x}[1 - B_k(x)]dx + [1 + \sigma_{k-1}\pi_{k-1}(z,s)]z_k(e^{\sigma_{k-1}u_{k-1}\theta_k}$$

$$-e^{\sigma_{k-1}\overline{\pi}_{k-1}(s+v_k)\theta_k}) \int_{\theta_k}^{\infty} e^{-(s+v_k+\sigma_{k-1})x}[1 - B_k(x)]dx + [1 + \sigma_{k-1}\pi_{k-1}(z,s)$$

$$+\sigma_{k-1}\pi_{k-1}(s+v_k) - v_k(z,s)]z_k(e^{-(\overline{\gamma}_{k-1}(s,z)-\gamma_{k-1}(s,z))\theta_k} \int_{\theta_k}^{\infty} e^{-\gamma_{k-1}(s,z)x}[1 - B_k(x)]dx$$

$$+\frac{\overline{\pi}_{k-1}(z,s)}{u_{k-1} - \overline{\pi}_{k-1}(s+v_k)}[\int_0^{\theta_k} (e^{-(s+v_k)x} - e^{-\overline{\gamma}_{k-1}(s,z)x})dB_k(x)$$

$$+(e^{\sigma_{k-1}u_{k-1}\theta_k} - e^{\sigma_{k-1}\overline{\pi}_{k-1}(s+v_k)\theta_k}) \int_{\theta_k}^{\infty} e^{-(s+v_k+\sigma_{k-1})x}dB_k(x)$$

$$+\sigma_{k-1}\pi_{k-1}(s+v_k)z_k(e^{\sigma_{k-1}u_{k-1}\theta_k} - e^{\sigma_{k-1}\overline{\pi}_{k-1}(s+v_k)\theta_k}) \int_{\theta_k}^{\infty} e^{-(s+v_k+\sigma_{k-1})x}[1 - B_k(x)]dx],$$

where $\overline{\gamma}_{k-1}(s,z) = \overline{\gamma}_{k-1}(s+v_k)$, $\gamma_{k-1}(s,z) = \gamma_{k-1}(s+v_k)$; $v_k(z,s)$ are defined from relations (21)–(23), $\pi_{k-1}(z,s)$ and $\overline{\pi}_{k-1}(z,s)$ are recurrently defined from below-mentioned formulas.

Theorem 4.
$$p(z,s) = \frac{1+\sigma\pi(z,s)}{s+\sigma-\sigma\pi(s)},$$

$$\sigma_k\overline{\pi}_k(z,s) = \sigma_{k-1}\overline{\pi}_{k-1}(z,s) + h_k(z,s)\frac{\overline{\gamma}_k(s,z) - \overline{\gamma}_{k-1}(s,z)}{z_k - h_k(s+v_k)},$$

$$\sigma_k\pi_k(z,s) = \sigma_{k-1}\pi_{k-1}(z,s) + h_k(z,s)[\sigma_{k-1}\pi_{k-1}(s+v_k) + \sigma_k\pi_k(s+v_{k+1})$$
$$+a_kz_kv_k(s+v_k)]\{z_k - h_k(s+v_k)\}^{-1} + a_kz_kv_k(z,s).$$

Remark 3. Similar to lemmas 8 and 9, the relations for schemes with "loss" of a request are obtained.

4 Probability of States

We'll consider that the system is in $\rightarrow j$ - state at the t moment, if at that moment takes place the switching to j class of priority; the system is in $*j$ - state, if at that moment takes place the service of the request of j class $(j = 1,\dots,r)$. We'll denote by $\rightarrow P_j(t)$ and $*P_j(t)$, respectively, the probabilities of those states, and by $\rightarrow p_j(s)$ and $*p_j(s)$ - their Laplace transforms.

4.1 Absolute DD Discipline

The following Lemmas 11–14 allow to calculate the Laplace transform of $\overrightarrow{j}h_k(s)$ and $\overset{*}{j}h_k(s)$ probabilities $\rightarrow j$ - and $*j$ - states on a separate k–cycle of service.

Lemma 11. For schemes I.1 for $j < k$

$$\overrightarrow{j}h_k(s) = \sigma_{k-1}[\pi_{k-1}(s)\overrightarrow{j}v_k(s) +_j \pi_{k-1}(s)]\int\limits_0^{\theta_k}[1 - B_k(x)]e^{-(s+\sigma_{k-1})x}dx$$

$$+\sigma_{k-1}\overrightarrow{j}\overline{\pi}_{k-1}(s)e^{-\sigma_{k-1}[\overline{\pi}_{k-1}(s)-\pi_{k-1}(s)v_k(s)]\theta_k}\int\limits_{\theta_k}^{\infty}[1-B_k(x)]e^{-(s+\sigma_{k-1}[1-\pi_{k-1}])x}dx,$$

and for $j = k$

$$\overrightarrow{k}h_k(s) = \int\limits_0^{\theta_k}[1 - B_k(x)]e^{-(s+\sigma_{k-1})x}dx\sigma_{k-1}\pi_{k-1}(s)\overrightarrow{k}v_k(s),$$

where, for each of schemes, $v_k(s)$ is determined from relations (3)–(5); $\overline{\pi}_{k-1}(s)$ and $\pi_{k-1}(s)$ - from (6)–(9); $\overrightarrow{j}v_k(s)$ - from [6]; $\overrightarrow{j}\pi_{k-1}(s)$ and $\overrightarrow{j}\overline{\pi}_{k-1}(s)$ - from lemma 15.

Lemma 12. For schemes I.2 for $j < k$

$$\vec{j}\, h_k(s) = \left\{ 1 - \sigma_{k-1}\pi_{k-1}(s)\nu_k(s) \int_0^{\theta_k} [1 - B_k(x)]e^{-(s+\sigma_{k-1})x}dx \right\}^{-1}$$

$$\times \left\{ \sigma_{k-1}[\pi_{k-1}(s)\vec{j}\,\nu_k(s) + \vec{j}\,\pi_{k-1}(s)] \int_0^{\theta_k} [1 - B_k(x)]e^{-(s+\sigma_{k-1})x}dx \right.$$

$$+\sigma_{k-1}\vec{j}\,\pi_{k-1}(s)e^{-\sigma_{k-1}[\overline{\pi}_{k-1}(s)-\pi_{k-1}(s)\nu_k(s)]\theta_k}$$

$$\times \int_{\theta_k}^{\infty} [1 - B_k(x)]e^{-(s+\sigma_{k-1}[1-\overline{\pi}_{k-1}(s)])x} \Big\},$$

and for $j = k$

$$\vec{k}\, h_k(s) = \frac{\sigma_{k-1}\pi_{k-1}(s)\vec{k}\,\nu_k(s) \int_0^{\theta_k} [1 - B_k(x)]e^{-(s+\sigma_{k-1})x}dx}{1 - \sigma_{k-1}\pi_{k-1}(s)\nu_k(s) \int_0^{\theta_k} [1 - B_k(x)]e^{-(s+\sigma_{k-1})x}dx},$$

where, $\vec{j}\,\overline{\pi}_{k-1}(s)$ and $\vec{j}\,\pi_{k-1}(s)$ are determined from lemma 15.

Lemma 13. For schemes I.1 for $j < k$

$$\overset{*}{j}h_k(s) = \sigma_{k-1}[\pi_{k-1}(s)\overset{*}{j}\nu_k(s) + \overset{*}{j}\,\pi_{k-1}(s)] \int_0^{\theta_k} [1 - B_k(x)]e^{-(s+\sigma_{k-1})x}dx$$

$$+\sigma_{k-1}\overset{*}{j}\overline{\pi}_{k-1}(s)e^{-\sigma_{k-1}[\overline{\pi}_{k-1}(s)-\pi_{k-1}(s)\nu_k(s)]\theta_k} \int_{\theta_k}^{\infty} [1 - B_k(x)]e^{-(s+\sigma_{k-1}[1-\overline{\pi}_{k-1}])x}dx,$$

and for $j = k$

$$\overset{*}{k}h_k(s) = \int_0^{\theta_k} [1 - B_k(x)]e^{-(s+\sigma_{k-1})x}dx$$

$$+e^{-\sigma_{k-1}[\overline{\pi}_{k-1}(s)-\pi_{k-1}(s)\nu_k(s)]\theta_k} \int_{\theta_k}^{\infty} [1 - B_k(x)]e^{-(s+\sigma_{k-1}[1-\overline{\pi}_{k-1}])x}dx,$$

where, $\overset{*}{j}\overline{\pi}_{k-1}(s)$ and $\overset{*}{j}\pi_{k-1}(s)$ are determined from lemma 16.

Lemma 14. For schemes I.2, for $j < k$

$$_j^* h_k(s) = \left\{ 1 - \sigma_{k-1}\pi_{k-1}(s)\nu_k(s) \int_0^{\theta_k} [1 - B_k(x)]e^{-(s+\sigma_{k-1})x}dx \right\}^{-1}$$

$$\times \left\{ \sigma_{k-1}[\pi_{k-1}(s)_j^*\nu_k(s) + {}_j^*\pi_{k-1}(s)] \int_0^{\theta_k} [1 - B_k(x)]e^{-(s+\sigma_{k-1})x}dx \right.$$

$$\left. + \sigma_{k-1}{}_j^*\pi_{k-1}(s)e^{-\sigma_{k-1}[\overline{\pi}_{k-1}(s)-\pi_{k-1}(s)\nu_k(s)]\theta_k} \int_{\theta_k}^{\infty} [1 - B_k(x)]e^{-(s+\sigma_{k-1}[1-\overline{\pi}_{k-1}(s)])x}dx \right\},$$

and for $j = k$

$$_k^* h_k(s) = \left\{ 1 - \sigma_{k-1}\pi_{k-1}(s)\nu_k(s) \int_0^{\theta_k} [1 - B_k(x)]e^{-(s+\sigma_{k-1})x}dx \right\}^{-1}$$

$$\times \left\{ \int_0^{\theta_k} [1 - B_k(x)]e^{-(s+\sigma_{k-1})x}dx + e^{-\sigma_{k-1}[\overline{\pi}_{k-1}(s)-\pi_{k-1}(s)\nu_k(s)]\theta_k} \right.$$

$$\left. \times \int_{\theta_k}^{\infty} [1 - B_k(x)]e^{-(s+\sigma_{k-1}[1-\overline{\pi}_{k-1}(s)])x}dx \right\}.$$

4.2   Probabilities of the $^{\rightarrow}j-$ and $^*j$–States of the System

In this section, there are obtained $^{\rightarrow}p_j(s)$ and $^*p_j(s)$ - Laplace transform probabilities $^{\rightarrow}j-$ and $^*j$–states of the system for Absolute DD discipline. First of all, we'll formulate (according to lemmas 15 and 16) probability $^{\rightarrow}j-$ and $^*j$–state on separate $\overline{\Pi}_k$–period and separate Π_k–period.

Lemma 15.

$$\sigma_{kj}{}^{\rightarrow}\overline{\pi}_k(s) = \sum_{i=j}^{k} \frac{{}_j^{\rightarrow}h_i(s)}{1 - h_i(s)}[\sigma_{i-1}\overline{\pi}_{i-1}(s) - \sigma_i\overline{\pi}_i(s) + a_i];$$

$$\sigma_{kj}{}^{\rightarrow}\pi_k(s) = \sum_{i=j}^{k} \left\{ {}_j^{\rightarrow}\nu_k(s)\gamma_{i-1}(s) + \frac{{}_j^{\rightarrow}h_i(s)}{1 - h_i(s)}Q_i(s) \right\}; \tag{27}$$

where

$$Q_i(s) = \gamma_{i-1}(s) + \sigma_{i-1}\pi_{i-1}(s + a_i) - \sigma_i\pi_i(s);$$

$$\gamma_{i-1}(s) = \sigma_{i-1}[\pi_{i-1}(s) - \pi_{i-1}(s + a_i)] + a_i.$$

Lemma 16.

$$\sigma_{kj}^{*}\overline{\pi}_k(s) = \sum_{i=j}^{k} \frac{_{j}^{*}h_i(s)}{1 - h_i(s)} \left[\sigma_{i-1}\overline{\pi}_{i-1}(s) - \sigma_i\overline{\pi}_i(s) + a_i\right]; \qquad (28)$$

$$\sigma_{kj}^{*}\pi_k(s) = \frac{_{j}^{*}h_j(s)}{1 - h_j(s)}Q_j(s) + \sum_{i=j+1}^{k} \left\{_{j}^{*}\nu_i(s)\gamma_{i-1}(s) + \frac{_{i}^{*}h_i(s)}{1 - h_i(s)}Q_i(s)\right\}.$$

Theorem 5. Laplace transform of the probability $^{\rightarrow}j$ - state of the system is determined from the expression

$$^{\rightarrow}p_j(s) = \frac{\sum_{i=j}^{r} \left\{_{j}^{\rightarrow}\nu_i(s)\gamma_{i-1}(s) + \frac{_{i}^{\rightarrow}h_i(s)}{1-h_i(s)}Q_i(s)\right\}}{s + \sigma - \sigma\pi(s)},$$

where $\sigma\pi(s) = \sigma_r\pi_r(s)$ is determined from relations (6)–(9).

Theorem 6. Laplace transform of the probability $^{*}j$ - state of the system is determined from the expression

$$^{*}p_j(s) = \frac{\frac{_{j}^{*}h_j(s)}{1-h_j(s)}Q_j(s) + \sum_{i=j+1}^{r} \left\{_{j}^{*}\nu_i(s)\gamma_{i-1}(s) + \frac{_{i}^{*}h_i(s)}{1-h_i(s)}Q_i(s)\right\}}{s + \sigma - \sigma\pi(s)}.$$

The proof of theorems 5 and 6 result from the formula $p_j(s) = \frac{\sigma_j\pi(s)}{s+\sigma-\sigma\pi(s)}$, which is fair for $^{\rightarrow}j$ and $^{*}j$ - state and for (27) and (28), lemmas 15 and 16.

Acknowledgments. This work is supported partially by 13.820.08.06 STCU.F/5854 project.

References

1. Jaiswal, N.K.: Priority Queues. Mir, Moscow (1973). (in Russian)
2. Bogunovic, N.: Process scheduling procedure for a class of real-time computer system. J. IEEE Trans. Ind. Electron. **34**(1), L29–34 (1987)
3. Volkovinski, M.I., Kabalevsky, A.N.: The service with mixed priorities in the systems with switching losses. Autom. Remote Control **11**, 16–22 (1975). (in Russian)
4. Mishkoy, GhK: On mixed priority service problem in systems with switching losses. Autom. Remote Control **2**, 15–21 (1978). (in Russian)
5. Dragalin, V.P., Mishkoy, GhK: The service with mixed priority and switchings. Bull. URSS Acad. Sci. **3**, 166–172 (1984). (in Russian)
6. Mishkoy, GhK: Generalized Priority Systems. Stiinta, Academy of Sciences of Moldova, Chisinau (2009). (in Russian)
7. Gnedenko, B.V., Danielian, E.A., Dimitrov, B.N., Klimov, G.P., Matveev, V.F.: Priority Queueing Systems. Moscow State University Press, Moscow (1973). (in Russian)

Multi-rate Loss Model for Optical Network Unit in Passive Optical Networks

Gelii Basharin$^{(\boxtimes)}$ and Nadezhda Rusina$^{(\boxtimes)}$

Peoples' Friendship University of Russia, Moscow, Russia
gbasharin@sci.pfu.edu.ru, rusina_nadezda@inbox.ru

Abstract. Nowadays, extensive research on passive optical network (PON) technology is being conducted. PON technology facilitates the use of inexpensive passive components and helps reduce the total length of optical fiber required. PON is a future optical technology that enables high-speed data transfer of multiservice traffic using optical fibers. This paper is concerned with multi-rate loss model for optical network units in PON like a queue system with a limited capacity buffer. The results obtained in the paper are applied to the calculation of the quality of service parameters of the network.

Keywords: Passive optical network (PON) · Optical line terminal (OLT) · Optical network unit (ONU) · Passive optical splitter/combiner (PO-SC) · Upstream · Wavelength division multiplexing (WDM) · Time division multiple access (TDMA) · Blocking probability

1   Introduction

Workaut on passive optical network (PON) technology began in 1995 and was undertaken by the Full Service Access Network (FSAN) working group, which was formed by major telecommunication service providers and system vendors such as British Telecom, France Telecom, Deutsche Telecom, NTT, KPN, Telefonica, and Telecom Italia. The organization's purpose was to determine the fundamental concepts for PON technology standardization and introduce it into the market. The initial architecture of PON was formulated by the FSAN working group. The International Telecommunication Union (ITU) did further work, and standardized two generations of PON.

PON is an optical access architecture [1–4], which supports transmitting various classes of the network traffic (voice, data, and video) between optical line terminal (OLT) and optical network units (ONUs) through a passive optical splitter/combiner (PO-SC).

In the present paper, we propose an upstream traffic multiservice model considering the functioning process of ONUs. The functioning of an ONU is modeled by the step Markov process with the transition rate of each ONU from ON to OFF-state and *vice versa*. These results are used in the blocking probability analysis of the model.

V. Vishnevsky et al. (Eds.): DCCN 2013, CCIS 279, pp. 219–228, 2014.
DOI: 10.1007/978-3-319-05209-0_19, © Springer International Publishing Switzerland 2014

2   PON Architecture and Principles of Operation

PON is a point-to-multipoint (PtMP) optical network that transfers different traffic classes between OLT and ONUs with no active elements in the signal path from source to destination (Fig. 1). A PON employs PO-SC to split an optical signal (power) from one fiber into several fibers and reciprocally, to combine optical signals from multiple fibers into one. The OLT resides in the central office (CO) and connects the optical access network to the metropolitan area network (MAN) or wide-area network (WAN), which is also known as backbone or long-haul network. The ONU is located either at the end-user location (fiber to the home (FTTH), fiber to the building (FTTB)), or at the curb (fiber to the curb (FTTC) architecture). PONs allow for long reach between the CO and customer premises, operating at distances of up to 20 km [4,5].

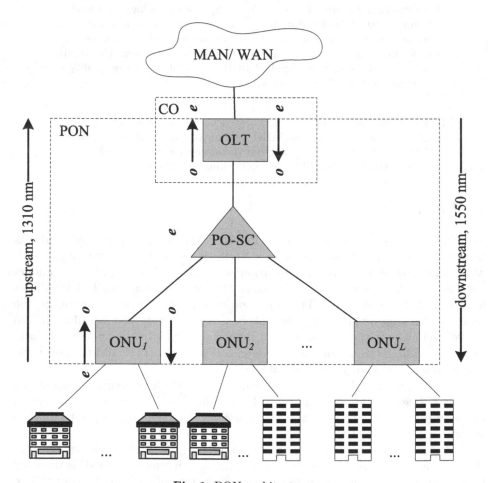

Fig. 1. PON architecture

A PO-SC broadcasts traffic from the OLT to ONUs (downstream direction) and transmits traffic from the ONUs to the OLT (upstream direction) through one fiber. PON employs different wavelengths for each direction to transmit downstream and upstream traffic through one fiber.

It can use wavelength division multiplexing (WDM) technology to transmit traffic in one direction. WDM PON employs a set of wavelengths for each direction

$$W_{\mathrm{up}} := \{\omega_{\mathrm{up},i}| \, 1 \le i < \infty\}, W_{\mathrm{down}} := \{\omega_{\mathrm{down},j}| 1 \le j < \infty\}. \tag{1}$$

Moreover, in WDM PON, the PO-SC is replaced by an array waveguide grating (AWG). This passive optical device (de)multiplexes a number of wavelengths in order for multiple wavelengths to be used either in the upstream or downstream direction.

It is economical to use one wavelength to transmit upstream traffic and another wavelength to transmit downstream traffic [4,5]

$$W_{\mathrm{up}} := \{1310\,\mathrm{nm}\}, W_{\mathrm{down}} := \{1550\,\mathrm{nm}\}. \tag{2}$$

According to this approach PON employs the time division multiple access (TDMA) scheme [4–6] to transmit upstream traffic in one direction. Although downstream traffic can be accepted by all ONUs, it is accepted only by the appropriate ONU owing to the packet header.

In TDMA PON technology (Fig. 2) each ONU uses its own frame, which involves several time-slots. ONU has to buffer user data until its frame becomes available. When its frame becomes available, ONU starts to transmit accumulated data at the full channel rate. Similarly, the OLT has to buffer user data until ONU's frame becomes available. When the frame becomes available, OLT starts to transmit accumulated data at the full channel rate.

Hence ONU may be in the ON-state, in other words, it is active and transmits data to and/or from the OLT through its earlier assigned frame, or the ONU may be in the OFF-state, in other words, it is in sleep period, when no data transmission to and/or from OLT occurs.

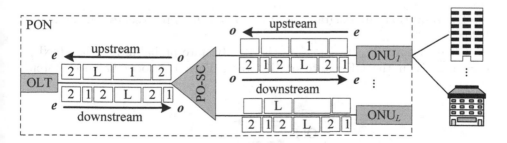

Fig. 2. TDMA PON

Looking ahead, the paper is concerned with upstream traffic transmission. The downstream direction can be analyzed similarly.

3   Functional Model of ONU in PON

Examine a WDM-TDMA passive optical sub-network [7] involving L ONUs. The upstream traffic transmission goes through several wavelengths

$$W_{up} := \{\omega_{up,i}|\ i = \overline{1,F}\}, F \le L. \tag{3}$$

The ONUs' function is described as a one-dimensional stationary Markov process $\mathbf{X}(t) = (X_l(t))_{l=\overline{1,L}}$, where $X_l(t)$ is the state of the ONU$_l$ at the moment $t > 0$. Then,

1. $X_l(t) = 1$ specifies the ON-state for ONU$_l$, $l = \overline{1,L}$;
2. $X_l(t) = 0$ specifies the OFF-state for ONU$_l$, $l = \overline{1,L}$.

The state vector of the system

$$\mathbf{n} := (n_l)_{l=\overline{1,L}}, n_l \in \{0,1\}. \tag{4}$$

The state space of the system

$$\Omega := \{\mathbf{n}|\ n_\bullet \in \{0,1,\ldots,F\}\}, n_\bullet := \mathbf{1}^T\mathbf{n} = \sum_{i=1}^{L} n_i. \tag{5}$$

Then,

$$|\Omega| = 1 + \sum_{f=1}^{F}\binom{L}{f} = \sum_{f=0}^{F}\binom{L}{f}. \tag{6}$$

For the same TDMA passive optical sub-network (the special case, when $F = 1$), when only one ONU can transmit upstream and downstream traffic in a time-slot or no single ONU can, then

$$|\Omega| = L + 1. \tag{7}$$

The transition rate diagram for ONU$_l$, $l = \overline{1,L}$, is represented in the figure below (Fig. 3).

Let us denote $P_{l,\text{off}}$ and $P_{l,\text{on}}$ as the steady-state probabilities when ONU$_l$, $l = \overline{1,L}$, is in the OFF-state or ON-state. Then the local balance system

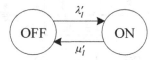

Fig. 3. State transition diagram for ONU$_l$, $l = \overline{1,L}$

of equations for the process, which describes the state of an ONU_l can be written as

$$P_{l,\text{off}} \cdot \lambda_l' = P_{l,\text{on}} \cdot \mu_l', l = \overline{1, L}. \tag{8}$$

Then steady-state probabilities can be calculated as

$$P_{l,\text{off}} = \tfrac{1}{1+\rho_l'} := 1 - \alpha_l , \\ P_{l,\text{on}} = \tfrac{\rho_l'}{1+\rho_l'} := \alpha_l , \tag{9}$$

where $\rho_l' := \tfrac{\lambda_l'}{\mu_l'}, l = \overline{1, L}$.

The examples of the process transitions $X(t)$ are represented in the figure below (Fig. 4).

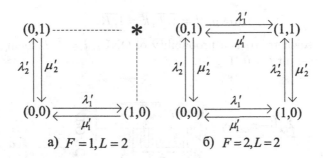

a) $F = 1, L = 2$ б) $F = 2, L = 2$

Fig. 4. The examples of the process transitions $X(t)$

The condition performs for the system

$$\alpha_\bullet := \sum_{l=1}^{L} \alpha_l \le F. \tag{10}$$

Then, for $F = 1$,

$$\alpha_\bullet \le 1. \tag{11}$$

4   Upstream Traffic Model

Let us consider the upstream traffic transmitting process in a TDMA passive optical sub-network involving L ONUs. Every ONU_l has a finite memory buffer of size R_l time-slots, $0 < R_l < \infty$, $l = \overline{1, L}$. The analyzed queue system supports K types of service classes. Service class k calls arrive at an ONU_l according to a Poisson process with mean arrival rate $\lambda_{l,k}$, $0 < \lambda_{l,k} < \infty$, $l = \overline{1, L}$, $k = \overline{1, K}$. Arrival streams are independent in total for each ONU_l. The service of a service class k call is realized by allocating a specific number of time slots, b_k, $k = \overline{1,K}$, in a frame for the entire duration of the service. In addition, the k call holds b_k time slots in the buffer until it is serviced, then it unbuffers immediately after the completion of service, thus idling the wavelength.

A new service class k call, $k = \overline{1,K}$, will be rejected by the ONU$_l$, $l = \overline{1,L}$, if there are no more than $R_l - b_k$ free time slots in the buffer. This k call leaves the system and has no effect on the stream arrival rate.

μ_k, $k = \overline{1,K}$ is the mean service time of service class k call in ONU$_l$, which is exponentially distributed. However, we have to take into consideration the functional process of ONU$_l$, $l = \overline{1,L}$, which transmits traffic through the assigned frames. It defines that ONU$_l$ can be in a state, when there is no transmission. Then the k call service rate, taking into consideration the ONU$_l$ functioning, will be

$$\alpha_l \cdot \mu_k, l = \overline{1,L}, k = \overline{1,K}. \tag{12}$$

Here and elsewhere, α_l is a probability of ONU$_l$, $l = \overline{1,L}$, being in an ON-state at some instant $t > 0$ (Fig. 5).

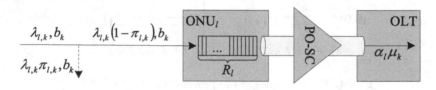

Fig. 5. Upstream traffic model for ONU$_l$, $l = \overline{1,L}$, and k call, $k = \overline{1,K}$

Then it is necessary to define the next functioning characteristic for the formal definition of the model

1. $\mathbf{R} := (R_l)_{l=\overline{1,L}}$ is the vector of the buffers size,
2. $\mathbf{\Lambda} := (\lambda_{l,k})_{l=\overline{1,L},k=\overline{1,K}}$ is the matrix of arrival rates,
3. $\mathbf{b} := (b_k)_{k=\overline{1,K}}$ is the vector of time slots amount necessary for k calls service,
4. $\boldsymbol{\mu} := (\mu_k)_{k=\overline{1,K}}$ is the vector of service rates,
5. $\boldsymbol{\alpha} := (\alpha_l)_{l=\overline{1,L}}$ is the vector of probabilities that ONUs are in an ON-state.

5   Stationary Markov Process and State Space Description

The queue system functioning is described by a two-dimensional stationary Markov process $\mathbf{Y} := (Y_{l,k}(t))_{l=\overline{1,L},k=\overline{1,K}}$, $t > 0$, where $Y_{l,k}(t)$ is the number of k calls in an ONU$_l$, at some instant $t > 0$.

The state matrix of the system

$$\mathbf{M} := (m_{l,k})_{l=\overline{1,L},k=\overline{1,K}}, m_{l,k} \in \left\{0, 1, \ldots, \left\lfloor \frac{R_l}{b_k} \right\rfloor \right\}. \tag{13}$$

The state space of the system

$$S := \{\mathbf{M} | 0 \leq \sum_{k=1}^{K} b_k m_{l,k} \leq R_l, l = \overline{1,L}\}. \tag{14}$$

The state subspace of k calls received by ONU_l

$$S_{l,k} := \{\mathbf{M} \in S | \sum_{k=1}^{K} b_k m_{l,k} \leq R_l - b_k\}, k = \overline{1,K}, l = \overline{1,L}. \tag{15}$$

The state subspace of k calls blocked by ONU_l

$$\bar{S}_{l,k} := \{\mathbf{M} \in S | \sum_{k=1}^{K} b_k m_{l,k} > R_l - b_k\}, k = \overline{1,K}, l = \overline{1,L}. \tag{16}$$

The transition rate diagram for any ONU_l, $l = \overline{1,L}$, is shown in the figure below (Fig. 6).

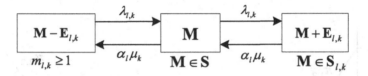

Fig. 6. The state transition scheme for ONU_l, $l = \overline{1,L}$, and k call, $k = \overline{1,K}$

The local balance system of equations for each k call, and any ONU_l is

$$p(\mathbf{M}) \cdot \alpha_l \cdot \mu_k = p(\mathbf{M} - \mathbf{E}_{l,k}) \cdot u(m_{l,k}) \cdot \lambda_{l,k}, \mathbf{M} \in S, l = \overline{1,L}, k = \overline{1,K}. \tag{17}$$

Then,

$$p(\mathbf{M}) = p(0) \cdot \prod_{l=1}^{L} \frac{1}{\alpha_l^{m_{l,\bullet}}} \prod_{k=1}^{K} \rho_{l,k}^{m_{l,k}}, \frac{1}{p(0)} = \sum_{\mathbf{M} \in S} \prod_{l=1}^{L} \frac{1}{\alpha_l^{m_{l,\bullet}}} \prod_{k=1}^{K} \rho_{l,k}^{m_{l,k}}, \tag{18}$$

where $\mathbf{M} \in S, \rho_{l,k} := \frac{\lambda_{l,k}}{\mu_k}, k = \overline{1,K}, m_{l,\bullet} := \sum_{k=1}^{K} m_{l,k}, l = \overline{1,L}$.
Then, the blocking probability of k calls for ONU_l

$$\pi_{l,k} = \sum_{\mathbf{M} \in \bar{S}_{l,k}} p(\mathbf{M}), k = \overline{1,K}, l = \overline{1,L}. \tag{19}$$

Table 1. The model parameters

Parameters	Value
K	2
$\mathbf{b}^T$	$(1, 2)$
L	2
$\mathbf{R}^T$	$(10, 10)$
α^T	$\alpha_1 + \alpha_2 = 1, \ 0.3 \leq \alpha_1 \leq 0.6$
Λ	$\begin{pmatrix} 0.8 & 0.16 \\ 1 & 0.3 \end{pmatrix}$
μ^T	$(4, 2)$
Blocking probability	$\pi_{l,k}, \ l, k = \overline{1, 2}$

6   Numerical Analysis Example

Let us examine the numerical analysis example of the upstream traffic mathematical model in TDMA PON. Parameter values are represented in the Table 1.

We analyze the chart $\pi_{l,k}, \ l, k = \overline{1, 2}$, from α_l. According to the diagram shown in figures below (Fig. 7, 8) we can improve the functioning of the ONU and reduce the blocking probability for each type of service class in every ONU configuring the parameter, which considers the ONU functioning process.

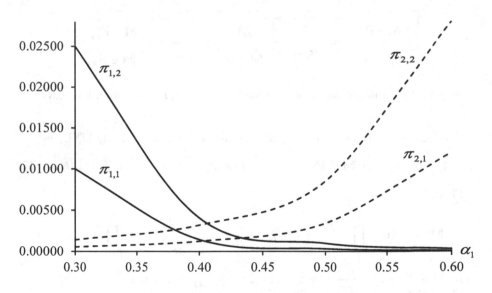

Fig. 7. The chart of blocking probabilities $\pi_{l,k}, \ l, k = \overline{1, 2}$, from α_1

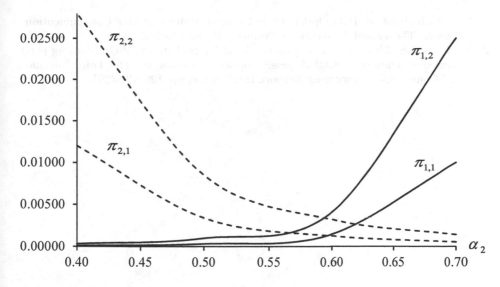

Fig. 8. The chart of blocking probabilities $\pi_{l,k}$, $l,k = \overline{1,2}$, from α_2

7   Conclusion

In this paper, novel formulas are presented to calculate blocking probabilities for the mathematical model of upstream traffic transfer in TDMA PON. In addition, there is the numerical analysis example in which the role of α_l, $l = \overline{1,L}$, is shown to select the optimal functioning behavior of the network.

The authors recommend the use of the concepts described in this paper for analysis of the upstream traffic blocking probability in OCDMA PON [8,9].

References

1. Greenfield, D.: The Essential Guide to Optical Networks. Prentice Hall, Upper Saddle River (2002)
2. Basharin, G.P.: Lectures on mathematical teletraffic theory. The 3rd publication, PFUR, Moscow (2009)
3. Naymov, V.A., Samyilov, K.Y., Yarkina, N.V.: Teletraffic theory of multiservice networks. Monography. PFUR, Moscow (2008)
4. Mukherjee, B.: Optical WDM Networks. Springer, Berlin (2006)
5. Semenov, J.A.: Telecommunications technology. http://book.itep.ru/4/41/pon.htm
6. Li, C.P., Neely, M.J.: Exploiting channel memory for multi-user wireless scheduling without channel measurement: capacity regions and algorithms. In: WIOPT, pp. 1–10 (2010)
7. Vardakas, J.S., Moscholios, I.D., Logothetis, M.D., Stylianakis, V.G.: An analytical approach for dynamic wavelength allocation in WDM-TDMA PONs servicing ON-OFF traffic. IEEE/OSA J. Opt. Commun. Netw. **3**(4), 347–358 (2011)

8. Yin, H., Richardson, D.J.: Optical Code Division Multiple Access Communication Networks. Theory and Applications. Springer, Berlin (2009)
9. Vardakas, J.S., Moscholios, I.D., Logothetis, M.D., Stylianakis, V.G.: Blocking performance of multi-rate OCDMA passive optical networks. In: The Third International Conference on Emerging Network Intelligence, pp. 125–130 (2011)

Investigation of the Queueing Network $GI - (GI|\infty)^K$ by Means of the First Jump Equation and Asymptotic Analysis

Anatoly Nazarov and Alexander Moiseev[✉]

Tomsk State University, Tomsk, Russian Federation
{nazarov.tsu,moiseev.tsu}@gmail.com

Abstract. Analysis of the open non-Markovian queueing network with renewal arrival process, Markovian routing, infinite servers count and general service time distribution is presented in the paper. Equation for characteristic function of the joint distribution of the number of customers in the nodes of the network is derived. Approximations for this characteristic function are presented under asymptotic condition of an infinite growth of the arrival rate. Both stationary and non-stationary cases are considered. It is shown that the multi-dimensional distributions under study can be approximated by the multi-dimensional Gaussian distributions. Expressions for parameters of these distributions are obtained.

Keywords: Queueing network · High-intensive arrival process · Renewal process

1 Introduction

Queueing network models [1–3] are widely applicable in analysis and design of different technical, economical, transport and other systems. One of the most important directions of application of these models is a solution of the problems related to telecommunication networks and distributed data processing systems. At present, nodes of such systems contain a large number of servers and data transmission channels between nodes a have high capacity. This allows to transfer and to process a large number of data packages per time unit. In terms of the Queueing Theory, such systems can be represented as open queueing networks with infinite-servers at each node and high-intensive [4] arrival process.

In the paper, asymptotic analysis of such queueing network is presented in a condition of infinite growth of the arrival rate. In [5], similar results were obtained for the same model by means of an original method called as *the multi-dimensional dynamic screening method*. In this paper we propose another approach for the queueing networks analysis based on constructing the so-called *first jump equations* and their asymptotic solution.

V. Vishnevsky et al. (Eds.): DCCN 2013, CCIS 279, pp. 229–240, 2014.
DOI: 10.1007/978-3-319-05209-0_20, © Springer International Publishing Switzerland 2014

2   Mathematical Model

Let the queueing network have K nodes, renewal arrival process, an infinite number of servers at each node and i.i.d. service times. Let $A(x)$ be a distribution function of independent time periods between consequent arrivals of customers, $B_k(x)$ $(k = \overline{1,K})$ be a distribution function of a service time at the node k, v_k be a probability that just coming customer occupies a server at the node k, $r_{k\nu}$ $(k,\nu = \overline{1,K})$ be a probability that, after customer service is complete at the node k, this customer occupies for further service a server at the node ν, r_{k0} $(k = \overline{1,K})$ be a probability that after customer service completion at the node k this customer leaves the network. It's obvious that $\sum\limits_{k=1}^{K} v_k = 1$ and $r_{k0} + \sum\limits_{\nu=1}^{K} r_{k\nu} = 1$ for $k = \overline{1,K}$. We denote a vector $\boldsymbol{v}^{\mathrm{T}} = \{v_1,\ldots,v_K\}$ and a matrix $\boldsymbol{r} = \{r_{k\nu}\}_{k,\nu=\overline{1,K}}$.

Let $i_k(t)$ be a number of customers that occupy servers of the network node k at the moment t. Denote a vector $\boldsymbol{i}^{\mathrm{T}}(t) = \{i_1(t),\ldots,i_K(t)\}$. The goal of our investigations is to obtain of the characteristics of the K-dimensional stochastic process $\boldsymbol{i}(t)$ at arbitrary moment t.

3   The First Jump Equation

Let us denote a probability $P(\boldsymbol{i},t) = \mathrm{P}\{\boldsymbol{i}(t) = \boldsymbol{i}\}$. To obtain equations for probabilities $P(\boldsymbol{i},t)$, we use a first jump separation technique which was presented in [6,7].

This method is described as follows. Let a customer come to an empty network at the moment $t_0 = 0$. This customer will be special for us and we call it as *a first customer*. Denote by $S_k(t)$ a probability that the first customer occupies a server at the network node k at the moment $t > 0$. So, $S_0(t) = 1 - \sum\limits_{k=1}^{K} S_k(t)$ is a probability that a first customer has left the network before the moment t. We will calculate probabilities $S_k(t)$ later, and lets suggest here that they are known.

Applying a technique by [6,7], we obtain the following system of equations for probabilities $P(\boldsymbol{i},t)$:

$$P(\boldsymbol{0},t) = S_0(t) \int\limits_0^t P(\boldsymbol{0},t-x)\, dA(x) + S_0(t)\left[1 - A(t)\right], \qquad (1)$$

$$P(\boldsymbol{e}_k,t) = S_0(t) \int\limits_0^t P(\boldsymbol{e}_k,t-x)\, dA(x) +$$

$$+ S_k(t)\left[1 - A(t)\right] + S_k(t) \int\limits_0^t P(\boldsymbol{0},t-x)\, dA(x), k = \overline{1,K}, \qquad (2)$$

where $\mathbf{0}$ is a zero vector, $\mathbf{e}_k$ is a vector with all zero components except a component number k which is equal to 1.

For each vector $\mathbf{i}$, which is not equal to $\mathbf{0}$ or to $\mathbf{e}_k$ $(k = \overline{1, K})$, we get a last equation of the system:

$$P(\mathbf{i}, t) = S_0(t) \int_0^t P(\mathbf{i}, t - x) \, dA(x) + \sum_{k=1}^K S_k(t) \int_0^t P(\mathbf{i} - \mathbf{e}_k, t - x) \, dA(x). \quad (3)$$

We call the system of Eqs. (1)–(3) as *the first jump equations* for probability distribution $P(\mathbf{i}, t)$ of the process $\mathbf{i}(t)$.

Consider a characteristic function $H(\mathbf{u}, t)$ of the distribution of the K-dimensional stochastic process $\mathbf{i}(t)$ at the moment t

$$H(\mathbf{u}, t) = \sum_{i_1=0}^\infty \cdots \sum_{i_K=0}^\infty e^{ju_1 i_1 + \cdots + ju_K i_K} P(i_1, \ldots, i_K, t) = \sum_{i=0}^\infty e^{j\mathbf{u}^T \cdot \mathbf{i}} P(\mathbf{i}, t).$$

Here vector $\mathbf{u}^T = \{u_1, \ldots, u_K\}$, $j = \sqrt{-1}$ is an imaginary unit.

Theorem 1. *The characteristic function $H(\mathbf{u}, t)$ satisfies the first jump equation as follows:*

$$H(\mathbf{u}, t) = \left[S_0(t) + \sum_{k=1}^K S_k(t) e^{ju_k} \right] \left[1 - A(t) + \int_0^t H(\mathbf{u}, t - x) \, dA(x) \right]. \quad (4)$$

Let's prove this statement. From (1)–(3), we obtain expression

$$H(\mathbf{u}, t) = S_0(t) \int_0^t e^{j\mathbf{u}^T \cdot \mathbf{0}} P(\mathbf{0}, t - x) \, dA(x) + S_0(t) e^{j\mathbf{u}^T \cdot \mathbf{0}} [1 - A(t)] +$$

$$+ \sum_{k=1}^K S_0(t) \int_0^t e^{j\mathbf{u}^T \cdot \mathbf{e}_k} P(\mathbf{e}_k, t - x) \, dA(x) + \sum_{k=1}^K S_k(t) e^{j\mathbf{u}^T \cdot \mathbf{e}_k} [1 - A(t)] +$$

$$+ \sum_{k=1}^K S_k(t) \int_0^t e^{j\mathbf{u}^T \cdot \mathbf{e}_k} P(\mathbf{0}, t - x) \, dA(x) + \sum_{i > \mathbf{e}_k} S_0(t) \int_0^t e^{j\mathbf{u}^T \cdot \mathbf{i}} P(\mathbf{i}, t - x) \, dA(x) +$$

$$+ \sum_{i > \mathbf{e}_k} \sum_{k=1}^K S_k(t) \int_0^t e^{j\mathbf{u}^T \cdot \mathbf{i}} P(\mathbf{i} - \mathbf{e}_k, t - x) \, dA(x).$$

Suggesting that $P(\mathbf{i}, t) = 0$ for vectors $\mathbf{i}$ with negative components, we can reduce this formula to the form (4). $\qquad \square$

4   Calculation of the Probabilities $S_k(t)$

Each customer entering into the network moves from one node to another until it leaves the system. So its transitions between network nodes form a realization of some semi-Markovian stochastic process $k(t)$ with transient states $1, \ldots, K$ and an absorbing state 0 (a customer leaves the network). Suppose that the process $k(t)$ starts at the moment $t_0 = 0$. So we have the following initial conditions:

$$S_0(0) = 0, \quad S_k(0) = v_k, k = \overline{1, K}.$$

Consider the following conditional probabilities:

$$G_{\nu k}(t) = \mathrm{P}\{k(t) = k | k(0) = \nu\}, \; \nu, k = \overline{1, K}.$$

Denote by $\boldsymbol{G}(t)$ a matrix with entries $G_{\nu k}(t)$ ($\nu, k = \overline{1, K}$), by $\boldsymbol{B}(t)$ a diagonal matrix with diagonal entries $\boldsymbol{B}_\nu(t)$ ($\nu = \overline{1, K}$) and let $\boldsymbol{I}$ is an identity matrix.

Lemma 1. *The conditional probabilities $G_{\nu k}(t)$ ($\nu, k = \overline{1, K}$) satisfy the following matrix integral equation:*

$$\boldsymbol{G}(t) = \boldsymbol{I} - \boldsymbol{B}(t) + \boldsymbol{r} \int\limits_0^t \boldsymbol{G}(t - x) \, d\boldsymbol{B}(x). \tag{5}$$

The proof of the statement we derive by applying the same technique of a first jump separation [6]. Denote by ξ_ν a first moment when the process $k(t)$ changes its state. We obtain the following equations:

$$G_{\nu\nu}(t) = \mathrm{P}\{k(t) = \nu, \xi_\nu > t | k(0) = \nu\} + \mathrm{P}\{k(t) = \nu, \xi_\nu \le t | k(0) = \nu\} =$$

$$= \mathrm{P}\{\xi_\nu > t\} + \sum_{l=1}^K \int\limits_0^t \mathrm{P}\{k(x) = l, \xi_\nu \in [x, x + dx) | k(0) = \nu\} \cdot \mathrm{P}\{k(t) = \nu | k(x) = l\} =$$

$$= 1 - B_\nu(t) + \sum_{l=1}^K r_{\nu l} \int\limits_0^t G_{l\nu}(t - x) \, dB_\nu(x), \tag{6}$$

$$G_{\nu k}(t) = \mathrm{P}\{k(t) = k, \xi_\nu \le t | k(0) = \nu\} = \sum_{l=1}^K r_{\nu l} \int\limits_0^t G_{l k}(t - x) \, dB_\nu(x), \; k \ne \nu \tag{7}$$

And we can rewrite the Eqs. (6)–(7) in the matrix form (5). $\qquad\square$

We solve the matrix integral Eq. (5) by using the matrix Fourier and Fourier-Stieltjes transforms in the form $\boldsymbol{G}^*(\alpha) = \int\limits_0^\infty e^{j\alpha t} \boldsymbol{G}(t) \, dt$, $\boldsymbol{B}^*(\alpha) = \int\limits_0^\infty e^{j\alpha t} \, d\boldsymbol{B}(t)$. Applying a Fourier transform to (GMatr), we get the equation

$$\boldsymbol{G}^*(\alpha) = \int\limits_0^\infty e^{j\alpha t} \left[\boldsymbol{I} - \boldsymbol{B}(t) \right] dt + \int\limits_0^\infty e^{j\alpha t} \boldsymbol{r} \int\limits_0^t \boldsymbol{G}(t - x) \, d\boldsymbol{B}(x) \, dt =$$

$$= \frac{1}{j\alpha} [\boldsymbol{B}^*(\alpha) - \boldsymbol{I}] + \boldsymbol{B}^*(\alpha)\boldsymbol{r}\boldsymbol{G}^*(\alpha).$$

Solution of this equation is given by

$$\boldsymbol{G}^*(\alpha) = [\boldsymbol{I} - \boldsymbol{B}^*(\alpha)\boldsymbol{r}]^{-1} [\boldsymbol{B}^*(\alpha) - \boldsymbol{I}] \frac{1}{j\alpha}. \tag{8}$$

Let $\boldsymbol{S}^{\mathrm{T}}(t) = \{S_1(t), \ldots, S_K(t)\}$.

Lemma 2. *The probabilities $\boldsymbol{S}(t)$ can be calculated by the following way:*

$$\boldsymbol{S}^{\mathrm{T}}(t) = \frac{1}{2\pi} \boldsymbol{v}^{\mathrm{T}} \int\limits_{-\infty}^{\infty} e^{-j\alpha t} [\boldsymbol{I} - \boldsymbol{B}^*(\alpha)\boldsymbol{r}]^{-1} [\boldsymbol{B}^*(\alpha) - \boldsymbol{I}] \frac{1}{j\alpha} \, d\alpha. \tag{9}$$

The proof. According to the total probability formula, we have the following expression:

$$\boldsymbol{S}^{\mathrm{T}}(t) = \boldsymbol{v}^{\mathrm{T}} \boldsymbol{G}(t).$$

Applying a Fourier transform to this expression, we obtain a formula

$$\boldsymbol{S}^{*\mathrm{T}}(\alpha) = \boldsymbol{v}^{\mathrm{T}} \boldsymbol{G}^*(\alpha),$$

where $\boldsymbol{S}^*(\alpha)$ is a Fourier transform for vector $\boldsymbol{S}(t)$. Using expression (8), we get a formula

$$\boldsymbol{S}^{*\mathrm{T}}(\alpha) = \boldsymbol{v}^{\mathrm{T}} [\boldsymbol{I} - \boldsymbol{B}^*(\alpha)\boldsymbol{r}]^{-1} [\boldsymbol{B}^*(\alpha) - \boldsymbol{I}] \frac{1}{j\alpha}.$$

After inversion of the Fourier transform, we obtain the final expression for the probability vector $\boldsymbol{S}(t)$ in the form (9).   □

5   Investigation of the Network Behavior under a Condition of the High Arrival Rate

Equation (4) and formula (9) provide a tool for analysis of the non-Markovian networks of type $GI - (GI|\infty)^K$. Here we consider asymptotic [8] investigation for such network under a condition of the high arrival rate.

Lets obtain asymptotic expressions for function $H(\boldsymbol{u}, t)$ under a condition of an infinite growth of arrival rate [4]. Rate of the arrival process is represented in a form $N\lambda$, where $N > 0$ is a parameter which gets large values (asymptotically $N \to \infty$), and value of λ is defined as

$$\lambda = \frac{1}{\int\limits_{0}^{\infty} [1 - A(x)] \, dx} = \frac{1}{a},$$

where a is an expected value of the random variable defined by distribution function $A(x)$.

It was shown [4] that time periods between customers arrival in the input process with rate $N\lambda$ are random variables defined by the distribution function $A(Nx)$. So, for the network with high arrival rate the Eq. (4) gets a form

$$H(\boldsymbol{u}, t) = \left[S_0(t) + \sum_{k=1}^{K} S_k(t) e^{ju_k}\right] \left[1 - A(Nt) + \int_0^t H(\boldsymbol{u}, t - x) \, dA(Nx)\right].$$

(10)

5.1   The First-Order Asymptotic Form

We make the following substitutions in the (10):

$$\frac{1}{N} = \varepsilon, \qquad \boldsymbol{u} = \varepsilon \boldsymbol{w}, \qquad H(\boldsymbol{u}, t) = F(\boldsymbol{w}, t, \varepsilon)$$

(11)

and get the equation

$$F(\boldsymbol{w}, t, \varepsilon) = \left[S_0(t) + \sum_{k=1}^{K} S_k(t) e^{jw_k\varepsilon}\right] \left[1 - A\left(\frac{t}{\varepsilon}\right) + \int_0^t F(\boldsymbol{w}, t - x, \varepsilon) \, dA\left(\frac{x}{\varepsilon}\right)\right].$$

(12)

Lets prove the following statement for an asymptotic approximation $F(\boldsymbol{w}, t) = \lim_{\varepsilon \to 0} F(\boldsymbol{w}, t, \varepsilon)$.

Theorem 2. *The expression for the function $F(\boldsymbol{w}, t)$ has the following form:*

$$F(\boldsymbol{w}, t) = \exp\left\{\lambda j \boldsymbol{w}^{\mathrm{T}} \int_0^t \boldsymbol{S}(\tau) \, d\tau\right\}.$$

(13)

The proof. Making a substitution $z = x/\varepsilon$ in the integral at the formula (12) we get

$$F(\boldsymbol{w}, t, \varepsilon) = \left[S_0(t) + \sum_{k=1}^{K} S_k(t) e^{jw_k\varepsilon}\right] \left[1 - A\left(\frac{t}{\varepsilon}\right) + \int_0^{\frac{t}{\varepsilon}} F(\boldsymbol{w}, t - z\varepsilon, \varepsilon) \, dA(z)\right].$$

Lets use the following expansions:

$$e^{jw_k\varepsilon} = 1 + jw_k\varepsilon + \mathrm{O}\left(\varepsilon^2\right),$$

$$F(\boldsymbol{w}, t - z\varepsilon, \varepsilon) = F(\boldsymbol{w}, t, \varepsilon) - z\varepsilon \frac{\partial F(\boldsymbol{w}, t, \varepsilon)}{\partial t} + \mathrm{O}\left(\varepsilon^2\right),$$

where $\mathrm{O}\left(\varepsilon^2\right)$ is infinitesimal which has order of ε^2. So we obtain

$$F(\boldsymbol{w}, t, \varepsilon) = \left[S_0(t) + \sum_{k=1}^{K} S_k(t) \left(1 + jw_k\varepsilon\right)\right] \times$$

$$\times \left[1 - A\left(\frac{t}{\varepsilon}\right) + \int\limits_0^{\frac{t}{\varepsilon}} \left\{ F(\boldsymbol{w},t,\varepsilon) - z\varepsilon\frac{\partial F(\boldsymbol{w},t,\varepsilon)}{\partial t} \right\} dA(z) \right] + O\left(\varepsilon^2\right).$$

Lets perform here an asymptotic transition $\varepsilon \to 0$:

$$F(\boldsymbol{w},t) = \lim_{\varepsilon \to 0} F(\boldsymbol{w},t,\varepsilon) =$$

$$= \lim_{\varepsilon \to 0} \left\{ \left[1 + \sum_{k=1}^K S_k(t)jw_k\varepsilon \right] \left[\int\limits_0^\infty F(\boldsymbol{w},t,\varepsilon)\,dA(z) \right. \right.$$

$$\left. \left. - \int\limits_0^\infty z\varepsilon\frac{\partial F(\boldsymbol{w},t,\varepsilon)}{\partial t}\,dA(z) \right] + O\left(\varepsilon^2\right) \right\} =$$

$$= \lim_{\varepsilon \to 0} \left\{ \left[1 + \sum_{k=1}^K S_k(t)jw_k\varepsilon \right] \left[F(\boldsymbol{w},t,\varepsilon) - \varepsilon\frac{\partial F(\boldsymbol{w},t,\varepsilon)}{\partial t}a \right] + O\left(\varepsilon^2\right) \right\}.$$

As a result, we obtain the following differential equation:

$$\frac{\partial F(\boldsymbol{w},t,\varepsilon)}{\partial t} = \lambda F(\boldsymbol{w},t)\sum_{k=1}^K S_k(t)jw_k.$$

Solving it with an initial condition $F(\boldsymbol{w},0) = 1$, we get the following expression for the function $F(\boldsymbol{w},t)$:

$$F(\boldsymbol{w},t) = \exp\left\{ \lambda\sum_{k=1}^K jw_k \int\limits_0^t S_k(\tau)\,d\tau \right\} = \exp\left\{ \lambda jw^{\mathrm{T}} \int\limits_0^t S(\tau)\,d\tau \right\}.$$

The theorem is proved.   □

Lets implement the inverse substitutions for (11) in the formula (13). We obtain an approximation

$$H(\boldsymbol{u},t) \approx \exp\left\{ \lambda jN\boldsymbol{u}^{\mathrm{T}} \int\limits_0^t S(\tau)\,d\tau \right\}.$$

when N has large value. So, for the network $GI - (GI|\infty)^K$ under a condition of high arrival rate, the average number of busy servers at the node k at the moment t can be approximated by value $\lambda N \int\limits_0^t S_k(\tau)\,d\tau$, where $S_k(t)$ is described in (9).

5.2   Second-Order Asymptotic Form

Lets make the following substitution in the Eq. (10)

$$H(\boldsymbol{u},t) = H_2(\boldsymbol{u},t)\exp\left\{ \lambda N\sum_{k=1}^K ju_k \int\limits_0^t S_k(\tau)\,d\tau \right\}. \tag{14}$$

We obtain

$$
H_2(\boldsymbol{u},t)\exp\left\{\lambda N\sum_{k=1}^{K}ju_k\int_0^t S_k(\tau)\,d\tau\right\}=\left[S_0(t)+\sum_{k=1}^{K}S_k(t)e^{ju_k}\right]\times
$$

$$
\times\left[1-A(Nt)+\int_0^t H_2(\boldsymbol{u},t-x)\exp\left\{\lambda N\sum_{k=1}^{K}ju_k\int_0^{t-x}S_k(\tau)\,d\tau\right\}dA(Nx)\right].
$$

(15)

Making here the substitutions

$$
\frac{1}{N}=\varepsilon^2,\qquad \boldsymbol{u}=\varepsilon\boldsymbol{w},\qquad H_2(\boldsymbol{u},t)=F_2(\boldsymbol{w},t,\varepsilon)
$$

(16)

we get the following equation

$$
F_2(\boldsymbol{w},t,\varepsilon)\exp\left\{\frac{\lambda}{\varepsilon^2}\sum_{k=1}^{K}j\varepsilon w_k\int_0^t S_k(\tau)\,d\tau\right\}=\left[S_0(t)+\sum_{k=1}^{K}S_k(t)e^{j\varepsilon w_k}\right]\times
$$

$$
\times\left[1-A\left(\frac{t}{\varepsilon^2}\right)+\int_0^t F_2(\boldsymbol{w},t-x,\varepsilon)\exp\left\{\frac{\lambda}{\varepsilon^2}\sum_{k=1}^{K}j\varepsilon w_k\int_0^{t-x}S_k(\tau)\,d\tau\right\}dA\left(\frac{x}{\varepsilon^2}\right)\right].
$$

(17)

Let's prove the following statement for an asymptotic approximation $F_2(\boldsymbol{w},t)=\lim_{\varepsilon\to0}F_2(\boldsymbol{w},t,\varepsilon)$.

Theorem 3. *The expression for the function $F_2(\boldsymbol{w},t)$ has the following form:*

$$
F_2(\boldsymbol{w},t)=\exp\left\{\lambda\sum_{k=1}^{K}\frac{(jw_k)^2}{2}\int_0^t S_k(\tau)\,d\tau+\frac{\kappa}{2}\sum_{k=1}^{K}\sum_{\nu=1}^{K}jw_kjw_\nu\int_0^t S_k(\tau)S_\nu(\tau)\,d\tau\right\},
$$

(18)

where $\kappa=\lambda^3\left(\sigma^2-a^2\right)$ and σ^2 is a variance of random variable with distribution function $A(x)$.

The proof. Lets make a substitution $z=x/\varepsilon^2$ in the integral on $dA(\cdot)$. The formula (17) gets a form:

$$
F_2(\boldsymbol{w},t,\varepsilon)=\left[S_0(t)+\sum_{k=1}^{K}S_k(t)e^{j\varepsilon w_k}\right]\times
$$

$$
\times\left[\left\{1-A\left(\frac{t}{\varepsilon^2}\right)\right\}\exp\left\{-\frac{\lambda}{\varepsilon}\sum_{k=1}^{K}jw_k\int_0^t S_k(\tau)\,d\tau\right\}+
$$

$$+ \int_0^{\frac{t}{\varepsilon^2}} F_2(\boldsymbol{w}, t - z\varepsilon^2, \varepsilon) \exp\left\{ -\frac{\lambda}{\varepsilon} \sum_{k=1}^{K} jw_k \int_{t-z\varepsilon^2}^{t} S_k(\tau)\, d\tau \right\} dA(z) \Bigg]. \tag{19}$$

Using an expansion $\int_{t-z\varepsilon^2}^{t} S_k(\tau)\, d\tau = z\varepsilon^2 S_k(t) + \mathrm{O}\left(\varepsilon^4\right)$, we get the relation

$$\exp\left\{ -\frac{\lambda}{\varepsilon} \sum_{k=1}^{K} jw_k \int_{t-z\varepsilon^2}^{t} S_k(\tau)\, d\tau \right\} = \exp\left\{ -\frac{\lambda}{\varepsilon} \sum_{k=1}^{K} jw_k \left[z\varepsilon^2 S_k(t) + \mathrm{O}\left(\varepsilon^4\right) \right] \right\} =$$

$$= \exp\left\{ -z\lambda \sum_{k=1}^{K} j\varepsilon w_k S_k(t) + \mathrm{O}\left(\varepsilon^3\right) \right\} =$$

$$= 1 - z\lambda \sum_{k=1}^{K} j\varepsilon w_k S_k(t) + \frac{z^2\lambda^2}{2} \left[\sum_{k=1}^{K} j\varepsilon w_k S_k(t) \right]^2 + \mathrm{O}\left(\varepsilon^3\right). \tag{20}$$

Further we consider a case when the functions $F_2(\boldsymbol{w}, t, \varepsilon)$ and $A(x)$ have the following features:

$$\int_{\frac{t}{\varepsilon^2}}^{\infty} F_2(\boldsymbol{w}, t - z\varepsilon^2, \varepsilon)\, dA(z) = \mathrm{o}\left(\varepsilon^2\right),$$

$$\left\{ 1 - A\left(\frac{t}{\varepsilon^2}\right) \right\} \exp\left\{ -\frac{\lambda}{\varepsilon} \sum_{k=1}^{K} jw_k \int_{0}^{t} S_k(\tau)\, d\tau \right\} = \mathrm{o}\left(\varepsilon^2\right), \tag{21}$$

where $\mathrm{o}\left(\varepsilon^2\right)$ is an infinitesimal of the order greater than ε^2. Substituting (21) and (20) into (19) and using the expansions

$$e^{j\varepsilon w_k} = 1 + j\varepsilon w_k + \frac{(j\varepsilon w_k)^2}{2} + \mathrm{O}\left(\varepsilon^3\right),$$

$$F_2(\boldsymbol{w}, t - z\varepsilon^2, \varepsilon) = F_2(\boldsymbol{w}, t, \varepsilon) - z\varepsilon^2 \frac{\partial F_2(\boldsymbol{w}, t, \varepsilon)}{\partial t} + \mathrm{o}\left(\varepsilon^2\right),$$

we get

$$F_2(\boldsymbol{w}, t, \varepsilon) = \left[S_0(t) + \sum_{k=1}^{K} S_k(t) + \sum_{k=1}^{K} S_k(t) j\varepsilon w_k + \sum_{k=1}^{K} S_k(t) \frac{(j\varepsilon w_k)^2}{2} \right] \times$$

$$\times \int_{0}^{\infty} \left\{ \left[F_2(\boldsymbol{w}, t, \varepsilon) - z\varepsilon^2 \frac{\partial F_2(\boldsymbol{w}, t, \varepsilon)}{\partial t} \right] \times$$

$$\times \left[1 - z\lambda \sum_{k=1}^{K} j\varepsilon w_k S_k(t) + \frac{z^2\lambda^2}{2} \left\{ \sum_{k=1}^{K} j\varepsilon w_k S_k(t) \right\}^2 \right] dA(z) \Bigg\} + o\left(\varepsilon^2\right) =$$

$$= \left[1 + \sum_{k=1}^{K} S_k(t) j\varepsilon w_k + \sum_{k=1}^{K} S_k(t) \frac{(j\varepsilon w_k)^2}{2} \right]$$

$$\left[F_2(\boldsymbol{w},t,\varepsilon) - F_2(\boldsymbol{w},t,\varepsilon)\lambda a \sum_{k=1}^{K} j\varepsilon w_k S_k(t) + \right.$$

$$\left. + F_2(\boldsymbol{w},t,\varepsilon)\frac{\lambda^2 a_2}{2} \left(\left\{ \sum_{k=1}^{K} j\varepsilon w_k S_k(t) \right\}^2 - \varepsilon^2 \frac{\partial F_2(\boldsymbol{w},t,\varepsilon)}{\partial t} a \right) \right] + o\left(\varepsilon^2\right),$$

where a_2 is a second initial moment of random variable with distribution function $A(x)$. As a result, we obtain the equation

$$\varepsilon^2 \frac{\partial F_2(\boldsymbol{w},t,\varepsilon)}{\partial t} a$$

$$= F_2(\boldsymbol{w},t,\varepsilon) \left[\sum_{k=1}^{K} S_k(t)\frac{(j\varepsilon w_k)^2}{2} + \left\{ \frac{\lambda^2 a_2}{2} - 1 \right\} \left\{ \sum_{k=1}^{K} j\varepsilon w_k S_k(t) \right\}^2 \right] + o\left(\varepsilon^2\right).$$

Dividing each part of this equation by ε^2 and making asymptotic transition $\varepsilon \to 0$, we get the following differential equation about function $F_2(\boldsymbol{w},t) = \lim_{\varepsilon \to 0} F_2(\boldsymbol{w},t,\varepsilon)$:

$$\frac{\partial F_2(\boldsymbol{w},t)}{\partial t} = F_2(\boldsymbol{w},t) \left[\lambda \sum_{k=1}^{K} \frac{(jw_k)^2}{2} S_k(t) + \frac{\kappa}{2} \sum_{k=1}^{K}\sum_{\nu=1}^{K} jw_k jw_\nu S_k(t)S_\nu(t) \right],$$

where $\kappa = \lambda^3\left(a_2 - 2a^2\right) = \lambda^3\left(\sigma^2 - a^2\right)$. Solving this equation under the initial condition $F_2(\boldsymbol{w},0) = 1$ we obtain

$$F_2(\boldsymbol{w},t) = \exp\left\{ \lambda \sum_{k=1}^{K} \frac{(jw_k)^2}{2} \int_0^t S_k(\tau)\,d\tau + \frac{\kappa}{2} \sum_{k=1}^{K}\sum_{\nu=1}^{K} jw_k jw_\nu \int_0^t S_k(\tau)S_\nu(\tau)\,d\tau \right\}.$$

The theorem is proved.   □

Lets make in (18) the inverse substitutions of (16) and (14). Supposing that N is large enough, we obtain an expression

$$H(\boldsymbol{u},t) \approx \exp\left\{ \lambda N \sum_{k=1}^{K} ju_k s_k(t) + \lambda N \sum_{k=1}^{K} \frac{(ju_k)^2}{2} s_k(t) \right.$$

$$\left. + \kappa N \sum_{k=1}^{K}\sum_{\nu=1}^{K} \frac{ju_k ju_\nu}{2} V_{k\nu}(t) \right\},$$

where $s_k(t) = \int_0^t S_k(\tau)\, d\tau$ and $V_{k\nu}(t) = \int_0^t S_k(\tau)S_\nu(\tau)\, d\tau$.

Using matrix notations $\boldsymbol{s}^{\mathrm{T}}(t) = \{s_1(t), \ldots, s_K(t)\}$, $\tilde{\boldsymbol{s}} = \mathrm{diag}\{s_1(t), \ldots, s_K(t)\}$ and $\boldsymbol{V}(t) = \{V_{k\nu}\}_{k,\nu=\overline{1,K}}$, an expression for an approximation $h(\boldsymbol{u}, t)$ of the characteristic function $H(\boldsymbol{u}, t)$ of the multi-dimensional distribution of the network states at the moment t is written in the form

$$h(\boldsymbol{u}, t) = \exp\left\{\lambda N j \boldsymbol{u}^{\mathrm{T}} \boldsymbol{s}(t) + \frac{1}{2} N j \boldsymbol{u}^{\mathrm{T}} \left[\lambda \tilde{\boldsymbol{s}}(t) + \kappa \boldsymbol{V}(t)\right] j \boldsymbol{u}\right\}. \tag{22}$$

So, the distribution of the states of the network $GI - (GI|\infty)^K$ under a condition of high arrival rate at the moment t can be approximated by the multi-dimensional normal distribution with a vector of means $\lambda N \boldsymbol{s}(t)$ and a covariance matrix $N[\lambda \tilde{\boldsymbol{s}}(t) + \kappa \boldsymbol{V}(t)]$.

To obtain an approximation $h(\boldsymbol{u}) = \lim_{t \to \infty} h(\boldsymbol{u}, t)$ for a characteristic function of the stationary distribution, lets make in (22) an asymptotic transition $t \to \infty$. The result is as follows

$$h(\boldsymbol{u}) = \exp\left\{\lambda N j \boldsymbol{u}^{\mathrm{T}} \boldsymbol{s} + \frac{1}{2} N j \boldsymbol{u}^{\mathrm{T}} \left[\lambda \tilde{\boldsymbol{s}} + \kappa \boldsymbol{V}\right] j \boldsymbol{u}\right\}, \tag{23}$$

where $\boldsymbol{s} = \int_0^\infty \boldsymbol{s}(\tau)\, d\tau$, $\tilde{\boldsymbol{s}} = \int_0^\infty \tilde{\boldsymbol{s}}(\tau)\, d\tau$, $\boldsymbol{V} = \int_0^\infty \boldsymbol{s}(\tau)\boldsymbol{s}^{\mathrm{T}}(\tau)\, d\tau$. So, under condition of N is large enough, the stationary joint distribution of the network states can be approximated by the multi-dimensional Gaussian distribution with a vector of means $\lambda N \boldsymbol{s}$ and a covariance matrix $N[\lambda \tilde{\boldsymbol{s}} + \kappa \boldsymbol{V}]$.

6   Conclusion

In the paper, we obtain the Eq. (4) and formula (9) which can be used for analysis of the behavior of non-Markovian queueing networks of the type $GI - (GI|\infty)^K$. Expressions (22) and (23) for the approximations of the characteristic function of the distribution of the network states under a condition of an infinite growth of the arrival process rate is derived. Both stationary and non-stationary cases are presented. It is shown that under such condition the distribution can be approximated by the multi-dimensional Gaussian distribution. Parameters of the approximations are obtained.

References

1. Jackson, J.R.: Networks of waiting lines. Oper. Res. 5(4), 518–521 (1957)
2. Walrand, J.: An Introduction to Queueing Networks. Prentice-Hall, Englewood Cliffs (1988)
3. Kelly, F.P.: Networks of queues. Adv. Appl. Probab. 8(2), 416–432 (1976)
4. Moiseev, A., Nazarov, A.: Investigation of high intensive general flow. In: Proceedings of the IV International Conference on Problems of Cybernetics and Informatics (PCT2012), Baku, pp. 161–163. IEEE (2012)

5. Nazarov, A., Moiseev, A.: Analysis of an open non-Markovian $GI - (GI|\infty)^K$ queueing network with high-rate renewal arrival process. Probl. Inf. Transm. **49**(2), 167–178 (2013)
6. Corolyuk, V.S.: Stochastic Models of Systems. Kluwer, Dordrecht (1999)
7. Bocharov, P.P., Pechinkin, A.V.: Queueing Theory. RUDN, Moscow (1995). (in Russian)
8. Nazarov, A.A., Moiseeva, S.P.: The Asymptotical Analysis Method in Queueing Theory. NTL, Tomsk (2006). (in Russian)

Performance Analysis and Monotone Control of a Tandem Queueing System

Dmitry Efrosinin[1](✉), Mais Farhadov[2], and Saule Kudubaeva[3]

[1] Johannes Kepler University Linz, Altenbergerstrasse 69, 4040 Linz, Austria
[2] Institute of Control Sciences, RAS, Profsoyuznaya Street 65,
Moscow 117997, Russia
[3] Kostanay State University, Beitursynova 47, 110000 Kostanay, Kazakhstan
dmitry.efrosinin@jku.at, mais.farhadov@gmail.com, saule_58@mail.ru
http://www.jku.at,www.ipu.ru,www.ksu.edu.kz

Abstract. A controllable tandem queueing system consists of two nodes in tandem of the type $M/M/n_i$ and a controller. Customers arrive to the controller, who allocates them between the nodes. After service completion at node 2 the controller can allocate the customer waiting at node 1 to node 2. With probability p after a service completion at node 1 a failure occurs. In this case the customer from node 1 joins node 2. With complement probability $1 - p$ the service completion at node 1 is successful. For the given cost structure we formulate an optimal allocation problem to minimize the long-run average cost per unit of time. Using dynamic-programming approach we show the existence of thresholds which divides the state-space into two contiguous regions where the optimal decision is to allocate the customers to node 1 or to node 2. Some monotonicity properties of the dynamic-programming value function are established.

Keywords: Tandem queue · Performance analysis · Long-run average cost · Dynamic-programming · Optimal allocation · Structural properties

1 Introduction

There is a huge number of models developed to study dynamics of different call-centers. The literature overview with references can be found in our papers [4] and [5]. The present paper deals with a two-node tandem queueing system introduced for the first time in [5]. This system is used to model a modern call-center with installed self-service facility operating on the basis of speech recognition technology. We remind that the node $i, i = \overline{1,2}$, specifies a multichannel queueing system of the type $M/M/n_i$ with n_i identical servers in each node. The customers from outside the system arrive according to a Poisson process with a

This work was funded by the COMET K2 Center "Austrian Center of Competence in Mechatronics (ACCM)", funded by the Austrian federal government, the federal state Upper Austria, and the scientific partners of the ACCM.

V. Vishnevsky et al. (Eds.): DCCN 2013, CCIS 279, pp. 241–255, 2014.
DOI: 10.1007/978-3-319-05209-0_21, © Springer International Publishing Switzerland 2014

parameter λ. The servers at node i have exponential service times with parameter μ_i. In [5] the system is supplied by a controller who allocates the customers based on a simple threshold control policy: when the number of customers in node 2 equal to or exceeds some threshold level, the customer goes to node 1, otherwise - to node 2. For the fixed threshold level and specified cost structure we have obtained explicitly the corresponding average cost which was minimized.

The model studied in this paper differs considerable form the previous case. Now we are looking for an optimal policy not only in a class of threshold policies defined above but on the set of all admissible stationary policies. We show that the optimal policy in this case has more complicated structure and is not of a simple threshold type and this policy provides the better values of performance characteristics. To calculate the policy we use a dynamic-programming approach. Several structural properties of a control policy are established as well.

The rest of the paper is organized as follows. Section 2 describes the mathematical model based on a controllable Markov process. In Sect. 3 optimization problem is formulated. Section 4 deals with optimality equations for the dynamic-programming value function and specifies the relationship to the control policy. Finally, some numerical examples are presented in Sect. 5.

2   Mathematical Model

Consider the system symbolically represented in Fig. 1 and described in introduction. Arriving customers can be sent by controller to one of two possible nodes, 1 or 2, according to a specified allocation control policy. The servers at node 1 and 2 have exponential service times with parameters μ_1 and μ_2. The servers at node 1 are assumed to be non-reliable, i.e. with probability p occurs a failure after service completion and the customer goes from the server of node 1 to node 2. With complement probability $1 - p$ the service completion is assumed to be successful. The interarrival and service times are assumed to be mutually independent.

Let $Q_i(t)$ denote the number of customers at node $i, i = 1, 2$. The system states at time t are described by a continuous-time Markov process

$$\{X(t)\}_{t \geq 0} = \{Q_1(t), Q_2(t)\}_{t \geq 0}.$$

The controllable model associated with a Markov process $\{X(t)\}_{t \geq 0}$ is a five-tuple

$$\{E, A, \{A(x), x \in E\}, \lambda_{xy}(a), c(x)\}.$$

- E is a *state space*,

$$E = \{x = (q_1, q_2); q_i \geq 0, \ i = 1, 2\}.$$

Further in the paper the notations $q_i(x)$ will be used to specify the certain components of the vector state $x \in E$.

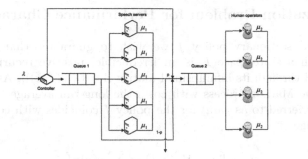

Fig. 1. Tandem queueing system as model of a call-center with self-services.

- $A = \{1, 2\}$ is an *action space* with elements $a \in A$, where $a = i$ means "to send a customer to node i", $i = 1, 2$. The *subsets* $A(x)$ *of control actions in state* x coincide with action space A for any $x \in E$.
- $\lambda_{xy}(a)$ is a *transition intensity* to go from state x to state y under a control action a. It is assumed that the model is *stable* and *conservative*, i.e.

$$\lambda_{xy}(a) \geq 0,\ y \neq x,\ \lambda_{xx}(a) = -\lambda_x(a) = -\sum_{y \neq x} \lambda_{xy}(a),\ \lambda_x(a) < \infty,$$

$$\lambda_{xy}(a) = \begin{cases} \lambda & y = x + e_a, \\ \min\{q_1(x), n_1\}p\mu_1 & y = x - e_1 + e_2,\ q_1(x) > 0, \\ \min\{q_1(x), n_1\}(1-p)\mu_1 & y = x - e_1,\ q_1(x) > 0, \\ \min\{q_2(x), n_2\}\mu_2 & y = x - e_2,\ q_1(x) = 0,\ q_2(x) > 0, \\ \min\{q_2(x), n_2\}\mu_2 & y = x - e_2 - e_1 + e_a,\ q_1(x) > n_1,\ q_2(x) > 0. \end{cases}$$

- $c(x)$ is an *immediate cost* in state x,

$$c(x) = \sum_{i=1}^{2} \Big[\min\{q_i(x), n_i\}c_{u,i} + (q_i(x) - n_i)c_{0,i}\mathbf{1}_{\{q_i(x)>n_i\}} \Big],$$

where $c_{0,i}$ – *waiting cost* per unit of time in node i, $c_{u,i}$ – *usage cost* of a server in node i per unit of time. If $c_{u,i} = c_{0,i} = 1$, $i = 1, 2$, then $c(x)$ represents the number of customers in state x.

We will next explain how the controller chooses its actions. According to the stationary Markov policy $f : E \to A$ whenever at a decision epoch the system state is $x \in E$, the controller chooses an action $f(x) = a \in A(x) \equiv A$ regardless of the past history of the system. We have two types of decision epochs:

- just after an arrival of a new customer the controller allocates it to one of the nodes;
- just after a service completion in node 2 the controller takes a customer from the queue of node 1, if it is not empty, and allocates it to node 2 or puts it back to node 1.

3   Optimization Problem for Performance Characteristics

For every fixed stationary policy f we wish to guarantee that the process $\{X(t)\}_{t\geq 0}$ with a state-space E is an irreducible, positive recurrent Markov process defined through its infinitesimal matrix $\Lambda = [\lambda_{xy}(f(x))]$. As it is known [15], for ergodic Markov process with costs the long-run average cost per unit of time (also referred to as *gain*) for the policy f coincides with corresponding assemble average,

$$g^f = \lim_{t\to\infty} \frac{1}{t} V^f(x,t) = \sum_{y\in E} c(y)\pi_y^f, \tag{1}$$

where

$$V^f(x,t) = \int_0^t \sum_{y\in E} \mathbb{P}^f[X(u) = y | X(0) = x] c(y) du \tag{2}$$

denotes the *total average cost up to time t* when the process starts in state x and $\pi_y^f = \mathbb{P}^f[X(t) = y]$ denotes a stationary probability of the process given policy f. Due to results in [5] the sum at the right hand side of (1) is finite if the following stability condition holds,

$$\frac{\lambda}{n_1 \mu_1} < 1, \quad \frac{p\lambda}{n_2 \mu_2} < 1$$

The policy f^* is said to be optimal when for any admissible policy f

$$g^{f^*} = \min_f g^f. \tag{3}$$

We assume that the gain g^{f^*} will be smaller or equal to the gain under optimal threshold policy studied in [5].

In many applications it is often needed to find a policy f^* which minimizes the long-run average cost per unit of time under the constraint on the sojourn time or the number of customers in the system (due to the Little's Law), namely

$$g^{f^*} = \min_f g^f \quad \text{subject to} \quad \bar{N}_1 \leq \alpha_1, \, \bar{N}_2 \leq \alpha_2, \tag{4}$$

where $\bar{N}_i$ is a mean number of customers at node i. The constrained Markov decision problem can be rewritten as an unconstrained one using Lagrange multipliers, see e.g. Altman [1], Beutler and Ross [8], Piunovskiy [11]. The application of Markov decision process with constraints to the problem of optimal allocation in queueing systems was illustrated e.g. by Yang et al. in [17,18]. For the system under study the Lagrange function $\mathcal{L}^f$ to be minimized is defined as

$$\mathcal{L}^f(\eta_1, \eta_2) = g^f + \sum_{i=1}^2 \eta_i(\bar{N}_i^f - \alpha_i) = \sum_{y\in E}\left[c(y) + \sum_{i=1}^2 \eta_i(l_i(y) - \alpha_i)\right]\pi_y^f, \tag{5}$$

where η_i is the Lagrange multiplier, $l_i(y)$ stands for the number of customers in state y at node i and $c(y) + \sum_{i=1}^{2} \eta_i(l_i(y) - \alpha_i)$ is a modified immediate cost in state $y \in E$. When η_i increases, then the value $\bar{N}_i$ decreases. Therefore, for each i there exist values η_i' and η_i'', such that $\bar{N}_i(\eta_i') > \alpha_i$ and $\bar{N}_i(\eta_i'') \le \alpha_i$, where $\eta_i'' = \eta_i' + \varepsilon_i$ for a small $\varepsilon_i \ge 0$.

The optimal policy f^* can be evaluated by means of a *Howard iteration algorithm* [7], which constructs a sequence of improved policies until the average cost optimal is reached. The key role in this algorithm is played by the *dynamic programming value function* $v : E \to \mathbb{R}_+$ which indicates a transition effect of an initial state x to the total average cost and satisfies a well-known asymptotic relation,

$$V^f(x,t) = g^f t + v^f(x) + o(1), \ x \in E, \ t \to \infty. \tag{6}$$

The functions V^f, v^f and g^f further in the paper will be denoted by V, v and g without upper index f.

4   Optimality Equation and Monotone Control

The system will be uniformized as in Puterman [12] with the uniformization constant

$$\lambda + n_1 \mu_1 + n_2 \mu_2 = 1,$$

which can be obtained by time scaling. As it is well known, the optimal policy f and the optimal average cost g are solutions of the optimality equation

$$Bv(x) = v(x) + g, \tag{7}$$

where B is the *dynamic programming operator* acting on value function v. Note that multiplication of (7) by π_x^f and subsequent summation over all states $x \in E$ leads to the relation (5).

Theorem 1. *The dynamic programming operator B is defined as follows*

$$Bv(x) = c(x) + \sum_{i=1}^{2} \eta_i(l_i(x) - \alpha_i) + \lambda T_0 v(x) \tag{8}$$

$$+ \min\{q_1(x), n_1\}\mu_1 p T_{1,1} v(x) + \min\{q_1(x), n_1\}\mu_1(1-p)T_{1,2}v(x)$$

$$+ \min\{q_2(x), n_2\}\mu_2 T_2 v(x) + \sum_{i=1}^{2}(n_i - q_i(x))\mu_i v(x)1_{\{q_i(x) < n_i\}},$$

where $T_0, T_{1,1}, T_{1,2}, T_2$ – event operators, respectively, for a new arrival, service completion with and without failure on a server at node 1, service completion on a server at node 2 and last term stands for dummy transition,

$$T_0 v(x) = \min_{a \in A} v(x + e_a),$$

$$T_{1,1}v(x) = \begin{cases} v(x - e_1 + e_2) & q_1(x) > 0, \\ 0 & q_1(x) = 0. \end{cases}$$

$$T_{1,2}v(x) = \begin{cases} v(x - e_1) & q_1(x) > 0, \\ 0 & q_1(x) = 0. \end{cases}$$

$$T_2v(x) = \begin{cases} T_0v(x - e_1 - e_2) & q_1(x) > n_1, q_2(x) > 0, \\ v(x - e_2) & 0 \le q_1(x) \le n_1, q_2(x) > 0, \\ 0 & q_2(x) = 0. \end{cases}$$

The notation e_j is used for the vector with 1 in the jth position (beginning from 0th) and 0 elsewhere.

Proof. The optimality equation is obtained by analyzing the function $V(x,t)$ in some infinitesimal interval $[t, t+dt]$. It leads to the differential equation. Applying further the limit expression

$$\lim_{dt \to 0} \frac{V(x, t + dt) - V(x, t)}{dt} = 0$$

and taking into account Markov property of $\{X(t)\}_{t \ge 0}$ with asymptotic relation (6) ones get (8).

Corollary 1. *It follows from (8) that just after a new arrival in state x or just after a service completion at node 2 in state $x + e_1 + e_2$ holds a relation,*

$$f(x) = \arg \min_{a \in A} \{v(x + e_a)\}. \tag{9}$$

Remark 1. Using (8) and (9) we can solve optimization problem for some other criteria:

If $c_{u,i} = c_{0,i} = 1, i = 1, 2$, then (3) is equivalent to minimization of the mean number of customers in the system $\bar{N}$ or mean sojourn time $\bar{T}$,

$$f^* = \arg \min_f \bar{N}^f = \arg \min_f \bar{T}^f.$$

If $c_{u,i} = 0, c_{0,i} = 1, i = 1, 2$, then (3) is equivalent to minimization of the mean number of customers in the queue $\bar{Q}$ or mean waiting time $\bar{W}$,

$$f^* = \arg \min_f \bar{Q}^f = \arg \min_f \bar{W}^f.$$

If $c(x) = 1, c(y) = 0, y \ne x$, then (3) is equivalent to minimization of the steady-state probability π_x^f of the given state $x \in E$,

$$f^* = \arg \min_f \pi_x^f.$$

The optimal equation (8) can be modified in such way to minimize the mean busy period $\bar{L}$ or system utilization $\bar{U}$,

$$f^* = \arg \min_f \bar{L}^f = \arg \min_f \bar{U}^f.$$

As is show in Efrosinin [3], the mean busy period can be evaluated by solving the system of equations for the mean first passage time to the empty state $\bar{L}(x)$ given initial state $x \in E$,

$$\bar{L}(x) = 1 + \lambda T_0 \bar{L}(x) + \min\{q_1(x), n_1\} \mu_1 p T_{1,1} \bar{L}(x) \tag{10}$$
$$+ \min\{q_1(x), n_1\} \mu_1 (1 - p) T_{1,2} \bar{L}(x)$$
$$+ \min\{q_2(x), n_2\} \mu_2 T_2 \bar{L}(x) + \sum_{i=1}^{2} (n_i - q_i(x)) \mu_i \bar{L}(x) 1_{\{n_i > q_i(x)\}}.$$

Note that using (8) the above performance measures can be calculated for any arbitrary policy f as well.

The relation (9) shows that the structural and monotone properties of the optimal control policy f can derived by analyzing the monotonicity properties of the value function v. Such properties for other types of controlled queues in a tandem were studied also by Koole [9], Liang and Kulkarni [10], Veatch and Wein [16]. It was shown that the value function has some monotonicity properties like nondecreasing and superconvexity. To prove such inequalities it is necessary to solve (7). Since the solution of the optimality equation in analytic form is hardly available, normally it can be solved recursively defining $v_{n+1} = Bv_n$ for some arbitrary initial v_0. Due to the limit relation

$$\lim_{n \to \infty} B^n v_0(x) = v(x) \tag{11}$$

we get an optimal solution for the value function. For existence and convergence solutions and optimal policies we refer to Aviv and Federgruen [2], Puterman [12], Sennott [14].

Now some monotonicity properties of the value function for the system under study will be presented and proved, but before we have to make a remark.

Remark 2. We have succeed in proving the structural results only for the simplified model. First, we assume a pure admission control problem, if the allocation can be performed only at moments of new arrivals. Second, we assume single server nodes, $n_i = 1, i = \overline{1,2}$, and absolutely reliable server at node 1, i.e. $p = 0$. The third assumption needed for the proof concerns the relation for the costs,

$$c_{u,i} - c_{0,i} \leq 0. \tag{12}$$

Nevertheless, we strongly believe that the presented properties are true for the original model as well, which was confirmed by numerical results.

Theorem 2. *The value function v satisfies the conditions for any $x \in E$:*
C1. Non-decreasing condition

$$v(x) - v(x + e_j) \leq 0, \ j = 1, 2.$$

C2. Superconvexity condition

$$v(x + e_1) - v(x + e_2) - v(x + 2e_1) + v(x + e_1 + e_2) \leq 0,$$
$$v(x + e_2) - v(x + e_1) - v(x + 2e_2) + v(x + e_1 + e_2) \leq 0$$

C3. Supermodularity condition

$$v(x + e_1) - v(x) - v(x + e_1 + e_2) + v(x + e_2) \leq 0,$$
$$v(x + e_2) - v(x) - v(x + e_1 + e_2) + v(x + e_1) \leq 0.$$

C4. Convexity condition

$$2v(x + e_j) - v(x) - v(x + 2e_j) \leq 0, \, j = 1, 2.$$

Note, that condition (C4) directly follows from conditions (C2) and (C3). The name of condition (C3) is borrowed from the terminology, given in [6].

Proof (Condition C1). The proof is by induction on n in v_n. Define $v_0(x) = 0$ for all states $x \in E$. This function obviously satisfies the condition (C1). Now, we assume (C1) for function $v_n(x)$, $x \in E$, and some $n \in \mathbb{N}$. One have to prove that $v_{n+1}(x)$ satisfies the non-decreasing property as well. Then for $j = 1$

$$v_{n+1}(x) - v_{n+1}(x + e_1) = c(x) - c(x + e_1) + \sum_{i=1}^{2} \eta_i(l_i(x) - l_i(x + e_1)) \qquad (I)$$

$$+ \lambda[T_0 v_n(x) - T_0 v_n(x + e_1)] \qquad (II)$$
$$+ \mu_1[T_{1,2} v_n(x) - T_{1,2} v_n(x + e_1)] \qquad (III)$$
$$+ \mu_2[T_2 v_n(x) - T_2 v_n(x + e_1)] \qquad (IV)$$

$$+ \sum_{i=1}^{2} \mu_i[v_n(x)1_{\{q_i(x)=0\}} - v_n(x + e_1)1_{\{q_i(x+e_1)=0\}}] \leq 0. \qquad (V)$$

The expression (I) in the right-hand side is obviously positive,

$$c(x) - c(x + e_1) + \sum_{i=1}^{2} \eta_i(l_i(x) - l_i(x + e_1)) = -c_{u,1}1_{\{q_1(x)=0\}} - c_{0,1}1_{\{q_1(x)>0\}} - \sum_{i=1}^{2} \eta_i \leq 0.$$

The term (II) is positive,

$$T_0 v_n(x) - T_0 v_n(x + e_1) = \min_{a \in A} v_n(x + e_a) - \min_{a \in A} v_n(x + e_1 + e_a) \leq 0,$$

since $v_n(x + 2e_1) \geq v_n(x + e_1)$ and $v_n(x + e_1 + e_2) \geq v_n(x + e_2)$ based on the induction hypothesis which is used further throughout the proof. If $q_1(x) > 0$ and $q_2(x) > 0$, the (III) and (IV) are non-positive due to (C1) at states $x - e_1$ and $x - e_2$. If $q_1(x) = 0$, then we sum up (III) and (V),

$$\mu_1 v_n(x) - T_{1,2} v_n(x + e_1) = \mu_1 v_n(x) - \mu_1 v_n(x) = 0.$$

If $q_2(x) = 0$, then we sum up (IV) and (V),

$$\mu_2[T_2 v_n(x) - T_2 v(x + e_1)] = v_n(x) - v_n(x + e_1) \leq 0,$$

due to (C1). Therefore, the condition (C1) holds by induction for any n and due to (11) $v(x)$ is a nondecreasing function. (C1) for $j = 2$ is proved in a similar manner.

Proof (Condition C2). The proof is done by induction on n in v_n. Define $v_0(x) = 0$ for all states $x \in E$. It is clear that this function satisfies condition (C2). Now suppose that properties (C1)–(C4) hold for $v_n, n \in \mathbb{N}$. Now we prove that it holds for $n + 1$ as well. Consider the first inequality (C2),

$$v_{n+1}(x + e_1) - v_{n+1}(x + e_2) - v_{n+1}(x + 2e_1) + v_{n+1}(x + e_1 + e_2)$$
$$= c(x + e_1) - c(x + e_2) - c(x + 2e_1) + c(x + e_1 + e_2) \tag{I}$$

$$+ \sum_{i=1}^{2} \eta_i(l_i(x + e_1) - l_i(x + e_2) - l_i(x + 2e_1) + l_i(x + e_1 + e_2))$$

$$+ \lambda[T_0 v_n(x + e_1) - T_0 v_n(x + e_2) - T_0 v_n(x + 2e_1) + T_0 v_n(x + e_1 + e_2)] \tag{II}$$
$$+ \mu_1[T_{1,2} v_n(x + e_1) - T_{1,2} v_n(x + e_2) - T_{1,2} v_n(x + 2e_1) + T_{1,2} v_n(x + e_1 + e_2)] \tag{III}$$
$$+ \mu_2[T_2 v_n(x + e_1) - T_2 v_n(x + e_2) - T_2 v_n(x + 2e_1) + T_2 v_n(x + e_1 + e_2)] \tag{IV}$$

$$+ \sum_{i=1}^{2} \mu_i[v_n(x + e_1)1_{\{q_i(x+e_1)=0\}} - v_n(x + e_2)1_{\{q_i(x+e_2)=0\}} \tag{V}$$
$$- v_n(x + 2e_1)1_{\{q_i(x+2e_1)=0\}} + v_n(x + e_1 + e_2)1_{\{q_i(x+e_1+e_2)=0\}}] \leq 0.$$

For (I) we have four subcases, namely $(q_1(x) > 0, q_2(x) > 0)$, $(q_1(x) > 0, q_2(x) = 0)$, $(q_1(x) = 0, q_2(x) > 0)$ and $(q_1(x) = q_2(x) = 0)$. In first two subcases this expression is equal to 0. In third and fourth subcases

$$c(x + e_1) - c(x + e_2) - c(x + 2e_1) + c(x + e_1 + e_2) = c_{u,1} - c_{0,1} \leq 0$$

due to the assumption (12) and

$$\sum_{i=1}^{2} \eta_i(l_i(x + e_1) - l_i(x + e_2) - l_i(x + 2e_1) + l_i(x + e_1 + e_2))$$

$$= \sum_{i=1}^{2} \eta_i(1 + l_i(x) - 1 - l_i(x) - 2 - l_i(x) + 2 + l_i(x)) = 0.$$

For (II) we have

$$T_0 v_n(x + e_1) - T_0 v_n(x + e_2) - T_0 v_n(x + 2e_1) + T_0 v_n(x + e_1 + e_2)$$
$$= \min_{a \in A} v_n(x + e_1 + e_a) - \min_{a \in A} v_n(x + e_2 + e_a)$$
$$- \min_{a \in A} v_n(x + 2e_1 + e_a) + \min_{a \in A} v_n(x + e_1 + e_2 + e_a).$$

Consider the following subcases. If $f_n(x + e_1) = f_n(x + e_2) = f_n(x + 2e_1) = f_n(x + e_1 + e_2) = a$, then the last expression is non positive due to the induction assumption (C2) at state $x + e_a$. If $f_n(x + e_2) = 2$ and $f_n(x + 2e_1) = 1$,

$$\min_{a \in A} v_n(x + e_1 + e_a) - v_n(x + 2e_2) - v_n(x + 3e_1) + \min_{a \in A} v_n(x + e_1 + e_2 + e_a)$$
$$\leq v_n(x + 2e_1) - v_n(x + 2e_2) - v_n(x + 3e_1) + v_n(x + e_1 + 2e_2) \leq 0,$$

which follows by summing up two inequalities (C2) at states $x + e_1$ and $x + e_2$,

$$v_n(x + e_1 + e_2) - v_n(x + 2e_2) - v_n(x + 2e_1 + e_2) + v_n(x + e_1 + 2e_2)$$
$$+ v_n(x + 2e_1) - v_n(x + e_1 + e_2) - v_n(x + 3e_1) + v_n(x + 2e_1 + e_2)$$
$$= v_n(x + 2e_1) - v_n(x + 2e_2) - v_n(x + 3e_1) + v_n(x + e_1 + 2e_2) \leq 0.$$

If $f_n(x + e_2) = 1$ and $f_n(x + 2e_1) = 2$,

$$\min_{a \in A} v_n(x + e_1 + e_a) - v_n(x + e_2 + e_1) - v_n(x + 2e_1 + e_2) + \min_{a \in A} v_n(x + 2e_1 + e_a)$$
$$\leq v_n(x + e_1 + e_2) - v_n(x + e_2 + e_1) - v_n(x + 2e_1 + e_2) + v_n(x + 2e_1 + e_2) = 0.$$

If $q_1(x) > 0$ and $q_2(x) > 0$, (III) and (IV) are non-positive due to (C2), respectively, at states $x - e_1$ and $x - e_2$. If $q_1(x) = 0$, then the sum (III) and (V),

$$\mu_1 [v_n(x) - v_n(x + e_2) - v_n(x + e_1) + v_n(x + e_2)] = \mu_1 [v_n(x) - v_n(x + e_1)] \leq 0$$

due to (C1). If $q_2(x) = 0$, we get for the sum (IV) and (V),

$$\mu_2 [v_n(x + e_1) - v_n(x) - v_n(x + 2e_1) + v_n(x + e_1)]$$
$$= \mu_2 [2v_n(x + e_1) - v_n(x) - v_n(x + 2e_1)] \leq 0,$$

which follows from (C4). The second part of (C2) can be proved in a similar manner. Hence, we conclude, by taking the limit $n \to \infty$, that the value function $v(x)$ preserves condition (C2).

Proof (Condition C3). The proof is done again by induction on n in v_n. Define $v_0(x) = 0$ for all states $x \in E$. Suppose that properties (C1)–(C4) hold for $v_n, n \in \mathbb{N}$. Consider the first inequality (C3),

$$v_{n+1}(x + e_1) - v_{n+1}(x) - v_{n+1}(x + e_1 + e_2) + v_{n+1}(x + e_2)$$
$$= c(x + e_1) - c(x) - c(x + e_1 + e_2) + c(x + e_2) \tag{I}$$

$$+ \sum_{i=1}^{2} \eta_i (l_i(x + e_1) - l_i(x) - l_i(x + e_1 + e_2) + l_i(x + e_2))$$

$$+ \lambda [T_0 v_n(x + e_1) - T_0 v_n(x) - T_0 v_n(x + e_1 + e_2) + T_0 v_n(x + e_2)] \tag{II}$$
$$+ \mu_1 [T_{1,2} v_n(x + e_1) - T_{1,2} v_n(x) - T_{1,2} v_n(x + e_1 + e_2) + T_{1,2} v_n(x + e_2)] \tag{III}$$
$$+ \mu_2 [T_2 v_n(x + e_1) - T_2 v_n(x) - T_2 v_n(x + e_1 + e_2) + T_2 v_n(x + e_2)] \tag{IV}$$

$$+ \sum_{i=1}^{2} \mu_i [v_n(x + e_1) 1_{\{q_i(x+e_1)=0\}} - v_n(x) 1_{\{q_i(x)=0\}} \tag{V}$$

$$- v_n(x + e_1 + e_2) 1_{\{q_i(x+e_1+e_2)=0\}} + v_n(x + e_2) 1_{\{q_i(x+e_2)=0\}}] \leq 0.$$

The expression (I) is equal to 0. For (II) we have

$$T_0 v_n(x + e_1) - T_0 v_n(x) - T_0 v_n(x + e_1 + e_2) + T_0 v_n(x + e_2)$$
$$= \min_{a \in A} v_n(x + e_1 + e_a) - \min_{a \in A} v_n(x + e_a)$$
$$- \min_{a \in A} v_n(x + e_1 + e_2 + e_a) + \min_{a \in A} v_n(x + e_2 + e_a).$$

Consider the following subcases. If $f_n(x + 1) = f_n(x) = f_n(x + e_1 + e_2) = f_n(x + e_2) = a$, then the last expression is non positive due to the induction assumption (C3) at state $x + e_a$. If $f_n(x) = 2$ and $f_n(x + e_1 + e_2) = 1$, then

$$\min_{a \in A} v_n(x + e_1 + e_a) - v_n(x + e_2) - v_n(x + 2e_1 + e_2) + \min_{a \in A} v_n(x + e_2 + e_a)$$
$$\leq 2v_n(x + e_1 + e_2) - v_n(x + e_2) - v_n(x + 2e_1 + e_2) \leq 0,$$

which follows form the first inequality of (C4) at state $x + e_2$. If $f_n(x) = 1$ and $f_n(x + e_1 + e_2) = 2$,

$$\min_{a \in A} v_n(x + e_1 + e_a) - v_n(x + e_1) - v_n(x + e_1 + 2e_2) + \min_{a \in A} v_n(x + e_2 + e_a)$$
$$\leq 2v_n(x + e_1 + e_2) - v_n(x + e_1) - v_n(x + e_1 + 2e_2) \leq 0,$$

due to the second inequality of (C4) at state $x + e_1$. If $q_1(x) > 0$ and $q_2(x) > 0$, (III) and (IV) are non-positive due to (C2). If $q_1(x) = 0$, we sum up (III) and (V),

$$\mu_1[v_n(x) - v_n(x) - v_n(x + e_2) + v_n(x + e_2)] = 0.$$

If $q_2(x) = 0$, we get for the sum (IV) and (V),

$$\mu_2[v_n(x + e_1) - v_n(x) - v_n(x + e_1) + v_n(x)] = 0.$$

The second part of (C3) can be proved in a similar manner. Hence, we conclude, by taking the limit $n \to \infty$, that the value function $v(x)$ preserves condition (C3).

This method was used for the first time in [13] for the queueing systems with heterogeneous servers. Condition (C2) shows that advantage to allocate a customer to node i decreases as the number of customers in this node increases.

Corollary 2. *Theorem 2 implies the existence of thresholds $q_1^*(q_2)$ for each fixed number q_2 of customers in node 2 to allocate them to node 1, when $q_1 < q_1^*(q_2)$, and to node 2, when $q_1 \geq q_1^*(q_2)$. The same is true for thresholds $q_2^*(q_1)$.*

Conjecture: We have strong reasons to believe that the optimal policy has additional property: In state x with $q_i(x) \leq n_i, i = 1, 2$, the controller allocates the customers to node 1 when

$$\frac{c_1}{n_1 \mu_1} < \frac{c_2}{n_2 \mu_2}$$

and to node 2 otherwise. This result was confirmed by numerical examples.

Table 1. Comparison of optimal allocation and simple threshold policies

$c_{u,1},\ c_{u,2},\mu_1,\mu_2$	q_2^*	$\bar{V}$	$\bar{V}_{opt}$	$\bar{N}$	$\bar{N}_{opt}$
0.5,1.0,0.5,0.6	(6,8)	1.5970	1.1354	1.5419	1.5063
0.5,1.0,1.5,0.6	(4,4)	1.4694	0.4235	1.4891	0.6694
0.5,1.0,0.5,1.6	(6,8)	0.5636	0.5625	0.5630	0.5627
0.5,1.0,1.5,1.6	(5,6)	0.5632	0.3316	0.5629	0.5625
2.5,1.0,0.5,0.6	(8,8)	1.6060	1.5470	1.5419	1.5063
2.5,1.0,1.5,0.6	(5,4)	1.5275	1.5017	1.4891	0.6694
2.5,1.0,0.5,1.6	(7,8)	0.5636	0.5633	0.5630	0.5627
2.5,1.0,1.5,1.6	(6,6)	0.5636	0.5630	0.5629	0.5625
0.5,3.0,0.5,0.6	(6,8)	4.5867	2.2512	1.5419	1.5063
0.5,3.0,1.5,0.6	(4,4)	4.3634	0.4919	1.4891	0.6694
0.5,3.0,0.5,1.6	(6,8)	1.6886	1.1994	0.5630	0.5627
0.5,3.0,1.5,1.6	(5,6)	1.6880	0.4308	0.5629	0.5625
2.5,3.0,0.5,0.6	(7,8)	4.6002	4.5055	1.5419	1.5063
2.5,3.0,1.5,0.6	(4,4)	4.4462	1.6876	1.4891	0.6694
2.5,3.0,0.5,1.6	(8,8)	1.6886	1.6881	0.5630	0.5627
2.5,3.0,1.5,1.6	(5,5)	1.6885	1.5341	0.5629	0.5625

5   Numerical Examples

Consider a queueing system with $n_1 = 2$ and $n_2 = 4$. The system parameters take the following values:

$$\lambda = 0.9,\ p = 0.01,\ c_{0,1} = c_{0,2} = 2.5. \tag{13}$$

To evaluate optimal policies we apply Howard iteration algorithm [7]. For numerical calculation we have to restrict the number of places in the buffers of the nodes. These numbers are taken enough large to make the system suit for infinite buffer case and to avoid the influence of the boundary states. The next example provides with comparison analysis of the optimal allocation and simple threshold policies. The results of calculation are summarized in Table 1. It can be seen that the optimal allocation policy is always superior in performance comparing to the simple threshold policy, calculated in Farhadov et al. [5].

Simple threshold policy with optimal level $q_2^* = 8$ is illustrated in Table 2. The structure of optimal allocation policy for the values

$$\mu_1 = 0.5,\ \mu_2 = 0.6,\ c_{u,1} = 0.5,\ c_{u,2} = 3.0. \tag{14}$$

is presented in Table 3. Other parameters take the same values as before.

Table 3 shows optimal threshold levels $q_1^*(q_2)$ for each value q_2. Moreover,

$$\frac{c_{u,1}}{n_1\mu_1} = 0.50 < 1.25 = \frac{c_{u,2}}{n_2\mu_2},$$

hence controller performs the allocation to node 1 in states x with $q_i(x) \leq n_i$, see the conjecture above. Table 4 illustrates the structure of an optimal allocation

Table 2. Simple threshold policy

$Q_1 \backslash Q_2$	0	1	2	3	4	5	6	7	8	9	10	11	12	13	14	15	16	...
0	2	2	2	2	2	2	2	2	1	1	1	1	1	1	1	1	1	...
1	2	2	2	2	2	2	2	2	1	1	1	1	1	1	1	1	1	...
2	2	2	2	2	2	2	2	2	1	1	1	1	1	1	1	1	1	...
⋮																		⋱
10	2	2	2	2	2	2	2	2	1	1	1	1	1	1	1	1	1	...
11	2	2	2	2	2	2	2	2	1	1	1	1	1	1	1	1	1	...
12	2	2	2	2	2	2	2	2	1	1	1	1	1	1	1	1	1	...
13	2	2	2	2	2	2	2	2	1	1	1	1	1	1	1	1	1	...
14	2	2	2	2	2	2	2	2	1	1	1	1	1	1	1	1	1	...
15	2	2	2	2	2	2	2	2	1	1	1	1	1	1	1	1	1	...
16	2	2	2	2	2	2	2	2	1	1	1	1	1	1	1	1	1	...
17	2	2	2	2	2	2	2	2	1	1	1	1	1	1	1	1	1	...
18	2	2	2	2	2	2	2	2	1	1	1	1	1	1	1	1	1	...
⋮																		⋱

Table 3. Optimal allocation policy for values (14)

$Q_1 \backslash Q_2$	0	1	2	3	4	5	6	7	8	9	10	11	12	13	14	15	16	...
0	1	1	1	1	1	1	1	1	1	1	1	1	1	1	1	1	1	...
1	1	1	1	1	1	1	1	1	1	1	1	1	1	1	1	1	1	...
2	2	2	2	1	1	1	1	1	1	1	1	1	1	1	1	1	1	...
⋮																		⋱
10	2	2	2	2	1	1	1	1	1	1	1	1	1	1	1	1	1	...
11	2	2	2	2	2	1	1	1	1	1	1	1	1	1	1	1	1	...
12	2	2	2	2	2	2	1	1	1	1	1	1	1	1	1	1	1	...
13	2	2	2	2	2	2	2	2	1	1	1	1	1	1	1	1	1	...
14	2	2	2	2	2	2	2	2	2	2	1	1	1	1	1	1	1	...
15	2	2	2	2	2	2	2	2	2	2	2	1	1	1	1	1	1	...
16	2	2	2	2	2	2	2	2	2	2	2	2	1	1	1	1	1	...
17	2	2	2	2	2	2	2	2	2	2	2	2	2	2	1	1	1	...
18	2	2	2	2	2	2	2	2	2	2	2	2	2	2	2	2	2	...
⋮																		⋱

policy for the following values of system parameters,

$$\mu_1 = 0.5, \ \mu_2 = 1.6, \ c_{u,1} = 0.5, \ c_{u,2} = 1.0. \tag{15}$$

Since the following inequality holds,

$$\frac{c_{u,1}}{n_1 \mu_1} = 0.50 > 0.15 = \frac{c_{u,2}}{n_2 \mu_2},$$

the controller performs the allocation to node 2 in states x with $q_i(x) \le n_i$.

In the next example $\lambda = 1.1$, other parameters take the values (13) and (14). For the optimal policy f_1^*, which minimizes the gain g, we obtain $g = 3.04$ and $\bar{N} = 2.06$ and for the optimal policy f_2^*, which minimizes the mean number of customers in the system $\bar{N}$, $g = 5.20$ and $\bar{N} = 1.86$. Consider the constrained problem (4), where $\bar{N} = \bar{N}_1 + \bar{N}_2 \le 1.93$. Solving this optimization problem, we get a new optimal policy f_3^* with $g = 4.62, \bar{N} = 1.92$. These optimal results were achieved for the Lagrange multiplier $\eta = \eta_1 + \eta_2 = 11.18$. To compare the results the optimal allocation policies f_1^*, f_2^* and f_3^* are shown together in Table 5.

Table 4. Optimal allocation policy for values (15)

$Q_1 \backslash Q_2$	0	1	2	3	4	5	6	7	8	9	10	11	12	13	14	15	16	...
0	2	2	2	2	1	1	1	1	1	1	1	1	1	1	1	1	1	...
1	2	2	2	2	1	1	1	1	1	1	1	1	1	1	1	1	1	...
2	2	2	2	2	1	1	1	1	1	1	1	1	1	1	1	1	1	...
⋮																		⋱
15	2	2	2	2	1	1	1	1	1	1	1	1	1	1	1	1	1	...
16	2	2	2	2	2	1	1	1	1	1	1	1	1	1	1	1	1	...
17	2	2	2	2	2	2	1	1	1	1	1	1	1	1	1	1	1	...
18	2	2	2	2	2	2	2	1	1	1	1	1	1	1	1	1	1	...
19	2	2	2	2	2	2	2	2	1	1	1	1	1	1	1	1	1	...
20	2	2	2	2	2	2	2	2	2	1	1	1	1	1	1	1	1	...
21	2	2	2	2	2	2	2	2	2	2	1	1	1	1	1	1	1	...
22	2	2	2	2	2	2	2	2	2	2	2	1	1	1	1	1	1	...
23	2	2	2	2	2	2	2	2	2	2	2	2	1	1	1	1	1	...
24	2	2	2	2	2	2	2	2	2	2	2	2	2	2	2	1	1	...
⋮																		⋱

Table 5. Optimal allocation policies f_1^*, f_2^* and f_3^*

$Q_1 \backslash Q_2$	0	1	2	3	4	5	6	7	8	9	10	11	12	0	1	2	3	4	5	6	7	8	9	10	11	12	0	1	2	3	4	5	6	7	8	9	10	11	12
0	1	1	1	1	1	1	1	1	1	1	1	1	1	2	2	2	2	1	1	1	1	1	1	1	1	1	2	2	2	2	1	1	1	1	1	1	1	1	1
1	1	1	1	1	1	1	1	1	1	1	1	1	1	2	2	2	2	1	1	1	1	1	1	1	1	1	2	2	2	2	1	1	1	1	1	1	1	1	1
2	2	2	2	2	1	1	1	1	1	1	1	1	1	2	2	2	2	1	1	1	1	1	1	1	1	1	2	2	2	2	1	1	1	1	1	1	1	1	1
⋮																																							
23	2	2	2	2	1	1	1	1	1	1	1	1	1	2	2	2	2	1	1	1	1	1	1	1	1	1	2	2	2	2	1	1	1	1	1	1	1	1	1
24	2	2	2	2	1	1	1	1	1	1	1	1	1	2	2	2	2	2	1	1	1	1	1	1	1	1	2	2	2	2	2	1	1	1	1	1	1	1	1
25	2	2	2	2	2	1	1	1	1	1	1	1	1	2	2	2	2	2	1	1	1	1	1	1	1	1	2	2	2	2	2	1	1	1	1	1	1	1	1
26	2	2	2	2	2	2	1	1	1	1	1	1	1	2	2	2	2	2	2	1	1	1	1	1	1	1	2	2	2	2	2	2	1	1	1	1	1	1	1
27	2	2	2	2	2	2	2	1	1	1	1	1	1	2	2	2	2	2	2	2	1	1	1	1	1	1	2	2	2	2	2	2	2	1	1	1	1	1	1
28	2	2	2	2	2	2	2	2	1	1	1	1	1	2	2	2	2	2	2	2	2	1	1	1	1	1	2	2	2	2	2	2	2	2	1	1	1	1	1
29	2	2	2	2	2	2	2	2	2	1	1	1	1	2	2	2	2	2	2	2	2	2	1	1	1	1	2	2	2	2	2	2	2	2	2	1	1	1	1
30	2	2	2	2	2	2	2	2	2	2	1	1	1	2	2	2	2	2	2	2	2	2	2	1	1	1	2	2	2	2	2	2	2	2	2	1	1	1	1
31	2	2	2	2	2	2	2	2	2	2	2	1	1	2	2	2	2	2	2	2	2	2	2	2	1	1	2	2	2	2	2	2	2	2	2	2	1	1	1
32	2	2	2	2	2	2	2	2	2	2	2	2	1	2	2	2	2	2	2	2	2	2	2	2	2	1	2	2	2	2	2	2	2	2	2	2	2	2	1
⋮																																							

6   Conclusion

In this paper we have studied a dynamic allocation problem for the tandem queueing system with two nodes of the type $M/M/n_i, i = 1, 2$. The optimal allocation policy for the long-run average criterion with possible constraints is calculated by means of the dynamic-programming approach. Some monotonicity properties of the value function are established. According to proposed results the following general conclusion can be made. Assuming a threshold policy it is necessary to understand that the corresponding optimal policy is optimal only in an appropriate class of threshold policies. Otherwise ones have to prove whether this structure is really optimal among available stationary policies. For this special queueing system it is not a case, since the optimal policy has a more complicated structure with certain threshold level $q_1^*(q_2)$ for each fixed number of customers q_2 in node 2. This optimal policy is superior in performance comparing to the simple threshold policy. It is also shown that the dynamic programming approach is quite appropriate for the calculation of different performance measures. Moreover it allows to analyze structural and monotonicity properties of the optimal control policy.

References

1. Altman, E.: Constrained Markov Decision Processes. Chapman and Hall, London (1999)
2. Aviv, Y., Federgruen, A.: The value-iteration method for countable state Markov decision processes. Oper. Res. Lett. **24**(5), 223–234 (1999)
3. Efrosinin, D.: Analysis of the busy period in threshold control system. Autom. Remote Control **71**(1), 87–104 (2010)
4. Farhadov, M.P., Petuchova, N.V., Efrosinin, D.V., Semenova, O.V.: Two-phase model with unbounded queues (in Russian). Problemy upravleniya **6**, 53–58 (2010)
5. Farhadov, M.P., Petuchova, N.V., Efrosinin, D.V., Semenova, O.V.: Modeling of a hybrid call-center with self-services and threshold-based control (in Russian). Upravlenie Bolshimi Sistemami. Special issue 30.1 Network Models in Control, pp. 352–370. IPU RAS, Moscow (2010)
6. Ghoneim, H.A., Stidham, S.: Control of arrivals to two queues in series. Eur. J. Oper. Res. **21**, 399–409 (1985)
7. Howard, R.: Dynamic Programming and Markov Processes. Wiley, New York (1960)
8. Beutler, F.J., Ross, K.W.: Optimal policies for controlled Markov chains with a constraint. J. Math. Anal. Appl. **112**, 236–252 (1985)
9. Koole, G.: Convexity in tandem queues. Prob. Eng. Inf. Sci. **18**(1), 13–31 (2004)
10. Liang, H.M., Kulkarni, V.G., et al.: Optimal routing control in retrial queues (chapter 14). In: Shanthikumar, J.G. (ed.) Applied Probability and Stochastic Processes, pp. 203–218. Kluwer Academic, New York (1999)
11. Piunovskiy, A.B.: Dynamic programming in a constrained Markov decision processes. Control Cybern. **35**(3), 645–660 (2006)
12. Puterman, M.L.: Markov Decision Process: Discrete Stochastic Dynamic Programming. Wiley, New York (1994)
13. Rykov, V.: Monotone control of queueing systems with heterogeneous servers. QUESTA **37**, 391–403 (2001)
14. Sennott, L.I.: Stochastic Dynamic Programming and the Control of Queueing Systems. Wiley, New York (1999)
15. Tijms, H.C.: Stochastic Models. An Algorithmic Approach. Wiley, New York (1994)
16. Veatch, M.N., Wein, L.M.: Monotone control of queueing networks. Queueing Syst. **12**, 391–408 (1992)
17. Yang, R., Bhulai, S., Mei, R.: Optimal resource allocation for multiqueue systems with a shared server pool. Queueing Syst. **68**, 133–163 (2011)
18. Yang, R., Bhulai, S., Mei, R.: Structural properties of the optimal resource allocation policy for single-queue systems. Ann. Oper. Res. **202**, 211–233 (2013)

Application of RFID-Technology to the Problem of Traffic Control through Roadway Intersections

Olga Semenova[✉] and Stanislav Lykov

Institute of Control Sciences, RAS, Profsoyuznaya Street 65, Moscow 117997, Russia
{olgasmnv,stasljr}@gmail.com

Abstract. This paper is devoted to an actual problem: the optimization of traffic light to control on roadway intersection. In order to reduce the total delay of vehicles, minimize the queues at junctions and optimize traffic flow through signalized intersection a model of stochastic polling systems is proposed. On order to determine the most effective service disciplines a comparative analysis of two different service disciplines was conducted. Exhaustive and gated service disciplines were analyzed, compared and simulated for the purpose to choose the most effective one. The viability and efficiency of the proposed algorithm using computer simulation are also demonstrated. Moreover the possibility of using RFID (Radio Frequency Identification) technology in order to reduce the total delay of vehicles was analyzed.

Keywords: RFID technology · Roadway intersection control · Polling systems

1 Introduction

The process of automobilization is spreading over of countries year by year with increasingly large number of vehicles and people involved in the roadway traffic area. All this causes inevitable growth of the traffic load. In cities, the situation becomes much worse especially on the roadway intersections, traffic jams multiply the transport delay, fuel consumption and consequently harmful substance emission.

In many countries, efforts are underway to develop and apply modern solutions for the effective transportation network control. In the paper, we investigate a possibility of the RFID-technology practical application to control the traffic on the signalized intersections, analyze the performance characteristics and carry out the comparative analysis of the traffic light switch control disciplines using a mathematical model of polling systems.

2 RFID-Technology

RFID-technology (Radio Frequency IDentification) is known from the 1960th. But only the last decades, the technology is becoming widely spread out and

V. Vishnevsky et al. (Eds.): DCCN 2013, CCIS 279, pp. 256–266, 2014.
DOI: 10.1007/978-3-319-05209-0_22, © Springer International Publishing Switzerland 2014

recently it is one of the most intensively developing branches in the automated object identification area. It is noteworthy that radio frequency identification makes possible the real-time automatical identification of the objects, allows to automatize the non-contact data collection and processing and to keep a temporal record of the tagged object events [1]. RFID-technology is based on three components [1,2]:

– radio-tags, or RFID-tags (transponders);
– a tool to interrogate and record tags (RFID-reader);
– server software decoding tag-data from the reader and transforming it in the form appropriate for the management control system.

Radio frequency identification process is described as follows [1]: an RFID reader transmits an encoded radio signal to interrogate the tag. The RFID tag receives the message and then responds with its identification and other information or record data received from the reader. RFID-tag is attached to the object to be identify becoming a source of the object information. RFID-tags allows multiple data reading and re-recording and is able to keep high data volume. The reading speed runs up to 1000 tags per second with 100 % accuracy within the interrogation area. The read distance can be extended for hundreds of meters depending on the reader and tag types.

3   Application of RFID-Technology to the Roadway Intersection

In the paper, we describe the application of RFID-technology in traffic control management systems. It should be noted that the technology is widely spread all over the world, particularly in USA, Canada and Europe. E.g, the patent US 2009/0231160 A1 [3] describes the algorithm of regulating the traffic flow at a roadway intersection having one or more traffic signals by using the RFID-technology in order to minimize the traffic delay and the total vehicle waiting time on the intersection. The algorithm as it is described in the patent is presented in Fig. 1.

The main idea of the algorithm is that each vehicle is tagged with a passive RFID-tag which is interrogated by a RFID-reader mounted in the vicinity of the traffic signal and connected with the processor. Thus the traffic control system has the complete information on the current intersection load allowing making a decision on each light signal duration and optimizing the intersection work (Fig. 2).

It should be noted that the algorithm is not optimal since it does not take into account the following factors: high-priority vehicles income (police, emergency ambulance, etc.), different speed of transit across the intersection, etc. In the present paper we use some ideas described in the patent US 2009/0231160 A1. The traffic management architecture considered here is similar to one in the patent but we use the other algorithm to control traffic light signals based on the polling service disciplines [4].

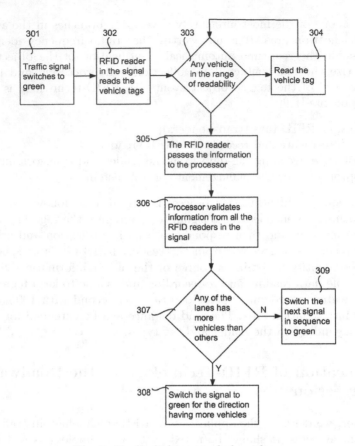

Fig. 1. Traffic light control algorithm.

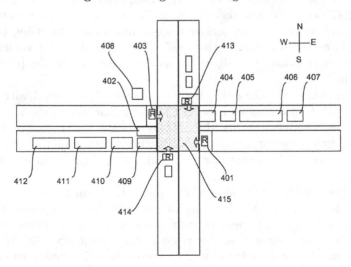

Fig. 2. Configuration of RFID-readers and passive tags.

4   Polling Model for the Roadway Intersection Performance Analysis

The algorithm to control traffic light signals is supposed to be a polling discipline (gated or exhaustive) to serve queues of vehicles at a roadway intersection. In the paper, we compare these polling disciplines to determine the most appropriate one for solving the traffic optimization problem. Note that the intersection can be considered as a polling system with cyclic polling order since the traffic lights are usually switched in a cyclic way.

Thus we model the queues of vehicles at the roadway intersection by a stochastic polling system with exhaustive or gated service [4] which is described as follows. The roadway intersection is supposed to be a server attending four queues (the simplest case). The i-th queue has the Poisson input of vehicles or rate λ_i. Service times at queue i are generally independent and identically distributed with distribution function $B_i(t)$ having the mean $b_i = \int_0^\infty t\,dB_i(t)$ and the second moment $b_2^{(i)}$, $i = \overline{1, N}$, where N is the number of queues.

The time of switching to the queue i, which is called the switchover time, has the distribution function $S_i(t)$ with the mean s_i and the second moment $s_i^{(2)}$. We denote by $\rho_i = \lambda_i b_i$ the traffic intensity to the queue i, and by $\rho = \sum_{i=1}^{N} \rho_i$ the total traffic intensity. Denote by s and $s^{(2)}$ the first and second moments of the server's total setup time in the cycle,

$$s = \sum_{j=1}^{N} s_j, \quad s^{(2)} = s^2 + \sum_{j=1}^{N}(s_j^{(2)} - s_j^2).$$

The service discipline at a queue is supposed to be exhaustive or gated. Under the exhaustive discipline, the server serves customers until the queue is emptied. Under the gated discipline, the server serves only those customers that presented in the queue at a polling moment (in our model it is time when the green light is being switched on for the queue).

One of the most important performance characteristics for the polling systems is the mean cycle time. It the time the server spends serving queues from 1 to N. For systems with exhaustive or gated cyclic polling, the mean cycle time involves the sum of time when server works at the queues (it takes the mean fraction of time ρ) and the total time of server switching between queues (it takes time s in the average). So, $C = \rho C + s$, that leads to the formula

$$C = \frac{s}{1 - \rho}.$$

For the system under consideration, the mean-value analysis can be applied [4] to calculate the mean length of vehicle queues and the mean delay at each queue. Mean value analysis allows calculating the mean queue length at an arbitrary epoch. The method implies that the mean duration of queue visits are known. The visit time of a queue is composed of the service time of the queue, i.e. the time the server spends servicing customers at the queue, plus the preceding setup time in case of exhaustive service or plus the succeeding setup time in case of gated service. By virtue of these two different definitions, a queue is empty exactly at the end of its visit time in case of exhaustive service, while the queue before the gate is empty at the beginning of a visit time in case of gated service (all customers waiting for service are then placed behind the gate). Since the method aims at the calculation of the performance characteristics at an arbitrary epoch, we need to know not only the mean queue visit time but its mean residual (or passed) time.

The mean residual and passed time of the customer service in the queue i are equal and defined by

$$R_{B_i} = \frac{b_i^{(2)}}{2b_i},$$

and the mean residual (or passed) setup time for the queue i is defined by

$$R_{S_i} = \frac{s_i^{(2)}}{2s_i}.$$

The probability that the server is working (serving a customer) at the queue i is equal to the traffic intensity to the queue, namely $\rho_i = \lambda_i b_i$.

Let V_i be the duration of the visit time for the queue i, $v_i = \mathbf{M}[V_i]$.

Since the server is working a fraction ρ_i of the time on queue i, the mean of a visit period of queue i reads, for exhaustive service,

$$v_i = \rho_i C + s_i,$$

and, for gated service,

$$v_i = \rho_i C + s_{i+1},$$

$i = \overline{1, N}$. Here we assume that $s_{N+1} = s_1$.

We define an (i, j)-period as the sum of j consecutive visit times starting in queue i, $j = \overline{1, N}$. The corresponding mean is given by

$$v_{i,j} = \sum_{n=i}^{i+j-1} v_n, \quad i, j = \overline{1, N}.$$

Note that in the case $j = 1$ and $j = N$, the value $v_{i,j}$ equals to the mean visit period v_i and the mean cycle time, respectively.

The fraction of the time $q_{i,j}$ the system is in an (i, j)-period equals

$$q_{i,j} = \frac{v_{i,j}}{C},$$

where, by definition, $q_{i,N} = 1$.

The mean of a residual (i, j)-period is given by

$$R_{V_{i,j}} = \frac{v_{i,j}^{(2)}}{2v_{i,j}},$$

with the remark that the second moments $v_{i,j}^{(2)}$ of the (i,j)-period length are still unknown.

Since the method deals with the mean values only, we will omit the symbol of expectation $\mathbf{M}$ within this Section.

Denote by L_{ij} the mean length of the queue i at an arbitrary epoch within a visit time of queue j, $i,j = \overline{1,N}$. The corresponding unconditional mean queue length L_i (regardless of the number of queue the server is working at) can be expressed in terms of $L_{i,j}$ as follows

$$L_i = \sum_{j=1}^{N} q_{j1} L_{ij}, \quad i = \overline{1,N}, \tag{1}$$

here $q_{j,1}$ means the probability that the server is attended to the queue j at an arbitrary epoch within a cycle, $j = \overline{1,N}$.

4.1 Exhaustive Service Discipline

Consider the polling system with exhaustive service and consider a tagged customer at the moment it arrives at queue i. Note that the state distribution seen by this tagged customer is identical to the equilibrium distribution. That is, this customer has to wait for the servicing of all customers L_i which were already waiting in this queue on its arrival. On the customer arrival, the server can be in one of the following states:

- with probability ρ_i it is working at queue i and the tagged customer has to wait for the residual service time of the customer in service;
- with probability s_i/C the server is switching to the queue i and the delay of the customer is increased by a residual setup time;
- with probability $1 - q_{i,1}$ the server is attended to some queue (not the queue i) and the service of the tagged customer is delayed until the server starts service again at queue i. The duration of this period is the residual time of the $(i+1, N-1)$-period plus the switchover time to the queue i.

Thus, the mean waiting time W_i in the queue i is the sum of the mean time intervals described above,

$$W_i = L_i b_i + \rho_i R_{B_i} + \frac{s_i}{C} R_{S_i} + (1 - q_{i,1})(R_{V_{i+1,N-1}} + s_i). \tag{2}$$

Application of the Little's Law,

$$L_i = \lambda_i W_i, \tag{3}$$

to the Eq. (2) yields

$$L_i = \frac{\lambda_i}{1 - \rho_i} \left(\rho_i R_{B_i} + \frac{s_i}{C} R_{S_i} + (1 - q_{i,1})(R_{V_{i+1,N-1}} + s_i) \right). \tag{4}$$

Note that in (4), the values $R_{V_{i+1,N-1}}$ remains unknown. To obtain them, we relate them to $L_{i,j}$, $i,j = \overline{1,N}$.

Firstly, as under the exhaustive policy no type-i customers are left at the end of a visit time of queue i, the following property can be obtained. The number of type-i customers present at an arbitrary moment within an $(i+1,j)$-period equals the number of Poisson arrivals during the passed time of this $(i+1,j)$-period. Since the passed time is in distribution equal to the residual time, the following equation holds

$$\sum_{n=i+1}^{i+j} \frac{q_{n,1}}{q_{i+1,j}} L_{i,n} = \lambda_i R_{V_{i+1,j}}, \quad i = \overline{1,N}, j = \overline{1,N-1}. \tag{5}$$

Substitution of (1) into the Eq. (4) yields

$$\sum_{n=1}^{N} q_{n,1} L_{i,n} = \frac{\lambda_i}{1 - \rho_i} \left(\rho_i R_{B_i} + \frac{s_i}{C} R_{S_i} + (1 - q_{i,1})(R_{V_{i+1,N-1}} + s_i) \right). \tag{6}$$

It is easily seen that Eqs. (5) and (6) represent a set of N^2 linear equations for the unknowns $L_{i,j}$ and $R_{V_{i,j}}$. In the remainder of this subsection, we derive additional equations by expressing $R_{V_{i,j}}$ in terms of $L_{i,j}$.

Consider the value $R_{V_{i,1}}$, the mean residual time of the queue i visit. At an arbitrary moment within a visit time of the queue i, $L_{i,i}$ type-i customers are waiting. Each of these customers increases the queue visit time by, in the average, $b_i/(1 - \rho_i)$ that is the mean busy period initiated by a customer in an M/G/1 queueing system corresponding to the queue i.

With the probability $\frac{\rho_i C}{v_{i,1}}$ ($\frac{s_i}{v_{i,1}}$), the server is serving a customer (switching to the queue) at an arbitrary epoch given that it is attended to the queue i. The customer being processed increase the queue visit time by the time, in average, $\frac{R_{B_i}}{1 - \rho_i}$. And the mean residual switchover time to the queue i is equal to $\frac{R_{S_i}}{1 - \rho_i}$. Thus, the mean residual time of the queue i visit is given by

$$R_{V_{i,1}} = \frac{1}{1 - \rho_i} \left(L_{i,i} b_i + \frac{\rho_i C}{v_{i,1}} R_{B_i} + \frac{s_i}{v_{i,1}} R_{B_i} \right), \quad i = \overline{1,N}. \tag{7}$$

Next, consider the value $R_{V_{i,2}}$. With probability $\frac{q_{i+1,1}}{q_{i,2}}$, the interval $R_{V_{i,2}}$ is simply equal to $R_{V_{i+1,1}}$. On the other hand, with probability $\frac{q_{i,1}}{q_{i,2}}$, this residual period equals $R_{V_{i,1}} + s_{i+1}$ plus the busy periods initiated by the type-$(i+1)$ customers arriving during $R_{V_{i,1}} + S_{i+1}$ and by the type-$(i+1)$ customers present at an arbitrary moment within a visit time of queue i. Thus, we have

$$R_{V_{i,2}} = \frac{q_{i,1}}{q_{i,2}} \left((R_{V_{i,1}} + s_{i+1}) \left(1 + \frac{\lambda_{i+1} B_{i+1}}{1 - \rho_{i+1}} \right) \right) + \left(1 - \frac{q_{i,1}}{q_{i,2}} \right) R_{V_{i+1,1}} =$$

$$= \frac{q_{i,1}}{q_{i,2}} \left(\frac{R_{V_{i,1}}}{1 - \rho_{i+1}} + \frac{s_{i+1} + L_{i+1,i} b_{i+1}}{1 - \rho_{i+1}} \right) + \left(1 - \frac{q_{i,1}}{q_{i,2}} \right) R_{V_{i+1,1}}.$$

The derivation of $R_{V_{i,j}}$ for general j, $j = \overline{1,N}$, can be made in the similar way

$$R_{V_{i,j}} = \frac{q_{i,1}}{q_{i,j}} \left(\frac{R_{V_{i,1}}}{\prod_{n=1}^{j-1}(1 - \rho_{i+n})} + \sum_{n=1}^{j-1} \frac{s_{i+n} + L_{i+n,i}b_{i+n}}{\prod_{m=n}^{j-1}(1 - \rho_{i+m})} \right) + \left(1 - \frac{q_{i,1}}{q_{i,j}}\right) R_{V_{i+1,j-1}}.$$
$$(8)$$

Finally, eliminating $R_{V_{i,j}}$ from Eqs. (5) and (6) with the help of Eqs. (7) and (8) renders a set of N^2 linear equations for equally many unknowns $L_{i,j}$. After solving these equations, the unconditional mean queue lengths L_i and mean waiting times W_i, $i = \overline{1,N}$, can be computed via and Little's Law (3).

Note that the residual cycle lengths $R_{V_{i,N}}$, $i = \overline{1,N}$, which are not required for the computation of the mean waiting times, satisfy the relation (8) as well. An important observation is that whereas the mean cycle lengths do not depend on the queue at which the cycle starts, the mean residual cycle lengths generally differ.

4.2   Gated Service Discipline

In case of gated service, all customers waiting in queue at the start of a visit time of this queue are placed behind a gate meaning that they are served in the current cycle. However, customers arriving during a visit time of their queue are placed before this gate and are, thus, only served in the next cycle. Thus, in the case $i = j$, the mean queue i length $L_{i,j}$ at an arbitrary epoch of its visit by the server is the sum of two auxiliary variables, i.e.,

$$L_{i,i} = \bar{L}_{i,i} + \tilde{L}_{i,i}, \quad i = \overline{1,N},$$

where $\bar{L}_{i,i}$ is the number of customers in the queue i which will be served in the current cycle (customers considered to be behind the gate), $\tilde{L}_{i,i}$ is the number of customers arrived to the queue i during the visit time, and they will be served in the next cycle (customers before the gate). The customer in service is excluded. In case $i \neq j$, all customers in queue i are obviously located before the gate, i.e.,

$$L_{i,j} = \tilde{L}_{i,j}, \quad i \neq j, \; i,j = \overline{1,N}.$$

The corresponding unconditional queue length L_i has mean

$$L_i = \tilde{L}_i + q_{i,1}\tilde{L}_{i,i} = \sum_{n=1}^{N} q_{n,1}\tilde{L}_{i,n} + q_{i,1}\bar{L}_{i,i}, \quad i = \overline{1,N}. \qquad (9)$$

The arguments for considering the case of gated service are similar to ones for the exhaustive service. Consider a tagged customer arriving to the queue i. So, the tagged customer has to wait for the service of all customers, on average,

$\tilde{L}_i$ which were already waiting before the gate upon its arrival. Furthermore, it has to wait until the first polling instant of queue i equalling a residual (i, N)-period, i.e., a residual cycle. By definition of the gated policy, this extra delay is incurred even in case the tagged type-i customer arrives in a visit time of queue i.

Consequently, the mean delay W_i of a type-i customer is given by

$$W_i = \tilde{L}_i b_i + R_{V_{i,N}}, \quad i = \overline{1, N},$$

which, in combination with Little's law (3), gives us the following relation

$$L_i = \rho_i \tilde{L}_i + \lambda_i R_{V_{i,N}}, \quad i = \overline{1, N}. \tag{10}$$

Note that in the case of gated service, we need to obtain the values $R_{V_{i,N}}$ which can be found in the similar way as shown in the Subsect. 4.1.

The gated policy, together with the definition of a visit time, clearly implies that the number of type-i customers before the gate at an arbitrary moment within an (i, j)-period is equal to the number of Poisson arrivals during the passed time of an (i, j)-period, which is in distribution again equal to a residual (i, j)-period. That is,

$$\sum_{n=i}^{i+j-1} \frac{q_{n,1}}{q_{i,j}} \tilde{L}_{i,n} = \lambda_i R_{V_{i,j}}, \quad i, j = \overline{1, N}. \tag{11}$$

Substituting the Eq. (9) into (10) results in

$$(1 - \rho_i) \sum_{i=1}^{N} q_{n,1} \tilde{L}_{i,n} + q_{i,1} \bar{L}_{i,i} = \lambda_i R_{V_{i,N}}, \quad i = \overline{1, N}. \tag{12}$$

Thus the Eqs. (11) and (12) form a system of $N(N + 1)$ equations for the unknowns $\tilde{L}_{i,j}$, $\bar{L}_{i,i}$? $R_{V_{i,j}}$. Now, we have to obtain more N^2 equations and express the values $R_{V_{i,j}}$ in terms of $\tilde{L}_{i,j}$ and $\bar{L}_{i,i}$.

Consider the value $R_{V_{i,1}}$ first. The $(i, 1)$-period lasts at least the sum of the service times of the customers behind the gate. With probability $\rho_i C / v_{i,1}$, a residual service time and a setup time for queue $i + 1$ is induced, while with probability $s_{i+1}/v_{i,1}$ only a residual setup time for queue $i + 1$ is generated. Thus, we have

$$R_{V_{i,1}} = \bar{L}_{i,i} b_i + \frac{s_{i+1}}{v_{i,1}} R_{S_{i+1}} + \frac{\rho_i C}{v_{i,1}} (R_{B_i} + s_{i+1}), \quad i = \overline{1, N}. \tag{13}$$

For the $(i, 2)$-period, the quantity $R_{V_{i,2}}$ equals $R_{V_{i+1,1}}$ with probability $\frac{q_{i+1,1}}{q_{i,2}}$. With probability $\frac{q_{i,1}}{q_{i,2}}$, however, this residual period equals $R_{V_{i,1}} + s_{i+2}$ plus the service times of the type-$(i+1)$ customers present at an arbitrary moment within a visit time of queue i and of the type-$(i + 1)$ customers arriving during $R_{V_{i,1}}$.

This yields

$$R_{V_{i,2}} = \frac{q_{i,1}}{q_{i,2}} \left(R_{V_{i,1}} + s_{i+2} + (\lambda_{i+1} R_{V_{i,1}} \tilde{L}_{i+1,i}) b_{i+1} \right)$$

$$+ \left(1 - \frac{q_{i,1}}{q_{i,2}}\right) R_{V_{i+1,1}} = \frac{q_{i,1}}{q_{i,2}} \left(R_{V_{i,1}}(1 + \rho_{i+1}) + s_{i+2} \tilde{L}_{i+1,i} b_{i+1} \right)$$

$$+ \left(1 - \frac{q_{i,1}}{q_{i,2}}\right) R_{V_{i+1,1}}.$$

The derivation of $R_{V_{i,j}}$ for general j is similar:

$$R_{V_{i,j}} = \frac{q_{i,1}}{q_{i,j}} \left(R_{V_{i,1}} \prod_{n=1}^{j-1}(1 + \rho_{i+n}) + \sum_{n=1}^{j-1}(s_{i+n+1} + \tilde{L}_{i+n,i} b_{i+n}) \right. \qquad (14)$$

$$\left. \times \prod_{m=n+1}^{j-1}(1 + \rho_{i+m}) \right) + \left(1 - \frac{q_{i,1}}{q_{i,j}}\right) R_{V_{i+1,j-1}}.$$

Equations (11) and (12) together with (13) and (14) form a system of $N(N+1)$ linear equations for equally many unknowns $\bar{L}_{i,i}$ and $\tilde{L}_{i,j}$. Together with relation (9) and Little's Law (3), the solution to these equations yields the unconditional mean queue lengths L_i and mean waiting times W_i, $i = \overline{1, N}$.

5   Numerical Results

In this section, we present the numerical results for the algorithm to calculate performance characteristics of the model considered. Let the mean service time be 5 s and the mean switchover time be 2 s. We compare exhaustive and gated service disciplines for various arrival rates. Figure 3 shows dependence of the weighted sum of mean waiting times T_{total}^{W} on the system load,

$$T_{total}^{W} = \sum_{1}^{4} \frac{\rho_i}{\rho} T_{total,i}^{W}$$

where ρ_i is the queue i load, $\rho = \sum_{i=1}^{4} \rho_i$ is the total system load. It follows from the figure that gated service is more appropriate to solve the traffic control problem at the roadway intersections since the weighted sum of the mean waiting times is less than one under the exhaustive service.

6   Conclusion

The paper is devoted to the actual problem of traffic optimization through roadway intersections and consider the possibility to use RFID-technology for solving traffic congestion problem and reducing traffic delay. The roadway intersection is considered to be a polling system with exhaustive or gated service and cyclic polling order since either the intersection light signals are switched in a cyclic way or the order of switching can be presented as a cyclic one.

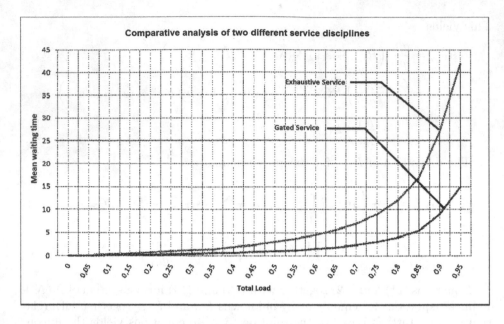

Fig. 3. Comparing of exhaustive and gated service disciplines.

Acknowledgments. This research is supported by the Russian Foundation for Basic Research (grant No. 13-07-00737).

References

1. Vishnevsky, V.M., Minnikhanov, R.N.: Automated system for driving offence control using RFID technology and the newest wireless tools. Inf. Probl. **1**, 52–65 (2012)
2. Mohammadia, S., Rajabi, A., Tavassolic, M.: Controlling of traffic lights using RFID technology and neural network. Adv. Mater. Res. **433–440**, 740–745 (2012)
3. Ramasubbu, S.S.: RFID intelligent traffic signaling. US 20090231160, A1 (2009)
4. Vishnevsky, V., Semenova, O.: Polling Systems: Theory and Applications for Broadband Wireless Networks. LAMBERT Academic Publishing, London (2012)

Author Index